Properties of Transition Metals and Their Compounds at Extreme Conditions

Properties of Transition Metals and Their Compounds at Extreme Conditions

Editors

Simone Anzellini
Daniel Errandonea

MDPI • Basel • Beijing • Wuhan • Barcelona • Belgrade • Manchester • Tokyo • Cluj • Tianjin

Editors
Simone Anzellini
Diamond House, Harwell
Science and Innovation Campus
UK

Daniel Errandonea
Universidad de Valencia
Spain

Editorial Office
MDPI
St. Alban-Anlage 66
4052 Basel, Switzerland

This is a reprint of articles from the Special Issue published online in the open access journal *Crystals* (ISSN 2073-4352) (available at: https://www.mdpi.com/journal/crystals/special_issues/transitionmetals_extremeconditions).

For citation purposes, cite each article independently as indicated on the article page online and as indicated below:

LastName, A.A.; LastName, B.B.; LastName, C.C. Article Title. *Journal Name* **Year**, *Volume Number*, Page Range.

ISBN 978-3-0365-2321-7 (Hbk)
ISBN 978-3-0365-2322-4 (PDF)

Contents

About the Editors

Simone Anzellini obtained a MSci in Physics from the University of RomaTre (Rome, IT) in 2010. In 2011, he started his PhD in physics at the CEA-DAM-DIF of Bruyer les Chatel (FR) under the supervision of Dr. P. Loubeyre and Dr. A. Dewaele. During this period, his main project was the characterization of the phase diagram of iron under extreme conditions of pressure and temperature. In 2014, he started working at Diamond Light Source as PDRA at the extreme condition beamline (I15). During this period, he mainly focussed on the design, building, and commissioning of a laser-heating system for in situ X-ray diffraction experiment in diamond anvil cells. The system is now open to the user community and allows the characterization of material under pressure in the Mbar range and temperatures up to 5000 K. Dr. Anzellini's research interests focus on materials under conditions of extreme pressure and temperature. In particular, he is interested in the characterisation of phase diagrams; melting lines; and equations of state of materials of geophysical interest, such as Fe, Ni, and Si. In pursuit of these interests, he performs experiments with diamond anvil cells combined with resistive-heating, laser-heating, or cryostat cooling, using XRD or XAS to characterize the structural, chemical, and textural evolution of the sample from atomic to macroscopic length scales. He is also interested in developing new techniques and methods to push the limits of the metrology and methodology used to explore such extreme conditions.

Daniel Errandonea is a full professor within the Department of Applied Physics of University of Valencia (Spain). He is an Argentinean-born physicist (married with two sons) who received a M.S. from the University of Buenos Aires (1992) and a Ph.D. from the University of Valencia (1998). He has authored/co-authored over 300 research articles in refereed scientific journals including Nat. Commun., Phys. Rev. Lett., and Advanced Science, which have attracted 9500+ citations. His work on materials under extreme conditions of pressure and temperature has implications for fundamental and applied research. Among other subjects, during the last decade, Prof. Errandonea has comprehensively explored phase transitions in ternary oxides, semiconductors, metals, and related materials. Some of his accomplishments include the determination of high-pressure phase transitions, high-pressure and high-temperature phase diagrams (including melting curves), and the study of their implications in the physical properties of materials. Prof. Errandonea is a fellow of the Alexander von Humboldt Foundation, winning the Van Valkenburg Award and IDEA Prize, among others. He is presently a member of the MALTA Consolider and the EFIMAT Teams, and serves on the executive committee of the International Association for the Advancement of High Pressure Science and Technology (AIRAPT).

Editorial

Properties of Transition Metals and Their Compounds at Extreme Conditions

Simone Anzellini [1,*] and Daniel Errandonea [2]

[1] Diamond Light Source Ltd., Harwell Science & Innovation Campus, Diamond House, Didcot OX11 0DE, UK
[2] Matter at High Pressure (MALTA) Consolider Team, Departamento de Física Aplicada-Instituto de Ciencia de Materiales, Universidad de Valencia, Edificio de Investigación, C/Dr. Moliner 50, Burjassot, 46100 Valencia, Spain; Daniel.Errandonea@uv.es
* Correspondence: simone.anzellini@diamond.ac.uk

check for
updates

Citation: Anzellini, S.; Errandonea, D. Properties of Transition Metals and Their Compounds at Extreme Conditions. *Crystals* **2021**, *11*, 1185. https://doi.org/10.3390/cryst11101185

Received: 22 September 2021
Accepted: 22 September 2021
Published: 29 September 2021

Publisher's Note: MDPI stays neutral with regard to jurisdictional claims in published maps and institutional affiliations.

The characterisation of the physical and chemical properties of transition metals and their compounds under extreme conditions of pressure and temperature has always attracted the interest of a wide scientific community. Their properties have numerous implications in fields ranging from solid-state physics, chemistry and materials science to Earth and planetary science.

In the last few decades, thanks to advancements in experimental techniques and computer simulations, the rate of new important discoveries in this field has significantly increased: from the prediction and the experimental observation of new ultra-hard materials to the combined characterisation of textural, structural, magnetic and chemical pressure-induced evolutions, and from the possible observation of topological transitions in the Fermi surface for valence electrons to newly predicted pressure-induced core level crossing transitions.

This Special Issue collects twelve contributions, starting with the paper of Minikayev et al. [1]. In this work the authors combine low temperature and high pressure to obtain detailed quantitative information on the evolution of the thermostructural and elastic properties of rock-salt-type crystals of PbTe and $Pb_{0.884}Cd_{0.116}Te$. The authors found a consistent image of influence on the partial substitution of Pb ions by Cd ions in the thermostructural properties of the PbTe lattice. Namely, the authors show how the lattice parameters, the thermal expansion coefficient and other thermostructural properties depend on the Cd content.

The following paper by Bettina Camin and Maximiliam Gille [2] is an interesting example of the direct application an "extreme conditions" research can have to everyday life tools. As a matter of fact, in the race towards a World with zero emissions, lightweight constructions and materials offer the opportunity to reduce CO_2 emission in the transport sector. However, the various components in vehicles are often exposed to temperatures that can be up to 40% higher than their melting temperatures. Therefore there is a risk of creep. In this paper, the authors are comparing the creep behaviour of hot extruded and heat treated ME21 magnesium alloy in both standardised miniature and standardised specimens. The obtained results show evidence of size effects in the creep parameters.

In the article by Littleton et al. [3], the authors report a characterisation of the electrical resistivity of solid and liquid Fe-FeS up to 5 GPa and the corresponding thermal conductivity. From the measured sharp changes in the measured electrical resistivity, the authors managed to delineate the eutectic temperatures of the studied Fe-FeS system as a function of pressure. The combination of the obtained results with thermal models provided an adiabatic heat flow of molten Fe-FeS eutectic composition indicating the possible presence of thermal convection in Fe-FeS under these pressure and temperature conditions. The obtained results provide important information for understanding the dynamo and thermal evolution of the core of Ganymede (the only moon in the solar system with its own magnetic field).

The characterisation of the pressure and temperature-induced modification of the properties of metallic alloys, such as morphology and microstructure, is extremely important for several technological applications. In the article by Zhang et al. [4], the authors probed the microstructural changes in the Si phase of the Al–20Si alloy at different pressure and temperature conditions, combining a multi-anvil apparatus and Scanning Electron Microscope images. In another original article, Baty et al. [5] performed an *ab initio* characterisation of the high pressure–high temperature phase diagram of copper. The results, obtained from a quantum molecular dynamic simulation based on the so called Z methodology, confirm the polymorphic nature of copper. The authors also discuss the consequent reliability of copper as a pressure standard for shock experiments.

In the subsequent contribution, Anzellini et al. [6] reported the first experimental characterisation of the phase diagram of iridium obtained with a combination of a laser-heated diamond anvil cell with synchrotron X-ray diffraction and density-functional theory calculations. Together with the solid and liquid phase boundary of iridium, the authors also report the corresponding thermal equation of state and compare it with the one of other transition metals.

In the next original contribution, Schiesaro et al. [7] characterise the pressure-induced modification of the local atomic structure of a Nb_3Sn superconductor. In particular, combining the standard analysis for X-ray absorption spectroscopy data with an evolutionary reverse Monte Carlo modelling, they evidenced a complex evolution in the Nb chains at the local atomic scale. The reported local effect appears to be correlated to anomalies evidenced by X-ray diffraction in other superconductors belonging to the same family.

In the following work by Mostafa et al. [8], the authors characterised the mechanical properties of a commercial purity aluminium modified with Zr micro-additive. The obtained results show a Zr-induced reduction in the grain size of the commercial purity aluminium, associated with an improvement of the microhardness number and resistance to fracture, as well as an improved Charpy impact energy and proof stress.

In the article by Liang et al. [9], the authors report the first high-pressure spectroscopy study on $Zn(IO_3)_2$, a member of the metal iodates, a family of materials characterised by non-linear optical properties. Using a combination of diamond anvil cells and synchrotron far-infrared spectroscopy, the authors observed three phase transitions all characterised by changes in the infrared spectra.

Finally, in the last original work, Diaz-Anichtchenko et al. [10] report a synchrotron X-ray diffraction characterisation of the pressure induced evolution of a Kagome staircase compound $(Ni_3V_2O_8)$, belonging to the large family of metal orthovanadates. The experimental results, obtained using diamond anvil cells, evidence an anisotropic response of the orthorombic structure to external compression, with the b-axis showing a compressibility almost 40% lower than the one of the other two axes. The authors also report a systematic comparison of the bulk moduli of various isomorphic metal orthovanadates.

In addition to the 10 original research articles described above, this Special Issue also include two review articles. In the first one, Tetsuya Komabayashi [11] shows an overview on the recent updates on the phase-relations in Fe-alloys as a function of pressure and temperature. Such phase-relations are of particular interest for the engineering and metallurgy fields—as they serve as recipes for creating specific phases with certain properties—and are of extreme importance for Earth science, as the major components of the Earth's core—subjected to extreme conditions of pressure and temperature—are Fe and a detailed characterisation of the phase domains of their (geophysically relevant) alloys at those extreme conditions can provide important information on the dynamics and evolution of our planet.

The characterisation of the melting line of transition metals under extreme conditions of pressure and temperature is one the most challenging (and debated) subject in the extreme conditions field. In his review paper, Paraskevas Parisiades [12] discusses the main static techniques used for these studies (with a strong focus on the diamond anvil cell), and explores the state-of-the-art in melting detection methods and analyses the possible reasons behind the discrepancy observed between melting lines obtained from various groups.

In summary, the articles presented in this Special Issue are a good representation of the broad interest—covering multiple scientific disciplines—in the present topic and they show the latest advancements in the investigation of transition metals under extreme conditions of pressure and temperature.

Acknowledgments: D.E. acknowledges financial support from Spanish Ministerio de Ciencia, Innovación y Universidades (Grants Nos. PID2019-106383GB-C41 and RED2018-102612-T), and Generalitat Valenciana (Prometeo/2018/123 EFIMAT).

Conflicts of Interest: The authors declare no competing interests.

References

1. Minikayev, R.; Safari, F.; Katrusiak, A.; Szuskiewicz, W.; Szczerbakow, A.; Bell, A.; Dynowska, E.; Paszkowicz, W. Thermostructural and Elastic Properties of PbTe and Pb0.884Cd0.116Te: A Combined Low-Temperature and High-Pressure X-ray Diffraction Study of Cd-Substitution Effects. *Crystals* **2021**, *11*, 1063. [CrossRef]
2. Camin, B.; Gille, M. The Effect of Specimen Size and Test Procedure on the Creep Behavior of ME21 Magnesium Alloy. *Crystals* **2021**, *11*, 918. [CrossRef]
3. Littleton, J.; Secco, R.; Yong, W. Thermal Convection in the Core of Ganymede Inferred from Liquid Eutectic Fe-FeS Electrical Resistivity at High Pressures. *Crystals* **2021**, *11*, 875. [CrossRef]
4. Zhang, R.; Zou, C.; Wei, Z.; Wang, H. Effect of High Pressure and Temperature on the Evolution of Si Phase and Eutectic Spacing in Al-20Si Alloys. *Crystals* **2021**, *11*, 705. [CrossRef]
5. Baty, S.; Burakovsky, L.; Errandonea, D. Ab Initio Phase Diagram of Copper. *Crystals* **2021**, *11*, 537. [CrossRef]
6. Anzellini, S.; Burakovsky, L.; Turnbull, R.; Bandiello, E.; Errandonea, D. P–V–T Equation of State of Iridium Up to 80 GPa and 3100 K. *Crystals* **2021**, *11*, 452. [CrossRef]
7. Schiesaro, I.; Anzellini, S.; Loria, R.; Torchio, R.; Spina, T.; Flükiger, R.; Irifune, T.; Silva, E.; Meneghini, C. Anomalous Behavior in the Atomic Structure of Nb3Sn under High Pressure. *Crystals* **2021**, *11*, 331. [CrossRef]
8. Mostafa, A.; Adaileh, W.; Awad, A.; Kilani, A. Mechanical Properties of Commercial Purity Aluminum Modified by Zirconium Micro-Additives. *Crystals* **2021**, *11*, 270. [CrossRef]
9. Liang, A.; Turnbull, R.; Bandiello, E.; Yousef, I.; Popescu, C.; Hebboul, Z.; Errandonea, D. High-Pressure Spectroscopy Study of Zn(IO3)2 Using Far-Infrared Synchrotron Radiation. *Crystals* **2021**, *11*, 34. [CrossRef]
10. Diaz-Anichtchenko, D.; Turnbull, R.; Bandiello, E.; Anzellini, S.; Errandonea, D. High-Pressure Structural Behavior and Equation of State of Kagome Staircase Compound, Ni3V2O8. *Crystals* **2020**, *11*, 910. [CrossRef]
11. Komabayashi, T. Phase Relations of Earth's Core-Forming Materials. *Crystals* **2021**, *11*, 581. [CrossRef]
12. Parisiades, P. A Review of the Melting Curves of Transition Metals at High Pressures Using Static Compression Techniques. *Crystals* **2021**, *11*, 416. [CrossRef]

Article

Thermostructural and Elastic Properties of PbTe and $Pb_{0.884}Cd_{0.116}Te$: A Combined Low-Temperature and High-Pressure X-ray Diffraction Study of Cd-Substitution Effects

Roman Minikayev [1,*], Fatemeh Safari [2], Andrzej Katrusiak [2], Wojciech Szuszkiewicz [1,3], Andrzej Szczerbakow [1], Anthony Bell [4,†], Elżbieta Dynowska [1] and Wojciech Paszkowicz [1,*]

[1] Institute of Physics, Polish Academy of Sciences, al. Lotników 32/46, 02-668 Warsaw, Poland; szusz@ifpan.edu.pl (W.S.); szczer@ifpan.edu.pl (A.S.); dynow@ifpan.edu.pl (E.D.)

[2] Faculty of Chemistry, Adam Mickiewicz University, ul. Umultowska 89b, 61-614 Poznań, Poland; fatemeh.safari@amu.edu.pl (F.S.); katran@amu.edu.pl (A.K.)

[3] Institute of Physics, College of Natural Sciences, University of Rzeszow, Pigonia 1, 35-310 Rzeszow, Poland

[4] Hamburger Synchrotronstrahlungslabor (HASYLAB) at Deutsches Elektronen-Synchrotron (DESY), Notkestr. 85, D-22607 Hamburg, Germany; anthony.bell@shu.ac.uk

* Correspondence: minik@ifpan.edu.pl (R.M.); paszk@ifpan.edu.pl (W.P.)

† Present address: 3 Materials and Engineering Research Institute, Sheffield Hallam University, Sheffield S1 1WB, UK.

check for updates

Citation: Minikayev, R.; Safari, F.; Katrusiak, A.; Szuszkiewicz, W.; Szczerbakow, A.; Bell, A.; Dynowska, E.; Paszkowicz, W. Thermostructural and Elastic Properties of PbTe and $Pb_{0.884}Cd_{0.116}Te$: A Combined Low-Temperature and High-Pressure X-ray Diffraction Study of Cd-Substitution Effects. *Crystals* **2021**, *11*, 1063. https://doi.org/10.3390/cryst11091063

Academic Editors: Daniel Errandonea and Simone Anzellini

Received: 10 July 2021
Accepted: 22 August 2021
Published: 3 September 2021

Publisher's Note: MDPI stays neutral with regard to jurisdictional claims in published maps and institutional affiliations.

Abstract: Rocksalt-type (Pb,Cd)Te belongs to IV–VI semiconductors exhibiting thermoelectric properties. With the aim of understanding of the influence of Cd substitution in PbTe on thermostructural and elastic properties, we studied PbTe and $Pb_{0.884}Cd_{0.116}Te$ (i) at low temperatures (15 to 300 K) and (ii) at high pressures within the stability range of NaCl-type PbTe (up to 4.5 GPa). For crystal structure studies, powder and single crystal X-ray diffraction methods were used. Modeling of the data included the second-order Grüneisen approximation of the unit-cell-volume variation, $V(T)$, the Debye expression describing the mean square atomic displacements (MSDs), $<u^2>(T)$, and Birch–Murnaghan equation of state (BMEOS). The fitting of the temperature-dependent diffraction data provided model variations of lattice parameter, the thermal expansion coefficient, and MSDs with temperature. A comparison of the MSD runs simulated for the PbTe and mixed (Pb,Cd)Te crystal leads to the confirmation of recent findings that the cation displacements are little affected by Cd substitution at the Pb site; whereas the Te displacements are markedly higher for the mixed crystal. Moreover, information about static disorder caused by Cd substitution is obtained. The calculations provided two independent ways to determine the values of the overall Debye temperature, θ_D. The resulting values differ only marginally, by no more than 1 K for PbTe and 7 K for $Pb_{0.884}Cd_{0.116}Te$ crystals. The θ_D values for the cationic and anionic sublattices were determined. The Grüneisen parameter is found to be nearly independent of temperature. The variations of unit-cell size with rising pressure (the NaCl structure of $Pb_{0.884}Cd_{0.116}Te$ sample was conserved), modeled with the BMEOS, provided the dependencies of the bulk modulus, K, on pressure for both crystals. The K_0 value is 45.6(2.5) GPa for PbTe, whereas that for $Pb_{0.884}Cd_{0.116}Te$ is significantly reduced, 33.5(2.8) GPa, showing that the lattice with fractional Cd substitution is less stiff than that of pure PbTe. The obtained experimental values of θ_D and K_0 for $Pb_{0.884}Cd_{0.116}Te$ are in line with the trends described in recently reported theoretical study for (Pb,Cd)Te mixed crystals.

Keywords: PbTe; substitutional disorder; thermal expansion; equation of state; bulk modulus; atomic displacement; low temperature; high pressure; compression; Debye temperature

1. Introduction

1.1. General Issues

IV–VI semiconductors of rocksalt-type (space group *Fm-3m*) are known to exhibit thermoelectric properties. One of the ways to modify the characteristics of such materials is fractional substitution of specific elements at the cationic or anionic sites of the crystal lattice, using a synthesis method allowing for conservation of the rocksalt-type structure. The achievable alloying level for such materials depends on the substituent and preparation method. Such substitution significantly affects the thermoelectric properties and others. We recognized that a determination of thermostructural and elastic properties can contribute to development of these materials, and that detailed studies involving the structure analysis under nonambient pressures and temperatures are lacking for mixed (IV,II)VI semiconductors. This observation gave us motivation to undertake the study for $Pb_{1-x}Cd_xTe$, which is one of most studied thermoelectric materials.

Thermostructural properties of a semiconductor are the subject of detailed studies in applied science, because they interfere with electric, optical and magnetic characteristics. Moreover, they are of importance in the design of semiconducting devices, as they affect the crystal and film growth processes and are meaningful for the first response of the material under mechanical or thermal load [1].

The family of thermoelectric materials includes tellurides, selenides, sulfides and some other compounds [2]. The thermoelectric properties of the IV–VI semiconductors of a NaCl structure have been extensively studied, both experimentally and theoretically. One of most important thermoelectrics in the family of Pb and Sn chalcogenides of a rocksalt structure is PbTe (mineral name altaite). Careful choice of the substituent and its amount incorporated into the PbTe lattice can significantly enhance the Seebeck coefficient [3,4]. PbTe exhibits polymorphism, and it is noteworthy that the orthorhombic high-pressure PbTe phase has also been reported to exhibit thermoelectric properties [5].

1.2. Cationic Substitutions

Experimental data on the effect of substitutions on the properties of PbTe have been reported for a number of systems containing various substituents. Scientific research is continuously expanding our knowledge about structural properties of such systems. For PbTe crystals modified by using various substituents, such studies have been performed, e.g., for Ba substituents [6] (experimental study), Cd or Mn [7], rare-earth metals [8], diverse substituents [9], and gold substituents [10] (theoretical studies). The mentioned theoretical study [9] convincingly demonstrates, how the choice of substituent (at the fixed level of 3.1%) incorporated into the PbTe crystal lattice, affects the physical properties and Seebeck coefficient describing the thermoelectric conversion efficiency. According to ref. [9], for most considered substituents incorporated into CdTe lattice, two important characteristics, the Debye temperature and bulk modulus, exhibit apparent decreasing trends. The prediction of the Seebeck coefficient value for tens of diversely substituted PbTe reported in ref. [9] provides a valuable basis and potential support for future work towards improvement of thermoelectrics.

1.3. PbX-CdX (X = Te, Se, S) Solid Solution

Thermoelectric properties of PbTe are improved when the Pb ions are partially replaced by Cd ions [11], forming a metastable $Pb_{1-x}Cd_xTe$ solid solution of the rocksalt type [12–15]. Therefore, this material attracts researchers' attention, being suitable for design of thermoelectric devices. Metastable single-phase NaCl-type crystals have been reported with maximal Cd content depending on the preparation conditions, particularly on applied quenching conditions from high temperature or high pressure or on the annealing method.

The lattice parameter of $Pb_{1-x}Cd_xTe$ decreases with the CdTe addition, as demonstrated experimentally [16–19] and theoretically [7]. The behavior is similar for PbSe and PbS matrices with the respective addition of CdSe and CdS. The linearity of lattice parameter variation in these three systems is illustrated in Figure 1, whereas the equations describing the lines are given in Table 1 (on the basis of the above references for PbTe, refs. [20–23] for PbSe and refs. [24–27] for PbS. (Some additional information is available in a study of the quaternary system [28].) The lines in Figure 1, representing the Vegard's rule (linear behavior of lattice parameter) are based on cited studies. The highest reported content, x_{max}, is shown for each system. Extrapolation to $x = 1$ gives the lattice parameters of hypothetical cadmium chalcogenides with a NaCl structure. The values of these parameters, of coefficients of Vegard's equation (in one case, of the equivalent Zen's equation), and the highest reported Cd content as well as the results of extrapolation are collected in Table 1, including the data from refs. [7,16,18–23,25–27].

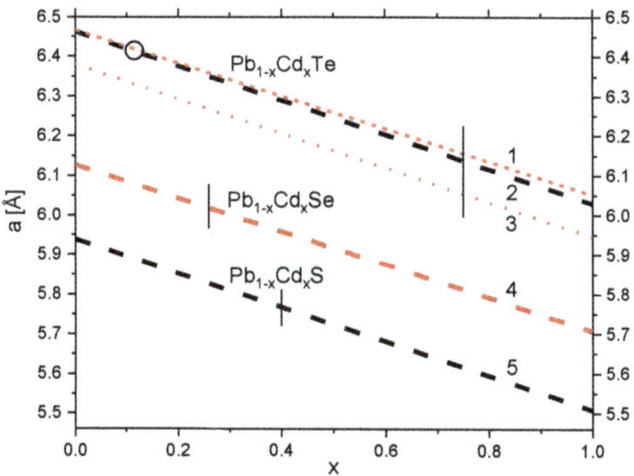

Figure 1. Reported variations of the metastable-rocksalt-phase lattice parameter of ternary Pb$_{1-x}$Cd$_x$Te, Pb$_{1-x}$Cd$_x$Se and Pb$_{1-x}$Cd$_x$S. The plots present the linear equations reported in ref. [18] (1), ref. [19] (2), ref. [7] (theoretical data) (3), ref. [20] (4), and ref. [25] (5). The highest values of Cd content in quenched samples among the reported ones, x_{max} (see Table 1), are marked with short vertical solid lines. The variations are extended above the achieved (in quenched crystals) Cd content towards $x = 1$, in order to indicate the extrapolated lattice parameters of rocksalt structures of the binary Cd chalcogenides. The open circle indicates the composition of the Pb$_{1-x}$Cd$_x$Te sample studied in this work.

Table 1. Linear equations for $a(x)$ or $V(x)$ (Vegard's rule and Zen's rule, respectively) for Pb$_{1-x}$Cd$_x$Te, Pb$_{1-x}$Cd$_x$Se and Pb$_{1-x}$Cd$_x$S solid solutions: equations, maximum reported Cd content, x_{max}, and extrapolated lattice parameters, a_{ex}, for rocksalt type CdTe, CdSe and CdS (at $x = 1$).

Compound	$a(x)$ [Å] or $V(x)$ [Å^3]	x_{max}	a_{ex} [Å]	Ref.	Year
Pb$_{1-x}$Cd$_x$Te	6.459–0.30x	0.20 (at 1139 K)	6.159	(a)	1964
	6.459–0.40x	0.144	6.059	(b)	1980
	6.466–0.414x	0.75	6.037 (*)	(c)	1989
	6.462–0.433(5)x	0.114	6.029	(d)	2009
	6.38–0.434x (&)	-	-	(e)	2012
Pb$_{1-x}$Cd$_x$Se	6.127–0.42x	0.26 (at 1213 K)	5.707	(f)	1965
	6.128–0.38x	0.03 (at 523 K)		(g)	1968
	-	0.18 (at 873 K)	-		
	-	~0.057 (at 673 K)	-	(h)	1973
	6.1263–0.3025x	0.04	-	(i)	2019
Pb$_{1-x}$Cd$_x$S	203.151–0.4389x ($)	0.016	-	(j)	1971
	5.9386–0.4302x	0.40	5.5084	(k)	2014
			5.412 (&)	(l)	2019
			5.435, 5.45, 5.72	(m)	2021

References: (a) [16], (b) [17], (c) [18], (d) [19], (e) [7], (f) [20], (g) [21], (h) [22], (i) [23], (j) [24], (k) [25], (l) [27], (m) [26]. More data can be found in a study of the quaternary system Pb$_{1-x}$Cd$_x$Se$_{1-y}$S$_y$ [28]. (*)–value for sample quenched from 2.5–3 GPa, 973–1473 K; this paper also gives 6.052 Å, derived from data for quenched Sn$_{1-x}$Cd$_x$Te. ($)–data refer to $V(x)$ dependence. (&)–theoretical data.

1.4. Knowledge on Thermostructural and Elastic Properties for PbTe and Pb$_{1-x}$Cd$_x$Te

Numerous studies on thermostructural and elastic properties of PbTe have been performed. However, some are not very detailed, and not many of them include the lowest (below 100 K) temperatures. Among the studies, the most detailed work is that of ref. [29], in which neutron powder diffraction was used to determine variations of the lattice parameter, thermal expansion coefficient (TEC), and mean square atomic displacements (MSDs) as a function of temperature. Other experimental studies refer to narrower temperature ranges or describe selected variables only. One of the frequently considered characteristics of thermoelectric materials is the degree of ordering [30]. Some recent studies focus on appearance of the cation disorder in PbTe and related chalcogenides [31–34]. The introduced disorder can affect the thermal conductivity of the crystal [31]. The substitutional disorder in the (Pb,Cd)Te alloys system has also been recently discussed in ref. [35]. However, the analysis of disorder based on temperature-dependent properties is still lacking for this ternary system.

The basic thermostructural data have been reported as functions of temperature for various temperature ranges:

- the lattice parameter $a(T)$ (experimental ones for PbTe, in refs. [29,31,36–41]), for Pb$_{1-x}$Cd$_x$Te in refs. [42–44], theoretical ones for PbTe in refs. [45–50]) (see Table 2),
- thermal expansion $\alpha(T)$ for PbTe (experimental ones in refs. [29,41,51,52] and theoretical ones in refs. [8,45–47,53,54] (see Table 3),
- atomic displacements for PbTe (experimental ones in refs. [29,31,34,37–41,50,55–57] and theoretical ones in refs. [39,49,50,58–60] (see Table 4).

Table 2. Temperature ranges for selected experimental and theoretical studies of the lattice parameter, $a(T)$, of PbTe and Pb$_{1-x}$Cd$_x$Te (x = 0.013, 0.020, 0.056, 0.096, 0.116). In the case of experimental studies, the ranges refer to the lower and upper limit of the experiment.

Mode	Compound	Temperature Range [K]	Method	Ref.	Year
Experiment	PbTe	0–400 ($)	n.a.	(a)	1971
	PbTe	120–298	SCXRD	(b)	1987
	PbTe	15–500	XRD/ND/PDF	(c)	2010
	PbTe	105–1000	SPXRD	(d)	2013
	PbTe	105–600	SPXRD	(e)	2016
	PbTe	10–500	ND	(f)	2016
	PbTe	125–293	SCXRDS	(g)	2018
	PbTe	50–600	NPD	(h)	2021
	PbTe	20–622	SCXRDS	(i)	2021
	PbTe	15–300	SPXRD	this work	2021
	Pb$_{0.987}$Cd$_{0.013}$Te,	300–~600, ~900–1073,	SPXRD	(j)	2009
	Pb$_{0.944}$Cd$_{0.056}$Te,	300–~430, ~970–1073,	"	"	"
	Pb$_{0.904}$Cd$_{0.096}$Te	300–~350	"	"	"
	Pb$_{0.98}$Cd$_{0.02}$Te	15–300	SPXRD	(k)	2011
	Pb$_{0.884}$Cd$_{0.116}$Te	15–300	SPXRD	this work	2021
Theory	PbTe	0–300	LDA, GGA	(l)	2009
	PbTe	4–550	PBEsol	(m)	2014
	PbTe	0–300	LDA, GGA	(n)	2014
	PbTe	100–800	QHA	(o)	2018
	PbTe	300–800 (*)	MD	(p)	2018
	PbTe	0–800 (&)	DFPT/LDA	(q)	2019

References: (a) [36], (b) [37], (c) [38], (d) [31], (e) [34], (f) [29], (g) [39], (h) [41], (i) [40], (j) [42,43], (k) [44], (l) [45], (m) [46], (n) [47], (o) [48], (p) [49], (q) [50]. (*)—$V(T)$ reported, (&)—relative values of lattice parameter are reported; "n.a."—stands for not available. ($)—values of lattice parameter are based on approximation curve deduced from earlier reported experiments. Abbreviations are explained at the end of this study.

Table 3. Temperature ranges for selected experimental and theoretical studies on the variation of the linear thermal expansion coefficient, $\alpha(T)$, of PbTe and $Pb_{1-x}Cd_xTe$ ($x = 0.116$). In the case of the experimental studies, the ranges refer to the lower and upper limit of the experiment.

Mode	Compound	Temperature Range [K]	Method	Ref.	Year
Experiment	PbTe	30-340	DM	(a)	1963
	PbTe	4–297	CM	(b)	1968
	PbTe	10–500	ND	(c)	2016
	PbTe	50–600	NPD	(d)	2021
	PbTe	15–300	SPXRD	this work	2021
	$Pb_{0.884}Cd_{0.116}Te$	15–300	SPXRD	this work	2021
Theory	PbTe	70–300	CDM	(e)	1966
	PbTe	0–300	LDA, GGA	(f)	2009
	PbTe	0–300	PBEsol	(g)	2014
	PbTe	0–300	LDA, GGA	(h)	2014
	PbTe	0–350	GGA	(i)	2015
	PbTe	0–300	FPBTF	(j)	2017
	PbTe	0–800	DFPT/LDA	(k)	2019

References: (a) [51], (b) [52], (c) [29], (d) [41], (e) [53], (f) [45], (g) [46], (h) [47], (i) [8], (j) [54], (k) [50]. Abbreviations are explained at the end of this study.

Table 4. Temperature ranges for selected earlier studies of the experimental and theoretical mean square displacements $<u^2>(T)$ of PbTe and $Pb_{1-x}Cd_xTe$ ($x = 0.116$). In the case of experimental studies, the ranges refer to the lower and upper limit of the experiment.

Mode	Compound	Temperature Range [K]	Method	Ref.	Year
Experiment	PbTe	78–400	PXRD, SCXRD	(a)	1973
	PbTe	100–300	SCXRD	(b)	1978
	PbTe	120–298	SCXRD	(c)	1987
	PbTe	15–500	XRD/ND/PDF	(d)	2010
	PbTe	105–1000	SPXRD	(e)	2013
	PbTe	8–500	SPXRD	(f)	2014
	PbTe	105–600	SPXRD	(g)	2016
	PbTe	10–500	ND	(h)	2016
	PbTe (*)	100–450	SCXRDS	(i)	2018
	PbTe	20–300	SCXRDS	(j)	2021
	PbTe	(40)–700	NPD	(k)	2021
	PbTe	15–300	SPXRD	this work	2021
	$Pb_{0.884}Cd_{0.116}Te$	15–300	SPXRD	this work	2021
Theory	PbTe	0–400	LKF	(l)	1968
	PbTe	0–700	MD (SME)	(m)	2014
	PbTe (*)	100–450	MD	(i)	2018
	PbTe	300–800	MD	(n)	2018
	PbTe	0–800	DFPT/LDA	(o)	2019

References: (a) [55], (b) [56], (c) [37], (d) [38], (e) [31], (f) [57], (g) [34], (h) [29], (i) [39], (j) [40], (k) [41], (l) [58], (m) [59], (n) [49], (o) [50]. (*)—the reported variation is for $u(T)$. Abbreviations are explained at the end of this study.

As for the properties of PbTe (a) describing the structural behavior under pressure or (b) describing it in the space of both variables, p and T, or (c) describing the variation of compressibility, elastic constants, heat capacity, Debye temperature and the Grüneisen parameter with temperature or pressure, the available data are scarce. For experimental data, see refs. [29,51,52,61–63], for theoretical data for PbTe, see refs. [8,45–47,53,64–69] and for $Pb_{1-x}Cd_xTe$, see refs. [66,68] (details are provided in Table 5).

Table 5. (Completing Tables 2–4). Pressure and/or temperature ranges for experimental and theoretical studies of structure-related variables for PbTe and $Pb_{1-x}Cd_xTe$ ($x = 0.116$), reported as functions of pressure and/or temperature: volume on pressure dependence, $V(p)$, bulk modulus, $K(T)$, elastic constants, $C(T)$, Debye temperature, $\theta_D(p,T)$ or $\theta_D(p)$, Grüneisen parameter, $\gamma(T)$ and $\gamma(p)$, thermal expansion, $\alpha(p,T)$, heat capacity, $c_p(T)$ or $c_p(T)$. For experimental studies, the ranges refer to the lower and upper limit of the experiment.

Mode	Compound	Variables	Pressure Range [GPa]	Temperature Range [K]	Method	Ref.	Year
Experiment	PbTe	$c_p(T)$, $c_v(T)$	-	20–260	CM	(a)	1954
	PbTe	$\gamma(T)$	-	30–340	CM+XRD	(b)	1963
	PbTe	$K(T)$, $C(T)$	-	4–297	CM	(c)	1968
	PbTe	$c_p(T)$	-	300–700	PTW	(d)	1983
	PbTe	$\theta_D(p)$, $\gamma(p)$	amb.–15, amb.–10.5	-	UIM	(e)	2013
	PbTe	$K(T)$, $\gamma(T)$, $c_v(T)$	-	10–300/300/260	ND	(f)	2016
	PbTe	$V(p)$, $K(p)$, $\gamma(T)$	amb.–4.5	-	SCXRD	this work	2021
	$Pb_{0.884}Cd_{0.116}Te$	$V(p)$, $K(p)$, $\gamma(T)$	amb.–4.5	-	SCXRD	this work	2021
Theory	PbTe	$\theta_D(T)$	-	0–200	CDM	(g)	1966
	PbTe	$K(T)$	-	0–300	LDA, GGA	(h)	2009
	PbTe	$K(p)$, $C(T)$	0–14	-	LDA	(i)	2012
	PbTe	$a(p,T)$, $\alpha(p,T)$, $K(p,T)$, $\theta_D(p,T)$, $c_v(T)$	0–10	0–300	LDA, GGA	(j)	2014
	PbTe	$K(T)$	-	0–600	PBEsol	(k)	2014
	PbTe	$c_v(T)$	-	0–400	GGA	(l)	2015
	PbTe	$K(T)$, $C(T)$	-	100–500, 0–500	LDY	(m)	2019
	PbTe (*)	$c_p(T)$, $c_v(T)$	-	~20–1000	THD	(n)	2019
	PbTe	$V(p)$	-	-	LDA	(o)	2020
	PbTe	$c_v(p)$	0–6	-	PBEsol	(p)	2021
	$Pb_{1-x}Cd_xTe$ (*)	$c_p(T)$, $c_v(T)$	-	~20–1000	THD	(n)	2019

References for experimental data: (a) [61], (b) [51], (c) [52], (d) [62], (e) [63], (f) [29], and for theoretical data: (g) [53], (h) [45], (i) [64], (j) [47], (k) [46], (l) [8], (m) [65], (n) [66,68], (o) [67], (p) [69]. (*)—this study refers to thin films; "amb" stands for ambient pressure. Abbreviations are explained at the end of this study.

1.5. Pb-Te and PbTe-CdTe System

The temperature phase diagram of Pb-Te system shows that the *Fm-3m* PbTe phase (with the lattice parameter $a = 6.460$ Å) is stable in the full range, up to $T_{max} = 1197$ K [70] (see refs. [71,72]). The off-stoichiometry range for PbTe is extremely narrow [70,73–76].

The PbTe–CdTe phase diagram [13] (for theoretical considerations, see ref. [66]) shows that the solubility, with equilibrium conditions at room temperature, of CdTe in PbTe is marginal. This is a consequence of the difference in their crystal structure—a rocksalt type for PbTe and a zinc-blende type for CdTe. However (as mentioned in the Introduction section), the $Pb_{1-x}Cd_xTe$ solid solution of the rocksalt type can be prepared in a metastable form. A diffraction study on the structure of $Pb_{1-x}Cd_xTe$ as a function of temperature has shown the decomposition process of metastable $Pb_{1-x}Cd_xTe$ ($x = 0.096$) during heating [43]; these results led to evaluation of the maximum achievable Cd content in the metastable solid solution [13]. Some theoretical calculations based on first principles have been presented in ref. [68].

Experimental investigations show that the transition to a high-pressure polymorph occurs at about 6–7 GPa [70,77–83], (see also theoretical studies [67,84,85]). The space group of this phase is *Pnma*, the lattice parameters at 7.5 GPa are $a = 11.91$ Å, $b = 4.20$ Å, $c = 4.51$ Å [70]. From about 18 GPa to at least 50 GPa, a CsCl type phase exists [83]. Recently, a topological transition at 4.8 GPa was reported in a density functional theory study [86]. Interestingly, the combination of the features of immiscibility and lattice-parameter similarity of the PbTe and CdTe components leads to the opportunity for the growth of heterostructures (which can be applied in the construction of room-temperature infrared detectors, for example [87]).

1.6. Aim

The purpose of this study is to systematically determine the influence of cadmium substitution on the thermostructural and elastic properties of PbTe. To achieve this goal, these properties (lattice parameter, thermal expansion, atomic displacements, bulk modulus and their variation with temperature or pressure) were studied and compared for two crystals, PbTe and $Pb_{0.884}Cd_{0.116}Te$. The literature data for PbTe are reviewed and taken into account in the comparative analysis of properties of these two crystals.

2. Materials and Methods

The PbTe and $Pb_{1-x}Cd_xTe$ single crystals were obtained by the self-selecting vapor growth (SSVG) method described in refs. [88,89]. High purity polycrystalline PbTe and CdTe compounds were used as reaction components. The conditions were similar to those used in earlier work [19]. To produce the PbTe–CdTe solid solution, the synthesis was performed using a mixture of PbTe and CdTe enclosed in a sealed quartz ampoule located in a furnace with a gradient of about 1 deg/cm at a temperature of about 850°C. The process of growth of homogeneous (Pb,Cd)Te crystals lasted two weeks. Further details of the growth procedure can be found in refs. [19,90]. The Cd content, $x = 0.116$, was derived from the $a(T)$ dependence reported in ref. [19].

Synchrotron-radiation techniques offer valuable experimental approaches for studies of materials; in particular, thermoelectric materials [91]. Here, we focus on the use of synchrotron radiation diffraction to extract the structural information on PbTe and $Pb_{1-x}Cd_xTe$. The in-situ low-temperature measurements were performed using synchrotron X-ray powder diffraction [92] at HASYLAB, Hamburg. The Debye–Scherrer geometry with monochromatic radiation ($\lambda = 0.5385$ Å) and an image plate detector [93] were applied. The incident beam size was 1×15 mm^2. The measurements were performed in the 2θ range of 7–58° (corresponding d-spacing range is 4.410–0.555 Å), and for samples mounted in glass capillaries (Hilgenberg) of 0.3 mm diameter, the X-ray powder diffraction patterns were recorded with a 0.004° (2θ) step.

The samples were prepared as a mixture of powdered $Pb_{1-x}Cd_xTe$ crystals and fine diamond powder (Sigma–Aldrich #48,359-1 synthetic powder), of ~1 µm monocrystalline grain size and purity of 99.9%). Addition of diamond powder served for both, (i) a diluent and (ii) an internal diffraction standard, avoiding the possible influence of wavelength instabilities (the use of such a standard has been proposed in ref. [94]). Low-temperature conditions (temperature range 15–300 K) were ensured by a closed-circuit He-cryostat. For the structural analysis, the Rietveld method [95,96] was applied using the refinement program, Fullprof.2k(v.7) [97]. In calculations, the pseudo-Voigt profile-shape function was assumed. The following parameters were refined: scale factor, lattice parameters, isotropic mean square displacement parameters, peak shape parameters, and systematic line-shift parameter. The background was set manually.

A Merrill-Bassett diamond-anvil cell (DAC) [98] was used in high-pressure experiments. The single-crystal sample was mounted inside the DAC chamber with a $MeOH:EtOH:H_2O$ (16:3:1) mixture as the pressure-transmitting medium. The pressure was calibrated with a Photon Control spectrometer by the ruby-fluorescence method [99], assuring a precision of 0.02 GPa. The experiments were conducted at a temperature of 296 K. High-pressure single-crystal X-ray diffraction data were collected at a four-circle KUMA X-ray diffractometer equipped with a graphite monochromator for the applied MoKα radiation. The gasket shadowing method was used for crystal centering and data collection [100]. The size of the diamond culets was 0.7 mm, the size of crystal for PbTe was $0.2 \times 0.05 \times 0.15$ mm^3, for $Pb_{0.884}Cd_{0.116}Te$ was $0.23 \times 0.05 \times 0.17$ mm^3 (only one crystal was loaded into DAC). UB-matrix determinations and data reductions were performed with the program CrysAlisPro [101]. The structures were solved by direct methods using the program ShelXS and refined by full-matrix least-squares on F^2 using the program ShelXT incorporated in Olex2 [102,103]. For high-pressure data analysis, the fitting procedures were conducted with the EoSFit7 program [104,105].

3. Results: Thermostructural and Elastic Properties of PbTe and $Pb_{0.884}Cd_{0.116}Te$

3.1. Effect of Cd Substitution on Temperature Variation of Unit Cell Size, Thermal Expansion Coefficient and Cationic and Anionic Mean Square Displacements

3.1.1. General Issues

The data provided by powder and single-crystal X-ray diffraction methods allowed the determination of the structural and elastic properties of PbTe and $Pb_{1-x}Cd_xTe$ ($x = 0.116$) in the 15–300 K temperature range and separately, in 0.1 MPa–4.5 GPa pressure range. Consequently, the properties measured at the same conditions for each of two crystals could be analyzed, leading to the understanding of the effect of Cd substitution on the crystal characteristics.

For pure PbTe, most of these properties were known in advance, but the information regarding temperatures below 105 K has been mostly based on the results of neutron powder diffraction of ref. [29]. The present study is one of few X-ray diffraction studies of the thermostructural properties of PbTe, covering an extended temperature range, and jointly analyzing all the three $a(T)$, $\alpha(T)$, and $<u^2>(T)$ experimental variations, completed by the $V(p)$ study. For the $Pb_{1-x}Cd_xTe$ system, the detailed investigations at non-ambient temperature and pressure have been almost completely lacking.

Phase analysis showed that the samples were single phase crystals. The analysis of powder diffraction data of PbTe and $Pb_{0.884}Cd_{0.116}Te$ by the Rietveld method yielded direct information on (i) the temperature dependencies of unit-cell size, $a(T)$, and (ii) mean square displacements, $<u^2>(T)$, of both, cations and anions. Illustrative examples of structure refinement plots for PbTe and $Pb_{1-x}Cd_xTe$ at temperatures 15 K and 300 K are given in Appendix A (Figure A1). Subsequent analysis of the $a(T)$ data led to the derivation of the temperature variation of the thermal expansion coefficient, $\alpha(T)$. The modeling of the temperature variations of the studied quantities allowed for independent determination also of other properties, for both the cationic and anionic sublattices of studied crystals.

3.1.2. Variation of Unit Cell Size of PbTe and $Pb_{0.884}Cd_{0.116}Te$ with Temperature

The experimental lattice parameter of PbTe varies in the 15–300 K range in a monotonic way (see Figure 2). The unit-cell volume, $V(T)$, variation was modeled by the second-order Grüneisen approximation, taking into account the Debye internal-energy function [106,107]:

$$V(T) = V_{(T=0)} + \frac{V_{(T=0)}E(T)}{Q - bE(T)} \tag{1}$$

where: Q and b are constants. $E(T)$ in Equation (1), expressed as

$$E(T) = \frac{9nk_BT}{(\theta_D/T)^3} \int_0^{\theta_D/T} \frac{x^3}{e^x - 1} dx \tag{2}$$

represents the Debye energy model of lattice vibrations, n is the number of atoms in the unit cell, k_B = Boltzmann constant, θ_D is the characteristic Debye temperature. The parameters Q, $V_{(T=0)}$ and b are obtained through fitting of experimental $V(T)$ data modeled by Equation (1) (refined parameters are quoted in Appendix C, Table A5). For PbTe, the lattice parameter increases by 0.50% over the whole temperature range. The run of the $a(T)$ (see inset in Figure 2b) is marginally different from the recent experimental data obtained in a wide temperature range (10–500 K) by neutron powder diffraction [29], and in the 105–300 K range by X-ray powder diffraction [31] (Figure 2a).

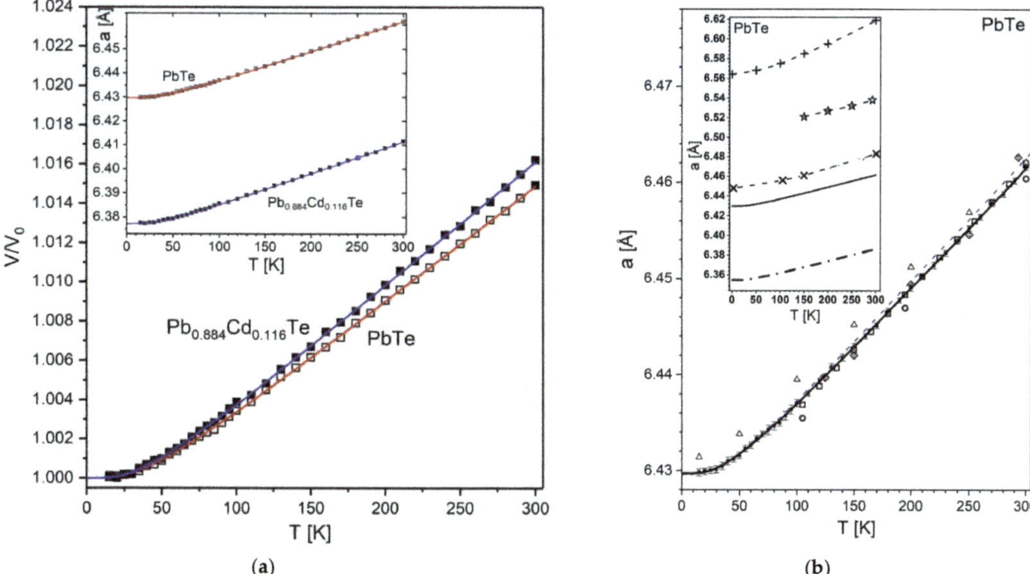

Figure 2. (**a**) Experimental variation of normalized unit-cell volume with temperature for PbTe (□) and $Pb_{0.884}Cd_{0.116}Te$ (■) (the main figure). The corresponding dependencies of the absolute values of lattice parameters are shown in the inset. The uncertainties are smaller than the symbol size. The fits of Equation (1) (second-order Grüneisen approximation) are shown in the figure and in the inset using red (PbTe) and blue ($Pb_{0.884}Cd_{0.116}Te$) solid lines, respectively. (**b**) Present temperature dependence of the PbTe lattice parameter, compared to selected literature data. Present experimental points (■) and the second-order Grüneisen approximation (thick solid line); experimental data of ref. [38] (△), ref. [31] (□), ref. [29] (thin dashed line), ref. [39] (◇), ref. [34] (O). The inset provides a comparison with theoretical data of ref. [45] (+ and dashed line), ref. [46] (✕ and dashed line), ref. [39] (☆ and dashed line), and ref. [50,108] (dot-dash line).

A comparison of both the experimental points and the fitted $a(T)$ curve to literature data is documented in Figure 2b (for refined parameters of all models see Appendix C, Table A5, whereas the numerical $a(T)$ data are quoted in Appendix B). The comparison based on this figure and on the recently reported experimental values near 0 K (data from refs. [29,44] quoted in Table 6) and near 300 K (data from refs. [26,29,31,34,39,41–43] quoted in Table 7), shows that the discrepancies between the present and earlier values of lattice parameter are quite small. Namely, near 0 K the discrepancy of the present a value, 6.42972(5) Å (the lattice parameter values given in this work with five decimal places refer to those obtained from fitted Equation (1) in this work and in the cited literature), with recent literature data, 6.42962 Å, is negligible (1×10^{-4} Å). As for the value at 300 K, our result of data fitting is 6.46148(87) Å. It agrees perfectly with the average of the high quality records for PbTe stored in the ICSD database [26] (the quality is based on ICSD-staff evaluation). There are five such records; their a(293 K) values are 6.462(1) Å, 6.459(1) Å, 6.461(1) Å, 6.461(1) Å, and 6.460(1) Å; the average is 6.46060(15) Å. After temperature correction from 293 to 300 K the average increases by 0.00088 Å (based on present $a(T)$ results) leading to the ICSD derived value at 300 K to be 6.46148(15) Å. This value is identical to the above-quoted present one. All these perfect agreements point out both, the high quality of the sample and precision of applied measurement approach, including the instrument calibration. This observation can justify recommendation of the present $a(T)$ run as a reference for the PbTe lattice parameter as a function of temperature; particularly in the near-RT temperatures, through interpolation of the data of Table A2 (Appendix B). The recommended a(300 K) value at 300 K is 6.46148(87) Å (thsi result is quoted together with other ones in Table 7).

Table 6. Present and recently reported values of experimental lattice parameter, a, near 0 K for PbTe and $Pb_{1-x}Cd_xTe$ (x = 0.013, 0.056, 0.116). For complete numerical data of $a(T)$ of this work, see Tables A1 and A2 (Appendix B).

Compound	T	a [Å]	Ref.	Year
PbTe	1	6.42962 (*)	(a)	2016
	0	6.42972(5) (*)	this work	2021
	15	6.42977(5) (*)	this work	2021
	15	6.4298(4)	this work	2021
$Pb_{0.98}Cd_{0.02}Te$	10	6.42114 (*)	(b)	2011
$Pb_{0.884}Cd_{0.116}Te$	0	6.37725(6) (*)	this work	2021
$Pb_{0.884}Cd_{0.116}Te$	15	6.37733(7) (*)	this work	2021
$Pb_{0.884}Cd_{0.116}Te$	15	6.3775(5)	this work	2021

References: (a) [29], (b) [44]. The values of the fitted model (Equation (1)) for present and literature data are starred.

Table 7. Present and selected recently reported values of the experimental lattice parameter, a, at room temperature for PbTe and $Pb_{1-x}Cd_xTe$ (x = 0.013, 0.056, 0.116). For complete numerical data of $a(T)$ of this work, see Tables A1 and A2 (Appendix B).

Compound	T [K]	a [Å]	Ref.	Year
PbTe	300	6.46179(3), 6.46201(4)	(a)	2013
	300	6.46255 (*)	(b)	2016
	300	6.46040(4), 6.46054(4)	(c)	2016
	293	6.4626(1)	(d)	2018
	300	6.4651(*)	(e)	2021
	300	6.459–6.462 ($), <$a$> = 6.46148(15)	(f)	2021
	300	6.4616(3)	this work	2021
	300	6.46148(87) (*)	this work	2021
$Pb_{0.987}Cd_{0.013}Te$	300	6.457(2)	(g)	2009
$Pb_{0.944}Cd_{0.056}Te$	300	6.437(2)	(g)	2009
$Pb_{0.884}Cd_{0.116}Te$	300	6.41133(116) (*)	this work	2021
$Pb_{0.884}Cd_{0.116}Te$	300	6.4116(4)	this work	2021

References: (a) [31], (b) [29], (c) [34], (d) [39], (e) [41], (f) [26], (g) [42,43]. ($)—range of five high-quality results (see text for details); the average is corrected for thermal expansion of PbTe (the source values refer to T = 293 K). The values of the fitted model (Equation (1)) for present and literature data are starred.

Also of interest is the compatibility of the experimental and theoretical data. Apparently, the shapes of the present (and other) experimental $a(T)$ variations are generally in line with earlier theoretical ones reported in refs. [39,45,46], whereas the absolute values differ by only 0.3% [46] to 2% [45]. For the best matching data of ref. [46], the increase of the a value across the 15–300 K range is only slightly larger than those experimentally observed (see Figure 2b).

The fitted Equation (1) describing the unit-cell size as a function of temperature perfectly approximates the experimental runs of both crystals, as shown in Figure 2a. The lattice parameter of $Pb_{0.884}Cd_{0.116}Te$ in the whole temperature range is reduced in respect to that of pure PbTe, and the reduction across the whole range is 0.53%, which is apparently larger than the value of 0.50% quoted at the beginning of this section for PbTe.

3.1.3. Variation of Thermal Expansion Coefficient with Temperature

For $Pb_{0.884}Cd_{0.116}Te$, the lattice parameter, according to the fitted model, increases from 6.37725(6) Å to 6.41133(116) Å. The difference in respect to PbTe in the slope of the cell-size dependence on temperature is visualized in Figure 2a, presenting the temperature variation of the cell volume for both studied crystals. The experimental dependence of the linear-thermal-expansion coefficient on temperature, $\alpha(T)$, was derived from the $V(T)$ Grüneisen approximation (Equation (1)), using the equation:

$$\alpha(T) = \alpha_V(T)/3 = (dV/dT)/(3V(T)) \tag{3}$$

For both materials, the general character of the $\alpha(T)$ variation is typical, with a nearly constant value up to 10 K, and with a pronounced increase observed up to ~100 K; above this temperature, the rise progressively becomes much smaller. In the ~170–300 K range, the variation of α with temperature is weak and nearly linear.

For PbTe, the TEC dependence on temperature obtained in the present work shows a fairly good matching with experimental results based on different earlier exploited techniques: dilatometry [51] and neutron powder diffraction [29] (see Figure 3). In particular, the resulting experimental TEC value of 19.6(6) MK^{-1} obtained in this work at 300 K matches very well the earlier reported experimental values of 19.80 MK^{-1} [51] and 19.91 MK^{-1} [29] (cf. Table 8) (this discrepancy is as low as 1.5%).

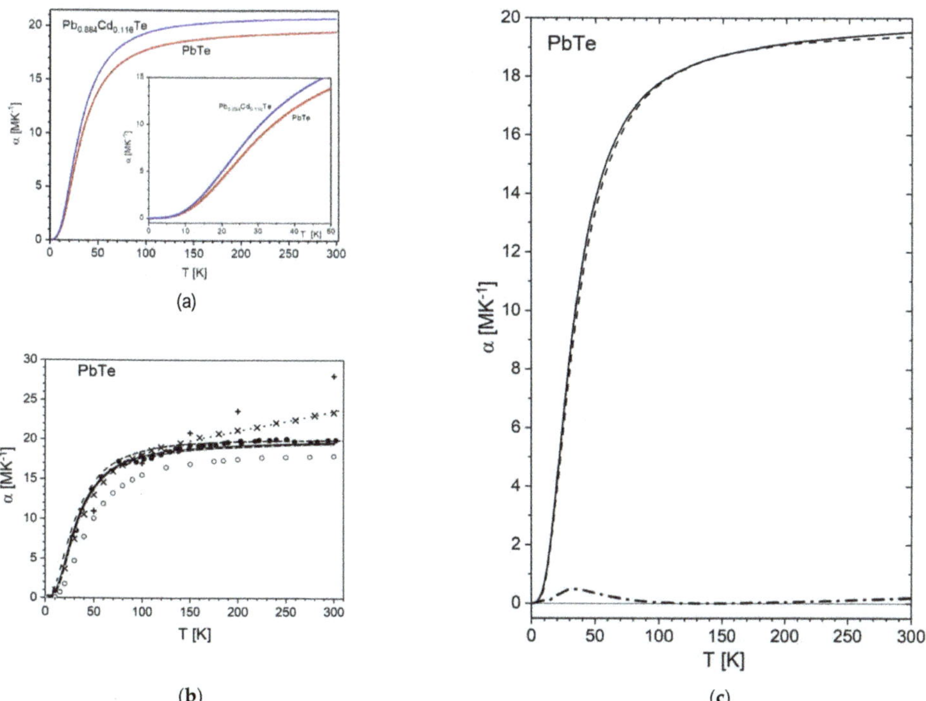

Figure 3. Variation of thermal expansion with temperature, $\alpha(T)$. (**a**) The linear thermal expansion coefficient $\alpha(T)$ for PbTe and Pb$_{0.884}$Cd$_{0.116}$Te as a function of temperature, was obtained from the experimental-data fitting using a second order Grüneisen approximation (Equation (1)). The inset includes the magnified low-temperature part (0–50 K) of the $\alpha(T)$ run. (**b**) Comparison of the present data for PbTe (black solid line) with previously reported experimental and theoretical data of PbTe. Selected data from literature (from ref. [51] (●), [29] (black dashed line) (experimental), (from ref. [45] (+); [46] (×), from ref. [8] (O) (theoretical)) are shown. (**c**) Comparison of the present experimental data (solid line, also shown as (strongly overlapping) dotted line in (**b**)), theory [60,108] (dashed line), difference curve (dash-dot line).

Table 8. Present and selected literature values of the experimental thermal expansion coefficient, α, at 300 K for PbTe and $Pb_{1-x}Cd_x$Te (x = 0.116). For full numerical data of $\alpha(T)$, see Table A2 (Appendix B).

Compound	T [K]	α [Å]	Ref.	Year
PbTe	300	19.94	(a)	1964
	300	19.91	(b)	2016
	300	19.36 (*)	(c)	2019
	300	18.12	(d)	2021
	300	19.6(6)	this work	2021
$Pb_{0.884}Cd_{0.116}$Te	300	20.7(8)	this work	2021

References: (a) [51], (b) [29], (c) [50,108], (d) [41]. (*)—theory.

A remarkable agreement of the present experimental thermal expansion data of PbTe is observed with the theory reported in ref. [50,108] (compare to the experimental and theoretical curves in Figure 3c). The agreement is visualized through the difference curve, and it is worth noting that the little bump of 2% height, observed at this curve would be twice as small if the temperature axis of the theoretical curve was shifted by only -0.7 K. The consistency with other theoretical data is not as perfect, but the trends of these results are generally compatible with the experiments described herein and others. In particular, the present data marginally differ from theoretical ones ref. [46] up to 100 K, whereas the discrepancy markedly increases at higher temperatures.

As for $Pb_{0.884}Cd_{0.116}$Te, the increase of the thermal expansion coefficient, $\alpha(T)$, in the studied temperature range is more pronounced than that observed for PbTe (Figure 3b). At 300 K, the coefficient reaches the value of 20.7(8) MK^{-1}, the increase in respect to PbTe being about 6.5% at this temperature (the rise is comparable at lower temperatures).

3.1.4. Variation of Mean Square Displacements with Temperature

The temperature dependencies of experimental mean square isotropic atomic displacement parameters for cations and anions of PbTe and $Pb_{0.884}Cd_{0.116}$Te display an apparent monotonically increasing behavior with rising temperature (Figure 4).

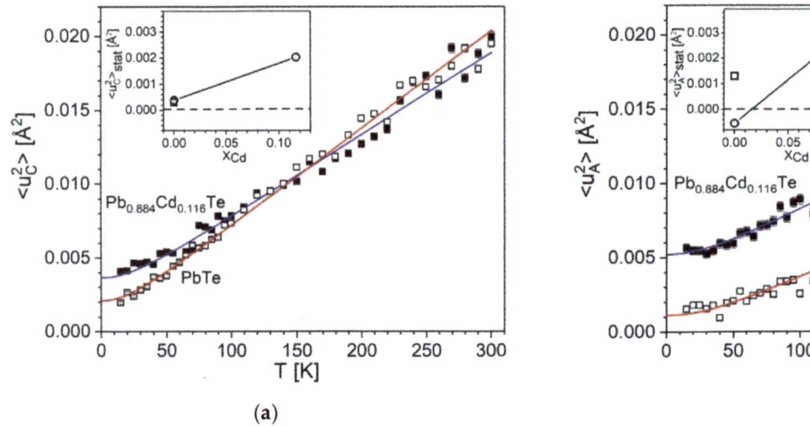
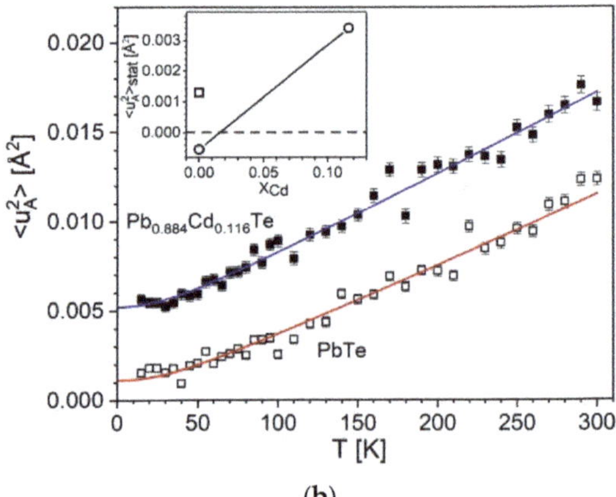

(a) (b)

Figure 4. Temperature dependence of mean square displacements for cations (**a**) and anions (**b**), for PbTe (□) and $Pb_{0.884}Cd_{0.116}$Te (■). The fitted Debye function is represented by red and blue solid lines, respectively. The insets show the static disorder term $<u^2>_{stat}$ as a function of Cd content, x_{Cd} (own results (○) with uncertainties smaller than symbol size, and results of ref. [29] (□)). The horizontal dashed line visualizes the zero-disorder level.

The experimental data were modeled using Equation (4)

$$<u^2>(T) = <u^2>_{dyn}(T) + <u^2>_{stat} \tag{4}$$

where the mean square displacement takes into account the temperature-dependent dynamic disorder term (Debye expression [109]) and the temperature-independent static disorder term $<u^2>_{stat}$ in the same way as that used in ref. [110]. The first term at the right side, $<u^2>_{dyn}(T)$, is the Debye function based on simplifying the assumption that takes into account the acoustic branches, whereas the optical branches are ignored:

$$\left\langle u^2 \right\rangle_{dyn}(T) = \frac{3\hbar^2 T}{mk_B\Theta_D^2}\left[\frac{T}{\Theta_D} \int_0^{\Theta_D/T} \frac{x}{e^x - 1}dx + \frac{\Theta_D}{4T} \right] \tag{5}$$

In the above expression, T stands for temperature, m for atomic mass, θ_D for the Debye temperature, k_B for the Boltzmann constant, and $\hbar$ for the reduced Planck constant. The second term in Equation (4), $<u^2>_{stat}$ is an empirical term attributed to the temperature-independent static disorder that can be connected in unsubstituted crystals, e.g., with the presence of point defects [111] (the presence of such defects is known to influence the electrical and other properties of thermoelectric crystals [112]), and in substituted crystals-with the presence of foreign atoms at the cationic or anionic sites.

For both crystals, PbTe and $Pb_{0.884}Cd_{0.116}Te$, the results of fitting of $<u^2>(T)$ defined by Equation (4) correctly describe the run of experimental points (the refined parameters of the model are provided in Table A5). The MSD values referring to temperatures near 0 K and near 300 K are compared with literature data in Table 9 (for values of fitted MSDs see Table A1, Appendix B).

Table 9. Present and selected reported experimental values of the mean square cationic and anionic displacements, $<u_C^2>(T)$ and $<u_A^2>(T)$, respectively, near 0 K and at room temperature for PbTe and $Pb_{1-x}Cd_xTe$ ($x = 0.116$). The (independent on temperature) static components, $<u_C^2>_{stat}$ and $<u_A^2>_{stat}$, are provided, where available. For detailed numerical data of the present study, see Tables A1 and A2, Appendix B.

Temperature	Compound	T [K]	$<u_C^2>(T)$ [Å²]	$<u_A^2>(T)$ [Å²]	$<u_C^2>_{stat}$ [Å²]	$<u_A^2>_{stat}$ [Å²]	Ref.	Year
low temperature	PbTe	15	0.0037	—			(a)	2010
	PbTe	0	0.0018 (*)	0.0018 (*)			(b) (#)	2014
	PbTe	1	0.00200	0.00315			(c)	2016
	PbTe	0	0.0021(1) (*)	0.0011(1) (*)	0.00038(4)	−0.00054(7)	this work (#)	2021
	$Pb_{0.884}Cd_{0.116}Te$	0	0.0036(1) (*)	0.0052(1) (*)	0.00203(6)	0.0034(1)	this work (#)	2021
room temperature	PbTe	300	0.0233(15)	0.0209(14)			(d)	1973
	PbTe	298	0.0204(3)	0.0141(3)			(e)	1987
	PbTe	300	0.0231	—			(a)	2010
	PbTe	300	0.0098(2)	0.01847(9)	small	small	(f)	2013
	PbTe	300	0.0238	0.0171			(b)	2014
	PbTe	300	0.0202	0.0136			(g)	2016
	PbTe	300	0.02155	0.01548	0.00031	0.00130	(c)	2016
	PbTe	300	0.0260(2)	0.0157(1)			(h)	2018
	PbTe	300	0.0204(2) (*)	0.0115(2) (*)	0.00038(4)	−0.00054(7)	this work	2021
	$Pb_{0.884}Cd_{0.116}Te$	300	0.0189(4) (*)	0.0172(3) (*)	0.00203(6)	0.0034(1)	this work	2021

References: (a) [38], (b) [57], (c) [29], (d) [55], (e) [37], (f) [31], (g) [34], (h) [39]. The values obtained from fiiting the model (Equation (4)) are starred. (#)—refers to value resulting from the model extrapolated to 0 K.

The run of each $<u^2>(T)$ curve shows (i) a characteristic nearly linear dependence at high temperatures, having a specific slope, and with (ii) a curvilinear behavior at low temperatures, characterized by a value of $<u^2>(T = 0)$. Each of these features has its own meaning. The given curve representing either the cationic or anionic site has its own characteristics determined by the fixed material parameter m, by the Debye temperature, θ_D, and by the disorder term, $<u^2>_{stat}$. Basically, $<u^2>_{stat}$ and θ_D are fittable parameters, and m could also be fitted if the composition was not well specified.

Examination of Figure 4 leads to following observations:

(1)   The fitted $<u^2>_{stat}(T)$ curves for PbTe and $Pb_{0.884}Cd_{0.116}Te$ behave differently. Namely:

(a) The MSDs at 0 K, $<u^2>(T = 0)$, increase significantly (by about 0.002–0.004 Å^2) with x rising from 0 to 0.116. We attribute this increase to the appearance of the static disorder expressed by the nonzero $<u^2>_{stat}$ term resulting from fitting Equation (4) (the values of $<u^2>_{stat}$ are quoted in Table 9). This effect is graphically presented in the insets of Figure 4a,b, where the variation of fitted $<u^2>_{stat}$ values is displayed. Appearance of marginally small negative fitted value for anionic site in PbTe (instead of zero that represents the lack of disorder) is attributed to be the effect of imperfections of fitted $<u^2>(T)$ data points. The quoted values (Table 9) show that the disorder in the anionic sublattice is considerably higher than that at the cationic site. Summarizing, an increase of the static disorder term, $<u^2>_{stat}$, in Equation (4), from approximately zero to a value of the order of 3×10^{-3} Å^2 is observed for the mixed crystal in respect to PbTe crystal. Namely, the rise is from $0.38(4) \times 10^{-3}$ Å^2 to $2.03(6) \times 10^{-3}$ Å^2 for cations, and from $-0.54(7) \times 10^{-3}$ Å^2 (a value marginally different from zero) to $3.4(1) \times 10^{-3}$ Å^2 for anions.

(b) At higher temperatures, the cationic MSDs are nearly equal for the two crystals, whereas the anionic ones differ markedly in the whole temperature range.

(c) The slope of the cationic $<u^2>(T)$ curve decreases with rising x, whereas the anionic one apparently increases. The property of Equation (4) is that the slope of $<u^2>(T)$ is governed at high temperatures by the Debye temperature (for high slope the Debye temperature is low and *vice versa*; the corresponding θ_D values are discussed in detail in Sections 3.3 and 4).

(2) The MSDs for the cationic and anionic sites behave differently for $x = 0$ than for $x = 0.116$.

(3) Comparison of Figure 4a,b shows that the cationic and anionic MSDs of Pb$_{0.884}$Cd$_{0.116}$Te are of comparable values in a broad temperature range. As this effect must depend on x, we expect that for $x < 0.116$, the $<u^2>$ values of anions are lower than those of cations, whereas for $x > 0.116$ (if the structure is stabilized), the anionic ones are higher.

A comparison of the MSDs for PbTe to literature data shows a similarity of runs with the detailed neutron-scattering based data [29,38] and with some other data based on X-ray diffraction [31,34,37] (see Figure 5). The differences in slopes of the quasilinear parts of experimental $<u^2>(T)$ runs can be connected with differences in the defect structure of studied single crystals and polycrystals.

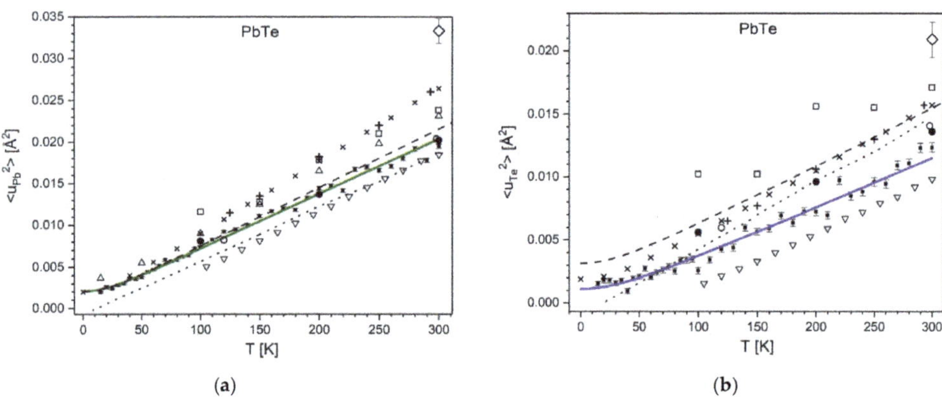

Figure 5. Temperature dependence of mean square displacements for PbTe for cations (**a**) and for anions (**b**). Experimental data: this work (solid line, ■), ref. [55] (◇), ref. [37] (**O**), ref. [38] (△), ref. [31] (▽), ref. [57] (□), ref. [34] (●), ref. [29] (dashed line), ref. [39] (+). Theoretical data: ref. [58] (✗), and ref. [39] (dotted line).

Among the theoretical MSD data, a better matching above 50 K with our experiments is found for the most recent molecular-dynamics-based data of ref. [39]. Near 0 K, the present experimental values match well the theoretical data of refs. [58,59], as shown in Figure 5.

3.2. Effect of Substitution of Cd in the PbTe Lattice on Variation of Unit-Cell Size and of Bulk Modulus with Pressure

The in-situ high-pressure X-ray diffraction study was performed under pressures ranging up to 4.5 GPa. The NaCl-type structure found for PbTe and $Pb_{0.884}Cd_{0.116}Te$ single crystals at ambient conditions (T = 295 K and p = 0.1 MPa) was conserved at the applied high-pressure conditions. The structure refinement yielded the lattice parameter monotonically varying with increasing pressure (for values see Table A3 in Appendix B).

In the analysis, the Birch–Murnaghan equation of state [104] was adopted. Its third-order variant is described by the following formula:

$$p(V) = 3K_0 f_E (1 + 2f_E)^{5/2} \left(1 + \frac{3}{4}(K' - 4)f_E + \frac{3}{2}\left((K' - 4)(K' - 3) + \frac{35}{9} \right) f_E^2 \right) \quad (6)$$

where p is the pressure, K_0 is the bulk modulus, and K' is the pressure derivative of the bulk modulus, $f_E = [(V_0/V)^{2/3} - 1]/2$ is the Eulerian strain (V is the volume under pressure p, and V_0 is the reference volume). When K' = 4, Equation (6) is reduced to a simpler, second order equation, applied in the present study (an equation of the second order has also been used in a recently reported experimental diffraction study of PbTe [83]).

The experimental relative unit-cell volume is well approximated for both crystals as a function of pressure by BMESO equation (see Figure 6; numerical data of the model are quoted with 0.5 GPa step in Table A4). The resulting bulk modulus value for PbTe K_0 is 45.6(2.5) GPa, which is consistent with previously reported values, in particular with those obtained from X-ray diffraction studies, 38.9 GPa [78,80] and 44(1) GPa [82] as well as with those from early ultrasonic wave velocity measurements of refs. [29,52,63,113], quoted in Table 10.

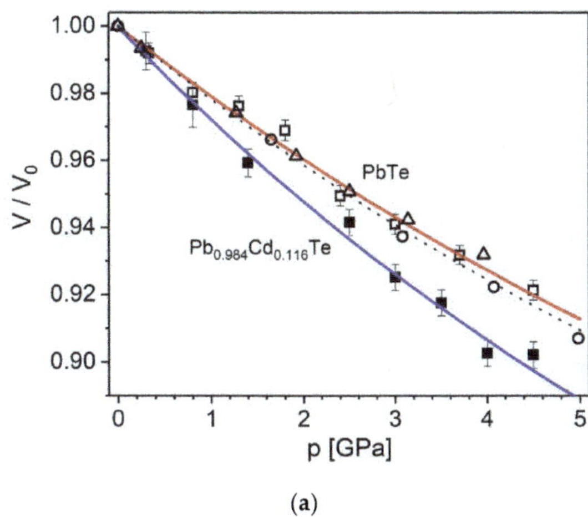
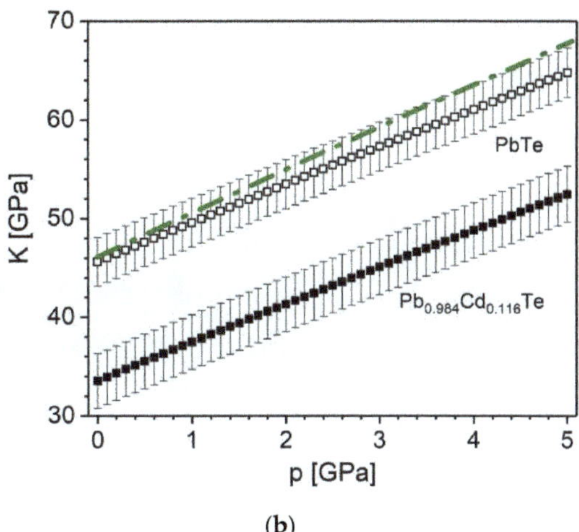

(a)   (b)

Figure 6. (a) Relative unit-cell volume as a function of pressure, for PbTe (□), and $Pb_{0.884}Cd_{0.116}Te$ (■). The solid lines correspond to the second-order BMEOS fit. Experimental literature data for PbTe are shown from ref. [83] (data (O) and fit (dotted line), the theoretical ones from ref. [64] (△ and green dashed line). (b) Bulk modulus, K, dependence on pressure for PbTe (■) and $Pb_{0.884}Cd_{0.116}Te$ (□) crystals (present data). as dash-dot line shows the theoretical data of [64].

Table 10. Fitted parameters of experimental equation of state for PbTe and $Pb_{1-x}Cd_xTe$ ($x = 0.116$) at various temperatures.

Type of Experiment	Compound	T [K]	BMEOS Parameters			Method and Remarks	Ref.	Year
			V_0 [Å^3]	K_0 [GPa]	K'			
X-ray diffraction	PbTe	RT	n.a.	38.9(1)	5.4	LEDPXRD	(a)	1984
	PbTe	RT	269.6(4)	44(1)	4 (fixed)	SPXRD (QHS)	(b)	2013
	PbTe	296	273.3(7)	45.6(2.5)	4 (fixed)	LSCXRD (t) (HS)	this work	2021
	$Pb_{0.884}Cd_{0.116}Te$	296	267.7(1.5)	33.5(2.8)	4 (fixed)	LSCXRD (t) (HS)	this work	2021
other	PbTe	0	n.a.	45.6(4)	-	UWVSC	(c)	1968
	PbTe	RT	n.a.	39.76	5.171	EC (s)	(d)	1981
		RT	n.a.	38.39	4.891	UWV (t)		
	PbTe	RT	n.a.	40.5(7)	3.8(2)	SV	(e)	2013
	PbTe	0	n.a.	44.89	-	UWVSC (*)	(f)	2016
		RT	n.a.	41.26	-			

References: (a) [78,80], (b) [83], (c) [52], (d) [113], (e) [63], (f) [29]. (s)—adiabatic; (t)—isothermal; (*)—calculation after PbTe data of ref. [52]. Abbreviations are explained at the end of this study.

The bulk modulus values provided by theoretical studies [10,45,46,64,114–117] based mostly on different approximations of the density functional theory fall into the range from 38.54 GPa to 51.7 GPa (for more details, see Table 10 (experimental data) and Table A6 (theoretical data from refs. [9,10,45–47,64–67,86,114,116–125]).

The room-temperature bulk modulus of $Pb_{0.884}Cd_{0.116}Te$ is found to be 33.5(2.8) GPa, providing the first experimental evidence that Cd substitution reduces the stiffness of the PbTe matrix. For both crystals, bulk modulus increases with pressure, in the range from 0.1 MPa to 4.5 GPa by about 50% (Figure 6, for numerical data see Table A4). For PbTe, the $K(p)$ dependence is in line with the theoretical one reported in ref. [64].

3.3. Effect of Cd Substitution on Values of Debye Temperature

Modeling three variations, $V(T)$, $<u^2>(T)$ and $V(p)$, namely the $V(T)$ variations using the second-order Grüneisen approximation (Equation (1)), the $<u^2>(T)$ variation involving the Debye expression (Equation (4)), and the $V(p)$ variations using the BMEOS (Equation (6)) led to determination of the Debye temperature, θ_D. In general, θ_D is frequently considered as a quantity depending on temperature, but for PbTe, the reported θ_D variations are weak and are observed mostly at cryogenic temperatures [53,61]. In most studies, including those based on diffraction, θ_D is considered a temperature-independent quantity. For compounds of the NaCl structure, different θ_D values are reported for the cation and anion sublattices. Such distinction is possible thanks to fitting of atomic displacements of the given (cationic or anionic) sublattice using Equation (4). Consequently, from the given experiment, we get a single overall θ_D value from fitting $V(T)$ and a pair of θ_D's from fitting of $<u_C^2>(T)$ and $<u_A^2>(T)$ (the corresponding symbols θ_{DV}, θ_{DUC}, and θ_{DUA} are used here, respectively, to highlight the distinction between these three θ_D definitions), whereas the overall θ_{DU} denotes the average of θ_{DUC} and θ_{DUA}.

The present overall θ_D values for PbTe (θ_{DV} and θ_{DU}) are almost identical (135.2(3.8) K and 135.9(7) K; the average is ~135.5 K). The Debye temperature for cation and anion sublattices in PbTe is $\theta_{DUC} = 102.8(3)$ K and $\theta_{DUA} = 169(1)$ K. For PbTe, there are a number of articles reporting the Debye temperature values. Selected literature data are collected in Table 11 (experimental X-ray diffraction and neutron diffraction based data from refs. [29,31,41,55,56]). Non-diffraction-based experimental data quoted in Table 12 are taken from refs. [13,52,61,63,126–132]). For theoretical data, see Table 13 providing the values from refs. [9,29,47,57,133,134]).

Table 11. Values of experimental Debye temperature θ_D determined by XRD/ND for PbTe and Pb$_{1-x}$Cd$_x$Te (x = 0.116) and earlier reported experimental values for PbTe: data θ_{DUC} and θ_{DUA} refer to values determined for cationic and anionic sublattices, respectively.

Compound	θ_{DUC} [K]	θ_{DUA} [K]	θ_{DU} [K] (*)	θ_{DV} [K]	Method	Ref.	Year
PbTe	95.5(2.0)	127(3)	111.3(3.0)	-	PXRD/SCXRD	(a)	1973
	-	-	-	107(2), 108(3)	LPXRD	(a)	1973
	-	-	-	111	SCXRD	(b)	1978
	-	-	87(1)	-	XRD	(c)	2013
	91(3) (&)	175(5) (&)	-	133(4) (&)	NPD+CM	(d)	2016
	99.6(2)	156.0(5)	127.8(4)	-	NPD	(d)	2016
	101.4	157.0	129.2	-	SCXRD	(d)	2016
	-	-	-	129(2)	NPD	(e)	2021
	102(1)	163(2)	132.5	-	NPD	(e)	2021
	102.0	161.4	131.7	-	NPD/PDXRDS	($)	2021
	-	-	-	135.2(3.8)	LPXRD	this work	2021
	102.8(3)	169(1)	135.9(7)	-	LPXRD	this work	2021
Pb$_{0.884}$Cd$_{0.116}$Te	-	-	-	130.1(4.4)	LPXRD	this work	2021
Pb$_{0.884}$Cd$_{0.116}$Te	115.1(5)	158(1)	136.6(8)	-	LPXRD	this work	2021

References: (a) [55], (b) [56], (c) [31], (d) [29], (e) [41]. (*)—average of the two (cationic and anionic) characteristic temperatures, θ_{DUC} and θ_{DUA}; ($)—weighted average of literature values collected in ref. [41] (only those are taken into account for which both cationic and anionic MSDs have been reported). (&)—for details of the applied method see ref. [29]. Abbreviations are explained at the end of this study.

Table 12. Experimental Debye temperature, θ_D, determined by non-diffraction methods for PbTe.

Compound	θ_D [K]	Method	Ref.	Year
PbTe	127 (at 20 K), 125 at 200 K	CM	(a)	1954
PbTe	176.7(5) (at 0 K) (*)	UWV	(b)	1968
PbTe	110	PM	(c)	1975
PbTe	168	HPM	(d)	1976
PbTe	140	SP	(e)	1979
PbTe	136	n.a.	(f)	1998
PbTe	105	TC	(g)	2006
PbTe	163	UPE	(h)	2011
PbTe	136	UWV	(i)	2012
PbTe	170(5)	DPS	(j)	2013
PbTe	143	SV	(k)	2013
PbTe	95	NS	(l)	2014
PbTe	128(1)	CM + NPD	(m)	2016

References: (a) [61], (b) [52], (c) [126], (d) [127], (e) [128], (f) [129], (g) [130], (h) [131], (i) [13], (j) [132], (k) [63], (l) [57], (m) [29]. (*)—The calculations from elastic constants involved the extrapolation to 0 K. Abbreviations are explained at the end of this study.

Table 13. Reported theoretical values of Debye temperature, θ_D, for PbTe and Pb$_{1-x}$Cd$_x$Te (x = 0.031, 0.116).

Compound	θ_D [K]	Method	Ref.	Year
PbTe	167 at 0 K 131 at 300 K (&)	CDM	(a)	1968
PbTe	177(1) (&)	NNI	(b)	1986
PbTe	152 (&)	GGA	(c)	2012
PbTe	141.5 (&) ($)	LDA, GGA	(d)	2014
PbTe	187.8 (&)	GGA	(e)	2015
Pb$_{0.969}$Cd$_{0.031}$Te	185.4	GGA	(e)	2015
Pb$_{0.884}$Cd$_{0.116}$Te	178.8	GGA	(*)	2021

References: (a) [53], (b) [133], (c) [134], (d) [47], (e) [9]. ($)—the value depends on T and p and varies between 140 and 170 K. (*)—evaluated in the present study from the data of ref. [9], using extrapolation assuming a linear variation of θ_D with x. (&)—used for calculation of the average, 157.9 K. Abbreviations are explained at the end of this study.

The earliest diffraction-based studies of PbTe have reported a relatively low overall Debye temperature of PbTe, about 110 K [55,56]. Our overall θ_{DU} and θ_{DV} values for the PbTe sublattices are in line with those determined in ref. [29] by both neutron powder and single crystal X-ray diffraction.

The variation of the direction of overall of (small) θ_D changes appearing with Cd substitution is indicated by θ_{DV} reduction by 5.1 K. A small reduction of Debye temperature for $Pb_{0.884}Cd_{0.116}Te$ ($\theta_{DV} = 130.1(4.4)$ K) in comparison with that for PbTe, $\theta_{DV} = 135.2(3.8)$ K, is observed.

The present overall θ_D values are also in line with the trends observed for those obtained by the non-diffraction methods in Table 12 (their average calculated for room-temperature data is 138.1 K, i.e., only 3 K larger than our value. The theoretical methods provided overall values with a higher average (Table 13) of 157.9 K, these data vary in an extended range.

The here-obtained cationic and anionic Debye temperature values are close to those obtained by neutron diffraction $\theta_{DUC} = 99.6(2)$ K and $\theta_{DUA} = 156.0(5)$ K [29] (the discrepancy is less than 8%). The results collected in Table 11 (the present one and those reported earlier) document that the cationic values determined in different laboratories are in very good agreement (between 95.5 and 102.8 K), whereas those for anions exhibit a larger scatter between 127 and 169 K. The θ_{DUC} and θ_{DUA} behave in an opposite way (the former rises, the latter decreases). Interestingly, the contribution of lighter Cd atoms at the Pb sites leads to a reduction of the difference between the cationic and anionic site from 66.2 K for the pure PbTe to 42.9 K in the mixed crystal.

Exploiting the data obtained in this work, the Grüneisen parameter, γ, variation with temperature was evaluated using the formula (see ref. [29] and refs. therein):

$$\gamma(T) = \alpha(T)K_0(T)V_m(T)/c_v(T) \tag{7}$$

where α is the thermal expansion coefficient, K_0 is the bulk modulus, and c_v describes the isochoric heat capacity. In calculations, the $\alpha(T)$ and $V_m(T)$ based on experimental results obtained in this work were used. The $K_0(T)$ variation reported in ref. [29] for PbTe was rescaled to the present K_0 at room temperature equal 45.6(2.5) GPa for the PbTe sample and to 33.5(2.8) GPa for the $Pb_{0.884}Cd_{0.116}Te$ sample (for $Pb_{0.884}Cd_{0.116}Te$ we adopted the rescaled $K_0(T)$ dependence of PbTe of ref. [29]). For PbTe, the temperature variation of the molar isochoric capacity $c_v(T)$ was taken from ref. [29], whereas for the Cd substituted sample the theoretical $c_v(T)$ data of $Pb_{0.88}Cd_{0.12}Te$ [68] were used. The dependencies obtained in this way are shown in Figure 7.

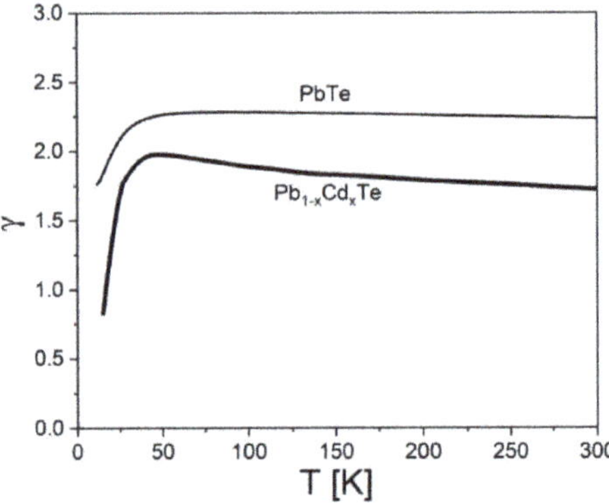

Figure 7. The temperature variation of the Grüneisen parameter for PbTe and $Pb_{0.884}Cd_{0.116}Te$ in the ranges 10–300 K and 15–300 K, respectively. The thin and the thick lines refer to PbTe and $Pb_{0.884}Cd_{0.116}Te$, respectively.

The obtained $\gamma(T)$ dependence for PbTe is comparable with those reported in refs. [51] with value of about 1.5 in the range 30–340 K and ref. [29] with values of about 2.1–2.2 in the range 50–260 K. The parameter γ is frequently considered a constant. Its experimental constant value determined by X-ray diffraction is reported to be 2.03 [29], whereas the sound velocity method has given a result of 0.95 [63] and the ultrasonic wave velocity method, 1.96 [13]. Theoretical values obtained by the density functional theory are 1.96– 2.18 [45], whereas the molecular dynamics yielded $\gamma = 1.66$ [117]. Interestingly, the present results and some of those referring to constant γ consistently suggest that its value is close to 2, whereas the roughly evaluated data on the mixed crystal indicate some decrease of gamma due to Cd substitution (see Figure 7).

The reliability of the γ values at the lowest temperatures, as calculated from Equation (7) depends on the accuracy of the very small, divided values of α and c_V, therefore the reduction of γ below ~50 K displayed in Figure 7 may be questioned.

4. Discussion

The results on the thermostructural and elastic properties of rocksalt-type crystals, PbTe and $Pb_{0.884}Cd_{0.116}Te$ solid solution, described in Section 3, are derived from X-ray diffraction data through fitting of Equations (1), (4) and (6). Temperature dependencies of the lattice parameter, $a(T)$, the thermal expansion coefficient, $\alpha(T)$, and the mean square displacements, $<u^2>(T)$, are determined for both crystals from X-ray diffraction powder diffraction data. These results for PbTe are consistent with recent literature data, in particular with the most detailed ones [29,31]. Moreover, the diffraction study of the equation of state, $V(p)$, provided the value of the PbTe bulk modulus dependence on pressure. The reliability of the present results is verified by the demonstrated close agreement of the $a(T)$, $\alpha(T)$ and $<u^2>(T)$ dependencies, as well as of the Debye temperature and bulk modulus variation, for PbTe with earlier experimental and theoretical data. It is also worth noting that the fitted model curves for $a(T)$, $<u^2>(T)$ and $V(p)$ dependencies match well the experimental points, therefore we do not expect occurrence of significant systematic errors which could add to the statistical errors quoted in Tables A3 and A4.

For $Pb_{0.884}Cd_{0.116}Te$, the obtained results are novel, they describe the thermal characteristics of this crystal and indicate the direction and magnitude of variation of the considered temperature-dependent properties with rising content of Cd at the cationic site. In other words, the earlier unknown effect of sharing the cationic sites by Pb and Cd atoms on thermal properties is revealed.

In Section 3, it is shown that the results on the PbTe lattice parameter, $a(T)$, are of high accuracy, as judged by the perfect agreement of PbTe data with the earlier-reported neutron powder diffraction data of ref. [29]. Moreover, the a value at 300 K is ideally equal to that derived from high-quality ICSD records [26]. Based on the analysis of the literature data, we show that the values of 6.42972(5) Å and 6.46148(87) Å are good candidates for the reference lattice parameter of PbTe at 0 and 300 K, respectively. The $Pb_{0.884}Cd_{0.116}Te$ sample shows a similar behavior with temperature. The $a(T)$ run for Cd-substituted PbTe crystal depends on the amount of substituent (as can be deduced from a comparison with earlier results for lower Cd content [42–44]). A related influence of substituent on the $a(T)$ runs is observed for Na and Eu substituted PbTe crystals [41]. In the above-cited results, which refer to temperatures exceeding the room temperature, the deviations from regular behavior indicate the decomposition of a metastable mixed crystal.

The here-obtained thermal expansion data for PbTe match other experimental data, especially those of ref. [29] (the discrepancy does not exceed (3%)). Unexpectedly, we found a surprisingly perfect agreement with data from ref. [50,108] in the whole studied temperature range. This agreement clearly suggests that both the present measurements and cited theory yielded accurate results for temperatures ranging up to 300 K. The fractional substitution of Cd atoms at the Pb site results in a discernible increase of the linear thermal expansion coefficient value. In particular, the value at 300 K is 19.6(6) MK^{-1} for PbTe and 20.7(8) MK^{-1} for $Pb_{0.884}Cd_{0.116}Te$: thus, the expansion rises by 6.2% at this temperature.

The investigation of mean square displacements (independently, the cationic and anionic ones) shows their nearly linear variation with rising temperature, except for the lowest temperatures (see Figures 4 and 5). This finding confirms, for the PbTe sample, the behavior known from earlier neutron diffraction and X-ray diffraction studies such as those described in refs. [29,31]. For $Pb_{0.884}Cd_{0.116}Te$, the cationic MSDs are comparable to those of PbTe, except in the region of the lowest temperatures. The Cd substitution causes apparent increase of the anionic MSDs. This increase is expected to be proportional to the Cd content.

In fitting the Equation (4), the $<u^2>_{stat}$ term describing the static disorder was determined, for both, cationic and anionic sublattices, in the unsubstituted and substituted crystal. As expected, the fitting for PbTe gave $<u^2>_{stat}$ a value close to zero, thus indicating that there is no significant disorder in this crystal (small values have also been reported in refs. [29,31]). We believe that the differences between the, values reported for pure PbTe by different groups can probably be attributed to differences in the defects' kind and density.

The observed increase of the $<u^2>_{stat}$ term after incorporating Cd into the PbTe lattice proves that alloying causes appearance of substitutional disorder in the mixed $Pb_{0.884}Cd_{0.116}Te$. crystal. We observe (see the insets in Figure 4) that the values of cationic and anionic $<u^2>_{stat}$ terms describing the static disorder are markedly different in the $Pb_{0.884}Cd_{0.116}Te$. crystal. Namely, the anionic disorder is significantly larger in this crystal.

The here-reported extraction the information on disorder for both, cationic and anionic sublattices, in a mixed PbTe based crystal, is an important novelty (previously, such calculations have been performed for pure PbTe, only). We notice a significant increase of the static disorder term, showing that information of this kind, extracted from analysis of carefully measured thermostructural properties, can be useful in future studies on IV-VI thermoelectric solid solutions and their application, because the disorder in solid solutions can affect the carrier mobility, electrical conductivity [32,35] and thermal conductivity [31,117] influencing the Seebeck coefficient. We evaluate that the opportunity for detection of disorder can concern low substituent fraction, even much less that $x = 0.1$ studied in the present work.

For both crystals, the in situ high-pressure single-crystal XRD experiment provided information on the lattice parameter variation for PbTe and $Pb_{0.884}Cd_{0.116}Te$, at pressures ranging up to 4.5 GPa. The observed pressure variation is in line with a theoretical result reported in ref. [64]. Modeling of the BMEOS led to determination of the bulk modulus and its pressure variation. At 0.1 MPa, the bulk modulus value is 45.6(2.5) for PbTe, well coinciding within error bars with the value 44(1) GPa reported in the most recent diffraction study [83]. The bulk modulus value significantly decreases with rising Cd content; in other words, the Cd substitution leads to a crystal of lower stiffness.

There are a number of theoretical works investigating the bulk modulus changes upon substitution of an element at the cationic site [8,9,123]; most typically, a reduction is predicted. In ref. [9], for 62 elements fractionally substituting Pb in PbTe, the resulting bulk modulus value is calculated; the same calculation is performed for nine substituents at an anionic Te site. For almost all of them, the bulk modulus is reduced; whereas for V, Nb, Ni and Bi, the K_0 value is larger than the calculated value of 46.61 GPa [9] for pure PbTe.

In ref. [9], the K_0 has been predicted to decrease from 46.61 GPa for PbTe to 46.42 GPa for $Pb_{0.969}Cd_{0.031}Te$. This leads to 45.90 GPa evaluated through extrapolation for $Pb_{0.884}Cd_{0.116}Te$. This evaluation differs from the experimental value obtained in the present study (33.5(2.8) GPa), but the direction of changes of K_0 with x is clear.

Experimental K_0 values for PbTe substituted with any cation are not available, except for the case of Ba substitution, where the mixing effect on K_0 consists of a 5% reduction of the PbTe value [6]. The assumption that $K_0(PbTe)$ equals 46.61 GPa leads to an evaluation of a (not explicitly reported) experimental value, 44.3 GPa for $Pb_{0.96}Ba_{0.04}Te$. Extrapolation of the theoretical value of 44.99 GPa for $Pb_{0.969}Ba_{0.031}Te$ quoted in ref. [9] leads to $K_0 = 44.5$ GPa for $Pb_{0.96}Ba_{0.04}Te$. The excellent agreement between the values of calculated 44.3 GPa and experimental 44.5 GPa points out the reliability of both, the cited experiment and the calculation.

The above-described fittings of $V(T)$ and $<u^2>(T)$ models led to the determination of values of Debye temperature, θ_D, for both crystals. Together with the Cd substitution, a small reduction of the overall Debye temperature, θ_{DU}, from 135.2(3.8) K to 130.1(4.4) K (i.e., a reduction of 5.1 K) is observed. Theoretical calculations predict reduction by 2.4 K for the composition of $x = 0.031$ [9]. Extrapolating this result to the composition of the mixed crystal studied in this work, ($x = 0.116$) gives a prediction of a 9 K (difference between $\theta_D = 187.8\,1$ K and 178.8 K quoted in Table 13) reduction of the theoretical overall θ_D. This theoretical result supports the observed trend of reduction of overall Debye temperature by increasing the cadmium content. Interestingly, the θ_D values reported by different authors for the cationic sublattice are in perfect agreement, whereas those for the anionic one are scattered. The influence of Cd substitution on Debye temperatures of cationic and anionic sublattices, described in Section 3.3, is not uniform; these values differ markedly for PbTe, but the difference is reduced for the Cd substituted crystal.

The observed influence of Cd substitution into PbTe lattice on the thermostructural and elastic properties studied can serve as a basis for evaluation of such features for crystals of different Cd content. They can also be useful in studies of more complex systems, such as those with dual cationic/anionic substitutions. As it is noted in ref. [30], various factors influence the Seebeck coefficient value. One of the ways to optimize this value consists of alloying with a selected element, which means a decrease or increase of the atomic order. It is equally possible to investigate other, more complex systems, for example those with less-conventional doubly substituted cationic systems, such as $Na_{0.03}Eu_{0.03}Pb_{0.94}Te$ [112]. The joint cationic/anionic substitution (Pb,Cu)(Se,Te) system has also been studied, providing another example of a dual system [135]. Mixed bi-cationic–bi-anionic systems such as $Na_{0.03}Eu_{0.03}Pb_{0.94}Te_{0.9}Se_{0.1}$ [112] are the subject of studies as well. It is noteworthy that upon replacing the Te anion by Se or S, the bond ionicity decreases (for the ionicity scale, see ref. [136]). Along the PbTe, PbSe, PbS series, some of the thermostructural/elastic properties (studied here for PbTe) vary monotonically; for example, the lattice parameter (see Figure 1 at $x = 0$), bulk modulus [46,85,114] and phase transition pressure [77,85].

5. Conclusions

PbTe of a rocksalt-type structure belongs to a family of thermoelectric materials. A modification of composition, typically by fractional substitution of an element such as Cd, at the Pb cation site, is known to improve the thermoelectric properties. In this work, the combined low-temperature–high-pressure study carried out here describes the effect of sharing the cationic sites by Pb and Cd atoms on the above-mentioned properties. These properties were derived for two samples, PbTe and $Pb_{0.884}Cd_{0.116}Te$, from X-ray diffraction data collected at varying temperature and pressure.

The dependencies of the lattice parameter, $a(T)$, the thermal expansion coefficient, $\alpha(T)$, and the mean square displacements, $<u^2>(T)$, are determined for both crystals. For PbTe, these results and thermal expansion are fully consistent with results of earlier X-ray diffraction, neutron diffraction, dilatometric and other experimental studies, as well as with those of multiple theoretical investigations, and this agreement supports the reliability of the data collected.

The experimental variation of the lattice parameter with temperature was modeled using the Grüneisen-approximation approach whereas the variation of mean square atomic displacements was modeled using the Debye expression. In addition, the equations of state were determined for pressures ranging up to 4.5 GPa, allowing conclusions to be drawn about the value of the bulk modulus and its variations under rising pressure and with varying Cd substitution.

The thermostructural and elastic properties for $Pb_{0.884}Cd_{0.116}Te$ crystal determined in the present study indicate the direction and magnitude of variation of the characteristics of $Pb_{1-x}Cd_xTe$ system with rising x. The stiffness of the alloy is smaller than that of pure PbTe, the thermal expansion is larger throughout the whole temperature range, and the atomic mean-square displacements change with Cd substitution in a complex way, indicating

(i) opposite variations of the Debye temperatures for both sublattices, as well as (ii) the appearance of substitutional disorder in the mixed crystal.

In summary, the study presents detailed quantitative information on the thermostructural and elastic properties of rocksalt-type crystals of PbTe and $Pb_{0.884}Cd_{0.116}Te$; such data are not yet available for alloys of the $Pb_{1-x}Cd_xTe$ system. The obtained results show a consistent image of influence of the partial substitution of Pb ions by Cd ions, in the PbTe lattice, on the thermostructural properties. Namely, the obtained results show how the lattice parameter, the thermal expansion coefficient, the atomic mean-square displacements and other thermostructural properties (compressibility, Debye temperature, Grüneisen parameter and others) depend on the cadmium content. In particular it was found, that the $Pb_{0.884}Cd_{0.116}Te$ lattice is less stiff than that of PbTe, whereas thermal expansion of the mixed crystal is discernibly larger. The described extension of the knowledge on the studied properties is expected to be profitable in a further work on the application of the fractionally substituted Cd lead telluride.

Author Contributions: Conceptualization, R.M. and W.P.; methodology, A.K., A.S., F.S., R.M., W.P. and W.S.; software, R.M.; investigation, A.B., A.K., A.S., E.D., F.S., R.M. and W.S.; writing—original draft preparation, R.M. and W.P.; writing—review and editing, R.M. and W.P.; supervision, W.P. All authors have read and agreed to the published version of the manuscript.

Funding: This work was partially supported by the National Science Centre (NCN, Poland) under grant UMO-2014/13/B/ST3/04393. The study of R.M. and A.S. has been partially supported by the National Science Centre for Development (Poland) through grant TERMOD No. TECHMAT-STRATEG2/408569/5/NCBR/2019.

Acknowledgments: The authors are grateful to José D. Querales-Flores (National Institute, Lee Maltings, Cork, Ireland) for providing the numerical data of ref. [50].

Abbreviations

APW	augmented plane-wave
BMEOS	Birch-Murnaghan Equation of State
CDM	crystal dynamics models
CM	calorimetry
DFPT	density functional perturbation theory
DFT	density functional theory
DM	dilatometry
DPS	double parton scattering (nuclear inelastic scattering)
EC	elastic constants
FP	full potential
FPBTF	first principles Boltzmann transport framework
GGA	generalized gradient approximation
GULP	computer program for the symmetry adapted simulation of solids, authored by Julian D. Gale
HPM	heat-pulse method
HS	hydrostatic conditions
HSEsolSOC	revised Heyd-Scuseria-Ernzerhof functional + spin-orbit coupling
LAPW	linearized augmented plane-wave
LDA	local density approximation
LDY	lattice dynamics calculations
LEDPXRD	laboratory energy-dispersive X-ray diffraction
LKF	Lin-Kleinman formalism
LSCXRD	laboratory single-crystal X-ray diffraction
MD	molecular dynamics

MSD	mean square displacement
n.a.	not available
ND	neutron diffraction
NNI	nearest-neighbor interaction model by Kagan and Maslow
NPD	neutron powder diffraction
NS	neutron scattering
PAW	projector augmented wave method
PBE	Perdew-Bucke-Ernzerhof exchange-correlation functional
PBEsol	Perdew–Bucke–Ernzerhof exchange-correlation functional revised for solids
PM	the paramagnetic resonance. θ_D estimated by the temperature-dependent hyperfine splitting constant $A(T)$
PTW	plane temperature waves method
QHA	quasiharmonic approximation
QHS	quasi-hydrostatic conditions
RT	room temperature
SCXRD	single crystal X-ray diffraction
SCXRDS	single crystal X-ray difraction at synchrotron
SO	soft-constraint based online
SME	slave mode expansion
SP	spectroscopy
SPXRD	synchrotron powder X-ray diffraction
SV	sound velocity method
TC	thermal conductivity
TEC	thermal expansion coefficient
THD	thermodynamic calculations
UPE	ultrasonic pulse-echo method
UIM	ultrasonic interferometry
UWV	ultrasonic wave velocity
UWVSC	ultrasonic wave velocity in single crystal
XRD/ND/PDF	X-ray diffraction and neutron diffraction, analyzed with pair distribution function (PDF) method

Appendix A

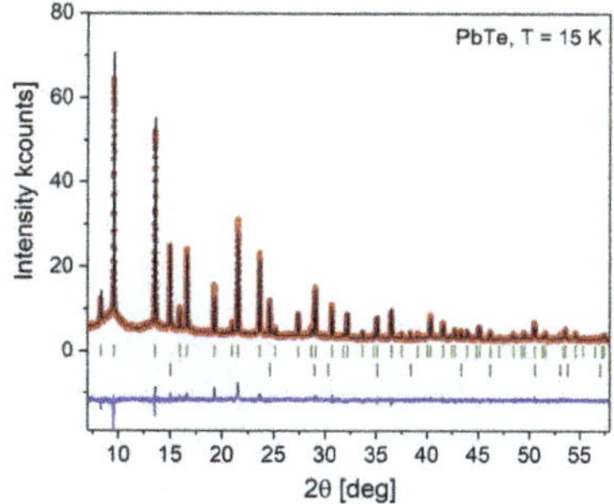

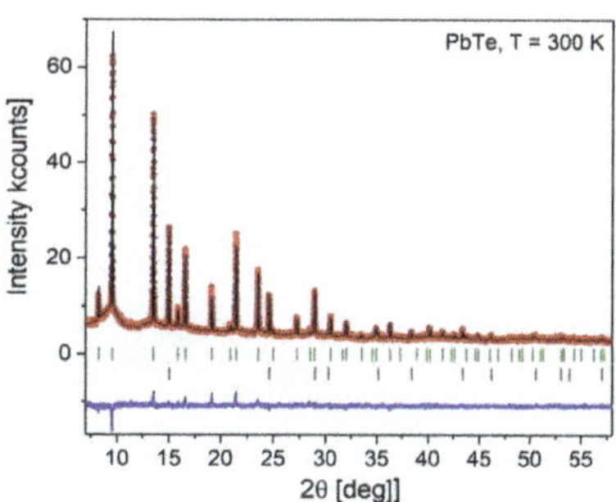

Figure A1. *Cont.*

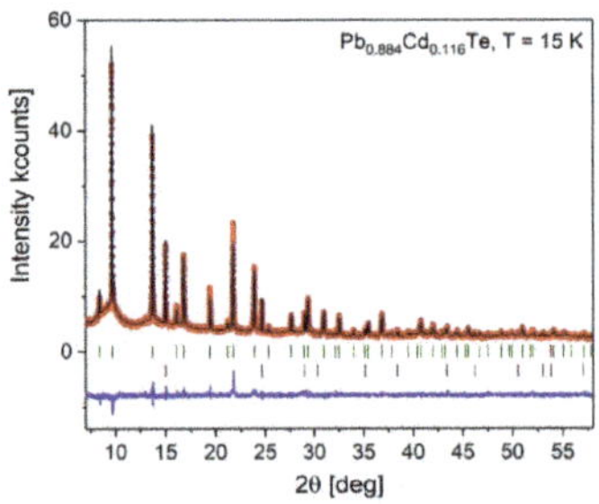

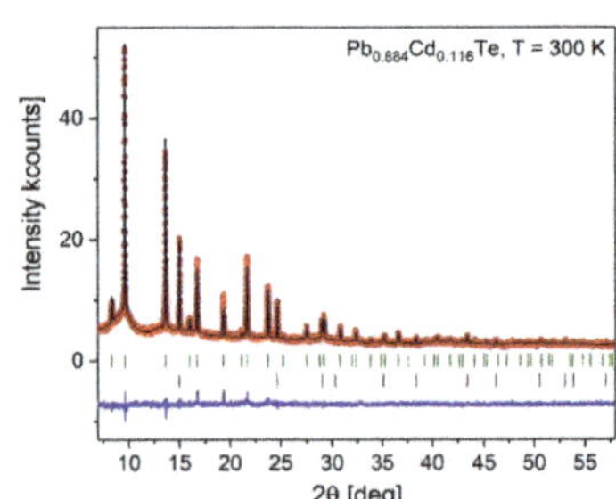

Figure A1. Rietveld refinement results for PbTe (**upper row**) and $Pb_{0.884}Cd_{0.116}Te$ (**bottom row**) at 15 K and 300 K. The experimental points are indicated by dots and the calculated patterns by the solid line. The positions of Bragg reflections of the sample are indicated as short vertical bars at the bottom (**upper row**), and those of the diamond calibrant (**lower row**). The difference patterns are displayed at the bottom of each subfigure.

Appendix B

Table A1. Present $a(T)$ and $<u^2>(T)$, obtained from Rietveld refinements for PbTe and $Pb_{0.884}Cd_{0.116}Te$.

T [K]	PbTe			$Pb_{0.884}Cd_{0.116}Te$		
	a [Å]	$<u_C^2>$ [Å^2]	$<u_A^2>$ [Å^2]	a [Å]	$<u_C^2>$ [Å^2]	$<u_A^2>$ [Å^2]
15	6.4298(4)	0.0020(1)	0.0016(2)	6.3775(5)	0.0041(2)	0.0057(3)
20	6.4299(3)	0.0027(1)	0.0018(2)	6.3773(4)	0.0041(2)	0.0055(3)
25	6.4301(3)	0.0024(1)	0.0018(2)	6.3776(5)	0.0047(2)	0.0055(3)
30	6.4301(4)	0.0028(1)	0.0016(2)	6.3777(5)	0.0046(2)	0.0052(3)
35	6.4305(4)	0.0030(1)	0.0018(2)	6.3783(5)	0.0047(2)	0.0055(3)
40	6.4310(4)	0.0037(1)	0.0010(2)	6.3787(7)	0.0046(2)	0.0059(3)
45	6.4312(3)	0.0036(1)	0.0019(2)	6.3792(6)	0.0053(2)	0.0058(3)
50	6.4316(3)	0.0038(1)	0.0021(2)	6.3793(6)	0.0054(2)	0.0060(3)
55	6.4323(3)	0.0044(1)	0.0027(2)	6.3800(7)	0.0053(2)	0.0066(3)
60	6.4327(3)	0.0047(1)	0.0021(2)	6.3805(5)	0.0047(2)	0.0067(3)
65	6.4333(3)	0.0052(1)	0.0024(2)	6.3810(7)	0.0054(2)	0.0064(3)
70	6.4338(3)	0.0059(1)	0.0026(2)	6.3817(6)	0.0056(2)	0.0071(3)
75	6.4342(5)	0.0057(2)	0.0029(2)	6.3823(5)	0.0072(2)	0.0072(3)
80	6.4347(5)	0.0058(2)	0.0025(2)	6.3829(6)	0.0071(2)	0.0074(3)
85	6.4350(3)	0.0062(1)	0.0034(2)	6.3833(5)	0.0069(2)	0.0084(3)
90	6.4358(4)	0.0064(2)	0.0034(2)	6.3840(4)	0.0078(2)	0.0077(3)
95	6.4364(5)	0.0072(2)	0.0035(3)	6.3848(4)	0.0075(2)	0.0087(3)
100	6.4371(4)	0.0074(2)	0.0026(3)	6.3855(4)	0.0079(2)	0.0089(3)
110	6.4381(3)	0.0083(2)	0.0034(2)	6.3863(5)	0.0085(2)	0.0079(4)
120	6.4394(3)	0.0092(2)	0.0043(3)	6.3875(4)	0.0094(2)	0.0092(4)
130	6.4408(4)	0.0095(2)	0.0044(3)	6.3891(4)	0.0095(2)	0.0094(3)
140	6.4418(4)	0.0100(2)	0.0060(3)	6.3903(4)	0.0099(2)	0.0097(4)
150	6.4429(3)	0.0111(2)	0.0057(3)	6.3915(4)	0.0102(2)	0.0104(4)
160	6.4440(3)	0.0117(2)	0.0059(3)	6.3931(5)	0.0115(2)	0.0115(4)
170	6.4451(3)	0.0120(2)	0.0069(3)	6.3941(5)	0.0108(2)	0.0129(4)
180	6.4466(3)	0.0118(2)	0.0064(3)	6.3953(5)	0.0117(2)	0.0103(4)
190	6.4477(3)	0.0133(2)	0.0073(3)	6.3969(5)	0.0121(2)	0.0129(4)
200	6.4491(4)	0.0144(2)	0.0072(3)	6.3981(5)	0.0127(2)	0.0132(4)
210	6.4503(3)	0.0147(2)	0.0070(3)	6.3997(4)	0.0132(2)	0.0131(4)
220	6.4514(4)	0.0142(2)	0.0097(3)	6.4007(5)	0.0137(3)	0.0138(4)
230	6.4527(3)	0.0167(2)	0.0085(3)	6.4020(4)	0.0156(3)	0.0137(4)
240	6.4539(3)	0.0170(3)	0.0088(3)	6.4035(3)	0.0170(3)	0.0135(4)
250	6.4552(3)	0.0166(2)	0.0096(3)	6.4045(3)	0.0173(3)	0.0153(4)
260	6.4564(3)	0.0171(2)	0.0094(3)	6.4062(3)	0.0160(3)	0.0148(4)
270	6.4578(3)	0.0180(3)	0.0109(4)	6.4071(3)	0.0192(3)	0.0160(5)
280	6.4588(3)	0.0192(3)	0.0111(4)	6.4086(4)	0.0171(3)	0.0165(5)
290	6.4602(3)	0.0178(3)	0.0123(4)	6.4100(4)	0.0188(3)	0.0176(5)
300	6.4616(3)	0.0195(3)	0.0124(4)	6.4116(4)	0.0200(3)	0.0166(5)

Table A2. Present values of $a(T)$, $\alpha(T)$ and $<u^2>(T)$, modeled by Equations (1) and (2) for PbTe and $Pb_{0.884}Cd_{0.116}Te$.

T [K]	PbTe				$Pb_{0.884}Cd_{0.116}Te$			
	a [Å]	α [MK^{-1}]	$<u_C^2>$ [Å^2]	$<u_A^2>$ [Å^2]	a [Å]	α [MK^{-1}]	$<u_C^2>$ [Å^2]	$<u_A^2>$ [Å^2]
0	6.42972(5)	0	0.0021(1)	0.0011(1)	6.37725(6)	0	0.0036(1)	0.0052(1)
10	6.42973(5)	0.70(7)	0.0022(1)	0.0012(1)	6.37726(6)	0.9(1)	0.0037(1)	0.0053(1)
20	6.42986(6)	4.2(3)	0.0025(1)	0.0013(1)	6.37742(7)	5.0(4)	0.0040(1)	0.0054(1)
30	6.43026(9)	8.5(4)	0.0030(1)	0.0015(1)	6.37788(11)	9.8(6)	0.0043(1)	0.0056(1)
40	6.43091(12)	11.7(4)	0.0035(1)	0.0017(1)	6.37861(15)	13.2(6)	0.0048(1)	0.0059(1)
50	6.43173(15)	13.9(4)	0.0041(1)	0.0020(1)	6.37952(18)	15.4(6)	0.0052(1)	0.0062(1)
60	6.43266(17)	15.3(4)	0.0047(1)	0.0023(1)	6.38054(22)	16.9(5)	0.0057(1)	0.0066(1)
70	6.43367(20)	16.2(4)	0.0053(1)	0.0027(1)	6.38165(25)	17.8(5)	0.0062(1)	0.0070(2)
80	6.43474(22)	16.9(4)	0.0059(1)	0.0030(1)	6.38281(28)	18.5(5)	0.0068(1)	0.0074(2)
90	6.43584(24)	17.4(4)	0.0066(1)	0.0033(1)	6.38400(31)	19.0(5)	0.0072(2)	0.0079(2)
100	6.43697(27)	17.8(4)	0.0072(1)	0.0037(1)	6.38523(35)	19.3(5)	0.0078(2)	0.0083(2)
110	6.43812(29)	18.0(4)	0.0079(1)	0.0041(1)	6.38647(38)	19.6(5)	0.0084(2)	0.0087(2)
120	6.43929(31)	18.3(4)	0.0085(1)	0.0045(1)	6.38773(41)	19.8(5)	0.0089(2)	0.0091(2)
130	6.44047(34)	18.5(4)	0.0092(1)	0.0049(1)	6.38900(44)	20.0(5)	0.0095(2)	0.0096(2)
140	6.44167(36)	18.6(4)	0.0098(1)	0.0052(2)	6.39028(47)	20.1(5)	0.0100(2)	0.0100(2)
150	6.44287(39)	18.7(4)	0.0105(1)	0.0056(2)	6.39157(51)	20.2(5)	0.0106(2)	0.0105(2)
160	6.44408(41)	18.8(4)	0.0111(1)	0.0060(2)	6.39287(54)	20.3(6)	0.0111(2)	0.0109(2)
170	6.44529(44)	18.9(4)	0.0118(1)	0.0064(2)	6.39417(58)	20.4(6)	0.0117(2)	0.0113(2)
180	6.44652(47)	19.0(4)	0.0125(1)	0.0068(2)	6.39547(62)	20.4(6)	0.0122(3)	0.0118(2)
190	6.44774(50)	19.1(4)	0.0131(1)	0.0072(2)	6.39678(66)	20.5(6)	0.0128(3)	0.0122(2)
200	6.44898(53)	19.1(5)	0.0138(1)	0.0076(2)	6.39809(70)	20.5(6)	0.0133(3)	0.0127(3)
210	6.45021(56)	19.2(5)	0.0144(1)	0.0080(2)	6.39941(74)	20.6(6)	0.0139(3)	0.0131(3)
220	6.45145(59)	19.3(5)	0.0151(1)	0.0084(2)	6.40073(78)	20.6(7)	0.0144(3)	0.0136(3)
230	6.45270(62)	19.3(5)	0.0158(1)	0.0088(2)	6.40205(82)	20.6(7)	0.0150(3)	0.0140(3)
240	6.45394(65)	19.3(5)	0.0164(1)	0.0091(2)	6.40337(87)	20.7(7)	0.0156(3)	0.0145(3)
250	6.45519(69)	19.4(5)	0.0171(1)	0.0095(2)	6.40469(91)	20.7(7)	0.0161(3)	0.0149(3)
260	6.45644(72)	19.4(5)	0.0177(1)	0.0099(2)	6.40602(96)	20.7(7)	0.0167(3)	0.0154(3)
270	6.45770(76)	19.5(6)	0.0184(1)	0.0103(2)	6.40734(100)	20.7(8)	0.0172(4)	0.0158(3)
280	6.45896(79)	19.5(6)	0.0191(1)	0.0107(2)	6.40867(105)	20.7(8)	0.0178(4)	0.0163(3)
290	6.46022(83)	19.5(6)	0.0197(2)	0.0111(2)	6.41000(110)	20.7(8)	0.0183(4)	0.0167(3)
300	6.46148(87)	19.6(6)	0.0204(2)	0.0115(2)	6.41133(116)	20.7(8)	0.0189(4)	0.0172(3)

Table A3. Present high-pressure $V(p)$ data from single-crystal structure refinement for PbTe and $Pb_{0.884}Cd_{0.116}Te$.

PbTe		$Pb_{0.884}Cd_{0.116}Te$	
p [GPa]	V [Å^3]	p [GPa]	V [Å^3]
0.33(2)	271.1(15)	0.30(2)	265.40(50)
0.80(2)	267.9(12)	0.80(2)	261.10(80)
1.30(2)	266.8(15)	1.40(2)	256.50(16)
1.80(2)	264.8(16)	2.50(2)	251.78(10)
2.40(2)	259.5(16)	3.00(2)	247.41(12)
3.00(2)	257.2(15)	3.50(2)	245.37(11)
3.70(2)	254.7(12)	4.00(2)	241.37(14)
4.50(2)	251.8(15)	4.50(2)	241.23(14)

Table A4. Present unit cell volume and bulk modulus as a function of pressure, modeled using Equation (6) for PbTe and $Pb_{0.884}Cd_{0.116}Te$.

p [GPa]	PbTe		$Pb_{0.884}Cd_{0.116}Te$	
	V [Å^3]	K [GPa]	V [Å^3]	K [GPa]
0	273.25	45.6(2.5)	273.25	33.5(2.8)
0.5	270.33	47.6(2.5)	270.33	35.5(2.8)
1.0	267.57	49.6(2.5)	267.57	37.5(2.8)
1.5	264.93	51.5(2.5)	264.93	39.4(2.8)
2.0	262.42	53.5(2.5)	262.42	41.3(2.8)
2.5	260.02	55.4(2.5)	260.02	43.2(2.8)
3.0	257.72	57.3(2.5)	257.72	45.1(2.8)
3.5	255.52	59.1(2.5)	255.52	47.0(2.8)
4.0	253.41	61.1(2.5)	253.41	48.8(2.8)
4.5	251.37	62.9(2.5)	251.37	50.7(2.8)
5.0	249.41	64.8(2.5)	249.41	52.5(2.8)

Appendix C

Table A5. Fitted values of the parameters of Equations (1), (4) and (6), obtained for PbTe and $Pb_{0.884}Cd_{0.116}Te$ crystals.

Function	Fitted Equation	Parameters for PbTe	Parameters for $Pb_{0.884}Cd_{0.116}Te$
$V(T)$	Equation (1)	$V_{(T=0)} = 265.813(6)$ Å^3, $Q = 2.86(3) \times 10^{-18}$, $b = 1.3(6)$, $\theta_D = 135.2(3.9)$ K	$V_{(T=0)} = 259.358(7)$ Å^3, $Q = 2.63(4) \times 10^{-18}$, $b = 0.4(7)$, $\theta_D = 130.1(4.4)$ K
$<u^2>(T)$, for cationic site	Equation (4)	$\theta_D = 102.8(3)$ K, $<u^2>_{stat} = 0.00038(4)$ Å^2	$\theta_D = 114.5(5)$ K, $<u^2>_{stat} = 0.00203(6)$ Å^2
$<u^2>(T)$, for anionic site	Equation (4)	$\theta_D = 169.2(1.1)$ K, $<u^2>_{stat} = -0.00054(7)$ Å^2	$\theta_D = 158.1(1.3)$ K, $<u^2>_{stat} = 0.0034(1)$ Å^2
$V(p)$	Equation (6)	$V_0 = 273.3(7)$ Å^3 $K_0 = 45.6(2.5)$ GPa $K' = 4$ (fixed)	$V_0 = 267.7(1.5)$ Å^3 $K_0 = 33.5(2.8)$ GPa $K' = 4$ (fixed)

Appendix D

Table A6. Reported theoretical bulk modulus and its derivative for PbTe and $Pb_{1-x}Cd_xTe$, $x = 0.031$, 0.116. In a number of papers (e.g., refs. [10,129]), multiple numerical approaches have been applied, so only selected representative values could be cited here. As a rule (with some exceptions), the experimental values refer to room temperature, whereas in the calculated ones (to 0 K), the temperature is marked if explicitly stated in the given reference.

Compound	K_0 [GPa]	K'	Method	Ref.	Year
PbTe	45	n.a.	LDA		
	48	n.a.	LDA	(a)	1983
	51.7	4.52	LAPW LDA	(b)	1997
			FP-LAPW LDA		
	51.44 (0 K)	5.50	FP-LAPW		
	40.30 (0 K)	4.27	GGA	(c)	2000
	49.82 (0 K)	5.76	FP-LAPW		
	39.5 (0 K)	3.92	LDA+SO		
			FP-LAPW GGA+SO		

Table A6. *Cont.*

Compound	K_0 [GPa]	K'	Method	Ref.	Year
	41.4	3.352	GGA		2002
	51.4	4.080	LDA	(d)	
	37.5 (0 K)	n.a.	FP-APW PBE	(e)	2007
	40.4 (0 K)	n.a.	LDA/GGA		
	50.3 (0 K)	n.a.	"	(f)	2009
	30.7	n.a.	"		
	39.05	4.32	FP-LAPW	(g)	2011
	46.0	4.27	LDA		2012
	46.1	4.53	LDA+SO	(h)	
	41.0	n.a.	MD (GULP)	(i)	2012
	39.1	n.a.	PAW PBE	(j)	2013
	47 (0 K)	n.a.	FP-LAPW	(k)	2014
	38.54 (300 K)	n.a.	PBEsol		2014
	~45.5 (0 K)	n.a.	"	(l)	
	34.04 (0 K)	n.a.	LDA, GGA	(m)	2014
	46.61 (0 K)	n.a.	PBEsol	(n)	2015
	44.1(&)	n.a.	HSEsolSOC	(o)	2016
	41.1	n.a.	FP-LAPW	(p)	2017
	36.19 (100 K)	n.a.	LDY		2019
	37.52 (300 K)	n.a.	"	(q)	
	48.242 (0 K)	5.576 (0 K)	LDA	(r)	2020
	43.6	4.6	GGA-PBE	(s)	2020
$Pb_{0.969}Cd_{0.031}Te$	46.42	n.a.	GGA	(n)	2015
$Pb_{0.884}Cd_{0.116}Te$	45.90	n.a.	GGA	($)	2021

References: (a) [118,119], (b) [115], (c) [116], (d) [114], (e) [120], (f) [45], (g) [122], (h) [64], (i) [117], (j) [123], (k) [124], (l) [46], (m) [47], (n) [9], (o) [10], (p) [125], (q) [65], (r) [67], (s) [86]. ($)—extrapolation of ref. [9] data reported for $Pb_{0.969}Cd_{0.031}Te$. (&)—the authors presented results also for seven other approaches. Abbreviations are explained at the end of this study.

References

1. Faber, K.T.; Malloy, K.J. (Eds.) *The Mechanical Properties of Semiconductors*; Academic Press: Boston, MA, USA, 1992; Volume 37.
2. Zhang, X.; Zhao, L.D. Thermoelectric materials: Energy conversion between heat and electricity. *J. Mater.* **2015**, *1*, 92–105. [CrossRef]
3. Heremans, J.P.; Jovovic, V.; Toberer, E.S.; Saramat, A.; Kurosaki, K.; Charoenphakdee, A.; Yamanaka, S.; Snyder, G.J. Enhancement of thermoelectric efficiency in PbTe by distortion of the electronic density of states. *Science* **2008**, *321*, 554–557. [CrossRef]
4. Jaworski, C.M.; Wiendlocha, B.; Jovovic, V.; Heremans, J.P. Combining alloy scattering of phonons and resonant electronic levels to reach a high thermoelectric figure of merit in PbTeSe and PbTeS alloys. *Energy Environ. Sci.* **2011**, *4*, 4155–4162. [CrossRef]
5. Wang, Y.; Chen, X.; Cui, T.; Niu, Y.; Wang, Y.; Wang, M.; Ma, Y.; Zou, G. Enhanced thermoelectric performance of PbTe within the orthorhombic *Pnma* phase. *Phys. Rev. B* **2007**, *76*, 155127. [CrossRef]
6. Dynowska, E.; Szuszkiewicz, W.; Szczepanska, A.; Romanowski, P.; Dobrowolski, W.; Lathe, C.; Slynko, E.V. High-Pressure Studies of Bulk $Pb_{1-x}Ba_xTe$ Mixed Crystals. In *Jahresbericht/Hamburger Synchrotronstrahlungslabor HASYLAB am Deutschen Elektronen-Synchrotron DESY= Annual Report*; HASYLAB: Hamburg, Germany, 2005; pp. 361–362.
7. Bukała, M.; Sankowski, P.; Buczko, R.; Kacman, P. Structural and electronic properties of $Pb_{1-x}Cd_xTe$ and $Pb_{1-x}Mn_xTe$ ternary alloys. *Phys. Rev. B* **2012**, *86*, 085205. [CrossRef]
8. Joseph, E.; Amouyal, Y. Enhancing thermoelectric performance of PbTe-based compounds by substituting elements: A first principles study. *J. Electron. Mater.* **2015**, *44*, 1460–1468. [CrossRef]
9. Joseph, E.; Amouyal, Y. Towards a predictive route for selection of doping elements for the thermoelectric compound PbTe from first-principles. *J. Appl. Phys.* **2015**, *117*, 175102. [CrossRef]
10. Śpiewak, P.; Kurzydłowski, K.J. Electronic Structure and Transport Properties of Doped Lead Chalcogenides from First Principles. *MRS Adv.* **2016**, *1*, 4003–4010. [CrossRef]
11. Sealy, B.J.; Crocker, A.J. A comparison of phase equilibria in some II-IV-VI compounds based on PbTe. *J. Mater. Sci.* **1973**, *8*, 1731–1736. [CrossRef]
12. Crocker, A.J.; Rogers, L.M. Valence band structure of PbTe. *J. Phys. Colloq.* **1968**, *29*, C4-129–C4-132. [CrossRef]
13. Pei, Y.; LaLonde, A.D.; Heinz, N.A.; Snyder, G.J. High thermoelectric figure of merit in PbTe alloys demonstrated in PbTe–CdTe. *Adv. Energy Mater.* **2012**, *2*, 670–675. [CrossRef]
14. Ahn, K.; Biswas, K.; He, J.; Chung, I.; Dravid, V.; Kanatzidis, M.G. Enhanced thermoelectric properties of p-type nanostructured PbTe–MTe (M = Cd, Hg) materials. *Energy Environ. Sci.* **2013**, *6*, 1529–1537. [CrossRef]

15. Sarkar, S.; Zhang, X.; Hao, S.; Hua, X.; Bailey, T.P.; Uher, C.; Wolverton, C.; Dravid, V.P.; Kanatzidis, M.G. Dual alloying strategy to achieve a high thermoelectric figure of merit and lattice hardening in p-type nanostructured PbTe. *ACS Energy Lett.* **2018**, *3*, 2593–2601. [CrossRef]

16. Rosenberg, A.J.; Woolley, J.C.; Nikolic, P.; Grierson, R. Solid solutions of CdTe and InTe in PbTe and SnTe. I. Crystal chemistry. *Trans. Metall. Soc. AIME* **1964**, *230*, 342–349.

17. Kulvitit, Y.; Rolland, S.; Granger, R.; Pelletier, C.M. Relation entre composition paramètre de maille et bande interdite des composés $Pb_{1-x}Cd_xTe$. *Rev. Phys. Appliquée* **1980**, *15*, 1501–1504. [CrossRef]

18. Marx, R.; Range, K.J. Homogene, abschreckbare Hochdruckphasen mit NaCl-Struktur in den systemen CdTe-SnTe und CdTe-PbTe. *J. Less Common Met.* **1989**, *155*, 49–59. [CrossRef]

19. Szot, M.; Szczerbakow, A.; Dybko, K.; Kowalczyk, L.; Smajek, E.; Domukhovski, V.; Łusakowska, E.; Dziawa, P.; Mycielski, A.; Story, T.; et al. Experimental and Theoretical Analysis of PbTe–CdTe Solid Solution Grown by Physical Vapour Transport Method. *Acta Phys. Polon. A* **2009**, *116*, 959–961. [CrossRef]

20. Wald, F.; Rosenberg, A.J. Solid solutions of CdSe and InSe in PbSe. *J. Phys. Chem. Solids* **1965**, *26*, 1087–1091. [CrossRef]

21. Crocker, A.J. Phase equilibria in PbTe/CdTe alloys. *J. Mater. Sci.* **1968**, *3*, 534–539. [CrossRef]

22. Sealy, B.J.; Crocker, A.J. Some physical properties of the systems $Pb_{1-x}Mg_xSe$ and $Pb_{1-x}Cd_xSe$. *J. Mater. Sci.* **1973**, *8*, 1247–1252. [CrossRef]

23. Qian, X.; Wu, H.; Wang, D.; Zhang, Y.; Wang, J.; Wang, G.; Wang, G.; Zheng, L.; Pennycook, S.J.; Zhao, L.D. Synergistically optimizing interdependent thermoelectric parameters of n-type PbSe through alloying CdSe. *Energy Environ. Sci.* **2019**, *12*, 1969–1978. [CrossRef]

24. Bethke, P.M.; Barton, P.B., Jr. Sub-solidus relations in the system PbS-CdS. *Am. Mineral. J. Earth Planet. Mater.* **1971**, *56*, 2034–2039.

25. Tan, G.L.; Liu, L.; Wu, W. Mid-IR band gap engineering of $Cd_xPb_{1-x}S$ nanocrystals by mechanochemical reaction. *AIP Adv.* **2014**, *4*, 067107. [CrossRef]

26. ICSD Database (FIZ, Karlsruhe), Records 620316, 620315, 620322. Available online: https://icsd.products.fiz-karlsruhe.de/ (accessed on 21 August 2021).

27. Bziz, I.; Atmani, E.H.; Fazouan, N.; Aazi, M.; Es-Smairi, A. Ab-initio study of structural, electronic and optical properties of CdS. In Proceedings of the 2019 7th International Renewable and Sustainable Energy Conference (IRSEC), Agadir, Morocco, 27–30 November 2019; 2019; pp. 1–6. [CrossRef]

28. Kiyosawa, T.; Takahashi, S.; Koguchi, N. Solid solubility range and lattice parameter of semiconducting $Pb_{1-x}Cd_xS_{1-y}Se_y$ for mid-infra-red lasers. *Phys. Status Solidi A* **1989**, *111*, K21–K25. [CrossRef]

29. Knight, K.S. Does Altaite Exhibit Emphanitic Behavior? a High Resolution Neutron Powder Diffraction Investigation of the Crystallographic and Thermoelastic Properties of PbTe Between 10 and 500 K. Does Altaite Exhibit Emphanitic Behaviour? *Can. Mineral.* **2016**, *54*, 1493–1503. [CrossRef]

30. Liu, Y.; Ibáñez, M. Tidying up the mess. *Science* **2021**, *371*, 678–679. [CrossRef] [PubMed]

31. Kastbjerg, S.; Bindzus, N.; Søndergaard, M.; Johnsen, S.; Lock, N.; Christensen, M.; Takata, M.; Spackman Brummerstedt, M.A.; Iversen, B. Direct evidence of cation disorder in thermoelectric lead chalcogenides PbTe and PbS. *Adv. Funct. Mater.* **2013**, *23*, 5477–5483. [CrossRef]

32. Wang, H.; LaLonde, A.D.; Pei, Y.; Snyder, G.J. The criteria for beneficial disorder in thermoelectric solid solutions. *Adv. Funct. Mater.* **2013**, *23*, 1586–1596. [CrossRef]

33. Knight, K.S. A high-resolution neutron powder diffraction investigation of galena (PbS) between 10 K and 350 K: No evidence for anomalies in the lattice parameters or atomic displacement parameters in galena or altaite (PbTe) at temperatures corresponding to the saturation of cation disorder. *J. Phys. Condens. Matter* **2014**, *26*, 385403. [CrossRef]

34. Christensen, S.; Bindzus, N.; Sist, M.; Takata, M.; Iversen, B.B. Structural disorder, anisotropic micro-strain and cation vacancies in thermo-electric lead chalcogenides. *Phys. Chem. Chem. Phys.* **2016**, *18*, 15874–15883. [CrossRef]

35. Szot, M.; Pfeffer, P.; Dybko, K.; Szczerbakow, A.; Kowalczyk, L.; Dziawa, P.; Minikayev, R.; Zajarniuk, T.; Piotrowski, P.; Gutowska, M.U.; et al. Two-valence band electron and heat transport in monocrystalline PbTe-CdTe solid solutions with Cd content up to 10 atomic percent. *Phys. Rev. Mater.* **2020**, *4*, 044605. [CrossRef]

36. Tsang, Y.W.; Cohen, M.L. Calculation of the temperature dependence of the energy gaps in PbTe and SnTe. *Phys. Rev. B* **1971**, *3*, 1254. [CrossRef]

37. Noda, Y.; Masumoto, K.; Ohba, S.; Saito, Y.; Toriumi, K.; Iwata, Y.; Shibuya, I. Temperature dependence of atomic thermal parameters of lead chalcogenides, PbS, PbSe and PbTe. *Acta Crystallogr. Sect. C Cryst. Struct. Commun.* **1987**, *43*, 1443–1445. [CrossRef]

38. Božin, E.S.; Malliakas, C.D.; Souvatzis, P.; Proffen, T.; Spaldin, N.A.; Kanatzidis, M.G.; Billinge, S.J. Entropically stabilized local dipole formation in lead chalcogenides. *Science* **2010**, *330*, 1660–1663. [CrossRef] [PubMed]

39. Sangiorgio, B.; Bozin, E.S.; Malliakas, C.D.; Fechner, M.; Simonov, A.; Kanatzidis, M.G.; Billinge, S.J.L.; Spaldin, N.A.; Weber, T. Correlated local dipoles in PbTe. *Phys. Rev. Mater.* **2018**, *2*, 085402. [CrossRef]

40. Holm, K.A.; Roth, N.; Zeuthen, C.M.; Iversen, B.B. Anharmonicity and correlated dynamics of PbTe and PbS studied by single crystal x-ray scattering. *Phys. Rev. B* **2021**, *103*, 224302. [CrossRef]

41. Male, J.P.; Hanus, R.; Snyder, G.J.; Hermann, R.P. Thermal Evolution of Internal Strain in Doped PbTe. *Chem. Mater.* **2021**, *33*, 4765–4772. [CrossRef]

42. Minikayev, E.D.; Dziawa, P.; Kamińska, E.; Szczerbakow, A.; Trots, D.; Szuszkiewicz, W. High-temperature studies of $Pb_{1-x}Cd_xTe$ solid solution: Structure stability and CdTe solubility limit, Synchrotron Radiation. *Nat. Sci.* **2009**, *8*, 8.

43. Minikayev, R.; Dynowska, E.; Kaminska, E.; Szczerbakow, A.; Trots, D.; Story, T.; Szuszkiewicz, W. Evolution of $Pb_{1-x}Cd_xTe$ solid solution structure at high temperatures. *Acta Phys. Pol. A* **2011**, *119*, 699–701. [CrossRef]

44. Minikayev, R.; Dynowska, E.; Story, T.; Szczerbakow, A.; Bell, A.; Trots, D.; Szuszkiewicz, W. Low-temperature expansion of metasY $Pb_{(1-x)}Cd_{(x)}Te$ solid solution. In Proceedings of the IX Krajowe Sympozjum Użytkowników Promieniowania Synchrotronowego (KSUPS-9), Warszawa, Polska, 26–27 September 2011.

45. Zhang, Y.; Ke, X.; Chen, C.; Yang, J.; Kent, P.R.C. Thermodynamic properties of PbTe, PbSe, and PbS: First-principles study. *Phys. Rev. B* **2009**, *80*, 024304. [CrossRef]

46. Skelton, J.M.; Parker, S.C.; Togo, A.; Tanaka, I.; Walsh, A. Thermal physics of the lead chalcogenides PbS, PbSe, and PbTe from first principles. *Phys. Rev. B* **2014**, *89*, 205203. [CrossRef]

47. Boukhris, N.; Meradji, H.; Korba, S.A.; Drablia, S.; Ghemid, S.; Hassan, F.E.H. First principles calculations of structural, electronic and thermal properties of lead chalcogenides PbS, PbSe and PbTe compounds. *Bull. Mater. Sci.* **2014**, *37*, 1159–1166. [CrossRef]

48. Xia, Y. Revisiting lattice thermal transport in PbTe: The crucial role of quartic anharmonicity. *Appl. Phys. Lett.* **2018**, *113*, 073901. [CrossRef]

49. Lu, Y.; Sun, T.; Zhang, D.B. Lattice anharmonicity, phonon dispersion, and thermal conductivity of PbTe studied by the phonon quasiparticle approach. *Phys. Rev. B* **2018**, *97*, 174304. [CrossRef]

50. Querales-Flores, J.D.; Cao, J.; Fahy, S.; Savić, I. Temperature effects on the electronic band structure of PbTe from first principles. *Phys. Rev. Mater.* **2019**, *3*, 055405. [CrossRef]

51. Novikova, S.I.; Abrikosov, N.K. Investigation of the thermal expansion of lead chalcogenides (Исследование теплового расширенияхалькогенидовсвинца). *Fiz. Tverd. Tela Sov. Phys. Solid State* **1963**, *5*, 1913–1916.

52. Houston, B.; Strakna, R.E.; Belson, H.S. Elastic constants, thermal expansion, and Debye temperature of lead telluride. *J. Appl. Phys.* **1968**, *39*, 3913–3916. [CrossRef]

53. Cochran, W.; Cowley, R.A.; Dolling, G.; Elcombe, M.M. The crystal dynamics of lead telluride. *Proc. R. Soc. Lond. Ser. A Math. Phys. Sci.* **1966**, *293*, 433–451. [CrossRef]

54. Murphy, A.R. Thermoelectric Properties of PbTe-Based Materials Driven Near the Ferroelectric Phase Transition from First Principles. Ph.D. Thesis, University College Cork, Cork, Irland, 2017; p. 75.

55. Ghezzi, C. Mean Square Vibrational Amplitudes in PbTe. *Phys. Status Solidi* **1973**, *58*, 737–744. [CrossRef]

56. Bublik, V.T. The mean square atomic displacements and enthalpies of vacancy formation in some semiconductors. *Phys. Status Solidi* **1978**, *45*, 543–548. [CrossRef]

57. Li, C.W.; Ma, J.; Cao, H.B.; May, A.F.; Abernathy, D.L.; Ehlers, G.; Hoffmann, C.; Wang, X.; Hong, T.; Huq, A.; et al. Anharmonicity and atomic distribution of SnTe and PbTe thermoelectrics. *Phys. Rev. B* **2014**, *90*, 214303. [CrossRef]

58. Keffer, C.; Hayes, T.M.; Bienenstock, A. PbTe Debye-Waller factors and band-gap temperature dependence. *Phys. Rev. Lett.* **1968**, *21*, 1676–1678. [CrossRef]

59. Keffer, C.; Hayes, T.M.; Bienenstock, A. Debye-Waller factors and the PbTe band-gap temperature dependence. *Phys. Rev. B* **1970**, *2*, 1966–1976. [CrossRef]

60. Chen, Y.; Ai, X.; Marianetti, C.A. First-principles approach to nonlinear lattice dynamics: Anomalous spectra in PbTe. *Phys. Rev. Lett.* **2014**, *113*, 105501. [CrossRef]

61. Parkinson, D.; Quarrington, J. The Molar Heats of Lead Sulphide, Selenide and Telluride in the Temperature Range 20°K to 260°K. *Proc. Phys. Soc.* **1954**, *A67*, 569–579. [CrossRef]

62. El-Sharkawy, A.A.; Abou El-Azm, A.M.; Kenawy, M.I.; Hillal, A.S.; Abu-Basha, H.M. Thermophysical properties of polycrystalline PbS, PbSe, and PbTe in the temperature range 300–700 K. *Int. J. Thermophys.* **1983**, *4*, 261–269. [CrossRef]

63. Jacobsen, M.K.; Liu, W.; Li, B. Sound velocities of PbTe to 14 GPa: Evidence for coupling between acoustic and optic phonons. *J. Phys. Condens. Matter* **2013**, *25*, 365402. [CrossRef]

64. Yang, Y.L. Elastic moduli and band gap of PbTe under pressure: Ab initio study. *Mater. Sci. Technol.* **2012**, *28*, 1308–1313. [CrossRef]

65. Tripathi, S.; Agarwal, R.; Singh, D. Nonlinear elastic, ultrasonic and thermophysical properties of lead telluride. *Int. J. Thermophys.* **2019**, *40*, 1–18. [CrossRef]

66. Naidych, B. Crystal Structure and Thermodynamic Parameters of Thin Film II-VI and IV-VI Alloys (Кристалічна структура та термодинамічні параметри тонкоплівкових конденсатівсистемII-VI, IV-VI). Ph.D. Thesis, Vasyl Stefanyk Precarpathian National University, Ivano-Frankivsk, Ukraine, 2019; p. 116. (In Ukrainian).

67. Öztürk, H.; Arslan, G.G.; Kürkçü, C.; Yamçıçıer, Ç. Structural phase transformation, intermediate states and electronic properties of PbTe under high pressure. *J. Electron. Mater.* **2020**, *49*, 3089–3095. [CrossRef]

68. Naidych, B.; Parashchuk, T.; Yaremiy, I.; Moyseyenko, M.; Kostyuk, O.; Voznyak, O.; Dashevsky, Z.; Nykyruy, L. Structural and thermodynamic properties of Pb-Cd-Te thin films: Experimental study and DFT analysis. *J. Electron. Mater.* **2021**, *50*, 580–591. [CrossRef]

69. Zhang, M.; Tang, G.; Li, Y. Hydrostatic Pressure Tuning of Thermal Conductivity for PbTe and PbSe Considering Pressure-Induced Phase Transitions. *ACS Omega* **2021**, *6*, 3980–3990. [CrossRef] [PubMed]

70. Lin, J.C.; Hsieh, K.C.; Sharma, R.C.; Chang, Y.A. The Pb-Te (lead-tellurium) system. *Bull. Alloy Phase Diagr.* **1989**, *10*, 340–347. [CrossRef]

71. Ngai, T.L.; Marshall, D.; Sharma, R.C.; Chang, Y.A. Thermodynamic properties and phase equilibria of the lead-tellurium binary system. *Mon. Für Chem. Chem. Mon.* **1987**, *118*, 277–300. [CrossRef]

72. Liu, Y.; Zhang, L.; Yu, D. Thermodynamic Descriptions for the Cd-Te, Pb-Te, Cd-Pb and Cd-Pb-Te Systems. *J. Electron. Mater.* **2009**, *38*, 2033–2045. [CrossRef]

73. Fujimoto, M.; Sato, Y. *P-T-x* phase diagram of the lead telluride system. *Jpn. J. Appl. Phys.* **1966**, *5*, 128–133. [CrossRef]

74. Sealy, B.J.; Crocker, A.J. The *P-T-x* phase diagram of PbTe and PbSe. *J. Mater. Sci.* **1973**, *8*, 1737–1743. [CrossRef]

75. Gierlotka, W.; Łapsa, J.; Jendrzejczyk-Handzlik, D. Thermodynamic description of the Pb–Te system using ionic liquid model. *J. Alloy. Compd.* **2009**, *479*, 152–156. [CrossRef]

76. Peters, M.C.; Doak, J.W.; Zhang, W.W.; Saal, J.E.; Olson, G.B.; Voorhees, P.W. Thermodynamic modeling of the PbX (X = S, Te) phase diagram using a five sub-lattice and two sub-lattice model. *Calphad* **2017**, *58*, 17–24. [CrossRef]

77. Shalvoy, R.B.; Fisher, G.B.; Stiles, P.J. Bond ionicity and structural stability of some average-valence-five materials studied by x-ray photoemission. *Phys. Rev. B* **1977**, *15*, 1680–1697. [CrossRef]

78. Chattopadhyay, T.; Werner, A.; Von Schnering, H.G. Pressure-Induced Phase Transition in IV–VI Compounds. *Mat. Res. Soc. Symp. Proc.* **1984**, *22*, 93–96. [CrossRef]

79. Chattopadhyay, T.; Werner, A.; Von Schnering, H.G.; Pannetier, J. Temperature and pressure induced phase transition in IV-VI compounds. *Rev. Phys. Appliquée* **1984**, *19*, 807–813. [CrossRef]

80. Chattopadhyay, L.; von Schnering, H.G.; Grosshans, W.A.; Holzapfel, W.B. High pressure X-ray diffraction study on the structural phase transitions in PbS, PbSe and PbTe with synchrotron radiation. *Phys. B+C* **1986**, *139–140*, 356–360. [CrossRef]

81. Ves, S.; Pusep, Y.A.; Syassen, K.; Cardona, M. Raman study of high pressure phase transitions in PbTe. *Solid State Commun.* **1989**, *70*, 257–260. [CrossRef]

82. Rousse, G.; Klotz, S.; Saitta, A.M.; Rodriguez-Carvajal, J.; McMahon, M.I.; Couzinet, B.; Mezouar, M. Structure of the intermediate phase of PbTe at high pressure. *Phys. Rev. B* **2005**, *71*, 224116. [CrossRef]

83. Li, Y.; Lin, C.; Li, H.; Liu, X.L.J. Phase transitions in PbTe under quasi-hydrostatic pressure up to 50 GPa. *High Press. Res.* **2013**, *33*, 713–719. [CrossRef]

84. Singh, R.K.; Gupta, D.C. High pressure phase transitions and elastic properties of IV-VI compound semiconductors. *Phase Transit. Multinatl. J.* **1995**, *53*, 39–51. [CrossRef]

85. Ahuja, R. High pressure structural phase transitions in IV–VI semiconductors. *Phys. Status Solidi* **2003**, *235*, 341–347. [CrossRef]

86. Aguado-Puente, P.; Fahy, S.; Grüning, M. GW study of pressure-induced topological insulator transition in group-IV tellurides. *Phys. Rev. Res.* **2020**, *2*, 043105. [CrossRef]

87. Chusnutdinow, S.; Schreyeck, S.; Kret, S.; Kazakov, A.; Karczewski, G. Room temperature infrared detectors made of PbTe/CdTe multilayer composite. *Appl. Phys. Lett.* **2020**, *117*, 072102. [CrossRef]

88. Stöber, D.; Hildmann, B.O.; Böttner, H.; Schelb, S.; Bachem, K.H.; Binnewies, M. Chemical transport reactions during crystal growth of PbTe and PbSe via vapour phase influenced by AgI. *J. Cryst. Growth* **1992**, *121*, 656–664. [CrossRef]

89. Szczerbakow, A. Crystal selection in the Self-selected Vapour Growth (SSVG) of PbSe in a vertical system. *Cryst. Res. Technol.* **1993**, *28*, K77–K80. [CrossRef]

90. Szczerbakow, A.; Durose, K. Self-selecting vapour growth of bulk crystals–Principles and applicability. *Prog. Cryst. Growth Character. Mater.* **2005**, *51*, 81–108. [CrossRef]

91. Xu, W.; Liu, Y.; Marcelli, A.; Shang, P.P.; Liu, W.S. The complexity of thermoelectric materials: Why we need powerful and brilliant synchrotron radiation sources? *Mater. Today Phys.* **2018**, *6*, 68–82. [CrossRef]

92. Knapp, M.; Baehtz, C.; Ehrenberg, H.; Fuess, H. The synchrotron powder diffractometer at beamline B2 at HASYLAB/DESY: Status and capabilities. *J. Synchrotron Radiat.* **2004**, *11*, 328–334. [CrossRef]

93. Knapp, M.; Joco, V.; Baehtz, C.; Brecht, H.H.; Berghaeuser, A.; Ehrenberg, H.; von Seggern, H.; Fuess, H. Position-sensitive detector system OBI for high resolution X-ray powder diffraction using on-site readable image plates. *Nucl. Instrum. Methods Phys. Res. Sect. A Accel. Spectrometers Detect. Assoc. Equip.* **2004**, *521*, 565–570. [CrossRef]

94. Paszkowicz, W.; Knapp, M.; Bähtz, C.; Minikayev, R.; Piszora, P.; Jiang, J.Z.; Bacewicz, R.J. Synchrotron X-ray wavelength calibration using a diamond internal standard: Application to low-temperature thermal-expansion studies. *Alloy. Comp.* **2004**, *382*, 107–111. [CrossRef]

95. Loopstra, B.O.; Rietveld, H.M. The structure of some alkaline–earth metal uranates. *Acta Crystallogr. Sect. B Struct. Crystallogr. Cryst. Chem.* **1969**, *25*, 787–791. [CrossRef]

96. Rietveld, H. A profile refinement method for nuclear and magnetic structures. *J. Appl. Crystallogr.* **1969**, *2*, 65–71. [CrossRef]

97. Rodríguez-Carvajal, J. Recent developments of the program FULLPROF. Commission on Powder Diffraction (IUCr) Newsletter; 2001; 26; 12–19. Available online: http://journals.iucr.org/iucr-top/comm/cpd/Newsletters/ (accessed on 21 August 2021).

98. Merrill, L.; Bassett, W.A. Miniature diamond anvil pressure cell for single crystal X-ray diffraction studies. *Rev. Sci. Instrum.* **1974**, *45*, 290–294. [CrossRef]

99. Mao, H.K.; Xu, J.; Bell, P.M. Calibration of the ruby pressure gauge to 800 kbar under quasi-hydrostatic conditions. *J. Geophys. Res.* **1986**, *91*, 4673–4676. [CrossRef]

100. Budzianowski, A.; Katrusiak, A. *High-Pressure Crystallography*; Katrusiak, A., McMillan, P.F., Eds.; Kluwer: Dordrecht, The Netherlands, 2004; pp. 101–112.

101. Xcalibur CCD System. *Crys Alis Pro Software System, Version 1.171.33*; Oxford Diffraction Ltd.: Wrocław, Poland, 2009.

102. Sheldrick, G.M. A short history of SHELX. *Acta Crystallogr. Sect. A* **2008**, *64*, 112–122. [CrossRef]

103. Dolomanov, O.V.; Bourhis, L.J.; Gildea, R.J.; Howard, J.A.K.; Puschmann, H. OLEX2: A Complete Structure Solution, Refinement and Analysis Program. *J. Appl. Crystallogr.* **2009**, *42*, 339–341. [CrossRef]

104. Angel, R.J.; Alvaro, M.; Gonzalez-Platas, J. EosFit7c and a Fortran module (library) for equation of state calculations. *Z. Für Krist. Cryst. Mater.* **2014**, *229*, 405–419. [CrossRef]

105. Gonzalez-Platas, J.; Alvaro, M.; Nestola, F.; Angel, R. EosFit7-GUI. *J. Appl. Crystallogr.* **2016**, *49*, 1377–1382. [CrossRef]

106. Wallace, D.C. *Thermodynamics of Crystals*; Dover: New York, NY, USA, 1998.

107. Vočadlo, L.; Knight, K.S.; Price, G.D.; Wood, I.G. Thermal expansion and crystal structure of FeSi between 4 and 1173 K determined by time-of-light neutron powder diffraction. *Phys. Chem. Miner.* **2002**, *29*, 132–139. [CrossRef]

108. Querales-Flores, J.; (Tyndall National Institute, University College Cork, Cork, Ireland). Personal Communication, 2021.

109. Willis, B.T.M.; Pryor, A.W. *Thermal Vibrations in Crystallography*; Cambridge University Press: London, UK, 1975.

110. Nakatsuka, A.; Shimokawa, M.; Nakayama, N.; Ohtaka, O.; Arima, H.; Okube, M.; Yoshiasa, A. Static disorders of atoms and experimental determination of Debye temperature in pyrope: Low-and high-temperature single-crystal X-ray diffraction study. *Am. Mineral.* **2011**, *96*, 1593–1605. [CrossRef]

111. Inagaki, M.; Sasaki, Y.; Sakai, M. Debye-Waller parameters of lead sulphide powders. *J. Mater. Sci.* **1987**, *22*, 1657–1662. [CrossRef]

112. Zhou, J.; Wu, Y.; Chen, Z.; Nan, P.; Ge, B.; Li, W.; Pei, Y. Manipulation of defects for high-performance thermoelectric PbTe-Based Alloys. *Small Struct.* **2021**, *2*, 2100016. [CrossRef]

113. Miller, A.J.; Saunders, G.A.; Yogurtcu, Y.K. Pressure dependences of the elastic constants of PbTe, SnTe and $Ge_{0.08}Sn_{0.92}Te$. *J. Phys. C Solid State Phys.* **1981**, *14*, 1569–1584. [CrossRef]

114. Lach-hab, M.; Papaconstantopoulos, D.A.; Meh, M.J. Electronic structure calculations of lead chalocgenides PbS, PbSe, PbTe. *J. Phys. Chem. Solids* **2002**, *63*, 833–841. [CrossRef]

115. Wei, S.H.; Zunger, A. Electronic and structural anomlies in lead chalcogenides. *Phys. Rev. B* **1997**, *55*, 13605–13609. [CrossRef]

116. Albanesi, E.A.; Okoye, C.M.I.; Rodriguez, C.O.; Blanca, E.L.P.; Petukhov, A.G. Electronic structure, structural properties, and dielectric functions of IV-VI semiconductors: PbSe and PbTe. *Phys. Rev. B* **2000**, *61B*, 16589. [CrossRef]

117. Kim, H.; Kaviany, M. Effect of thermal disorder on high figure of merit in PbTe. *Phys. Rev. B* **2012**, *86*, 045213. [CrossRef]

118. Rabe, K.M.; Joannopoulos, J.D. Ab initio relativistic pseudopotential study of the zero-temperature structural properties of SnTe and PbTe. *Phys. Rev. B* **1985**, *32*, 2302–2314. [CrossRef]

119. Rabe, K.M. Ab Initio Statistical Mechanics of Structural Phase Transitions. Ph.D. Thesis, Massachusetts Institute of Technology, Cambridge, MA, USA, 1987.

120. Hummer, K.; Grüneis, A.; Kresse, G. Structural and electronic properties of lead chalcogenides from first principles. *Phys. Rev. B* **2007**, *75*, 195211. [CrossRef]

121. Xu, L.; Zheng, Y.; Zheng, J.C. Thermoelectric transport properties of PbTe under pressure. *Phys. Rev. B* **2010**, *82*, 195102. [CrossRef]

122. Bencherif, Y.; Boukra, A.; Zaoui, A.; Ferhat, M. High-pressure phases of lead chalcogenides. *Mater. Chem. Phys.* **2011**, *126*, 707–710. [CrossRef]

123. Petersen, J.E. First Principles Study of Structural, Electronic, and Mechanical Properties of Lead Selenide and Lead Telluride. Master Thesis, Master Graduate College, Texas State University, San Marcos, TX, USA, 2013.

124. Gelbstein, Y.; Davidow, J.; Leshem, E.; Pinshow, O.; Moisa, S. Significant lattice thermal conductivity reduction following phase separation of the highly efficient $Ge_xPb_{1-x}Te$ thermoelectric alloys. *Phys. Status Solidi* **2014**, *251*, 1431–1437. [CrossRef]

125. Komisarchik, G.; Gelbstein, Y.; Fuks, D. Solubility of Ti in thermoelectric PbTe compound. *Intermetallics* **2017**, *89*, 16–21. [CrossRef]

126. Inoue, M.; Yagi, H.; Muratani, T.; Tatsukawa, T. EPR studies of Mn^{2+} in SnTe and PbTe crystals. *J. Phys. Soc. Jpn.* **1976**, *40*, 458–462. [CrossRef]

127. Bevolo, A.J.; Shanks, H.R.; Eckels, D.E. Molar heat capacity of GeTe, SnTe, and PbTe from 0.9 to 60 K. *Phys. Rev. B* **1976**, *13*, 3523–3533. [CrossRef]

128. Finkenrath, H.; Franz, G.; Uhle, N. Reststrahlen spectra of $PbSe_{1-x}Te_x$. *Phys. Status Solidi* **1979**, *95*, 179–184. [CrossRef]

129. Lead Telluride (PbTe). *Debye Temperature, Heat Capacities, Density, Melting Point, in Non-Tetrahedrally Bonded Elements and Binary Compounds I*; Madelung, O., Rössler, U., Schulz, M., Eds.; Springer: Berlin/Heidelberg, Germany, 1998; pp. 1–3.

130. Morelli, D.T.; Slack, G.A. High lattice thermal conductivity solids. In *High Thermal Conductivity Materials*; Springer: New York, NY, USA, 2006; pp. 37–68.

131. Wang, H.; Pei, Y.; LaLonde, A.D.; Snyder, G.J. Heavily doped p-type PbSe with high thermoelectric performance: An alternative for PbTe. *Adv. Mater.* **2011**, *23*, 1366–1370. [CrossRef]

132. Bauer Pereira, P.; Sergueev, I.; Gorsse, S.; Dadda, J.; Müller, E.; Hermann, R.P. Lattice dynamics and structure of GeTe, SnTe and PbTe. *Phys. Status Solidi* **2013**, *250*, 1300–1307. [CrossRef]

133. Kemerink, G.J.; Pleiter, F. Recoilless fractions calculated with the nearest-neighbour interaction model by Kagan and Maslow. *Hyperfine Interact.* **1986**, *30*, 187–198. [CrossRef]

134. Kong, F.; Liu, Y.; Wang, B.; Wang, Y.; Wang, L. Lattice dynamics of PbTe polymorphs from first principles. *Comput. Mater. Sci.* **2012**, *56*, 18–24. [CrossRef]

135. Zhou, C.; Yu, Y.; Lee, Y.L.; Ge, B.; Lu, W.; Cojocaru-Mirédin, O.; Im, J.; Cho, S.P.; Wuttig, M.; Shi, Z.; et al. Exceptionally High Average Power Factor and Thermoelectric Figure of Merit in n-type PbSe by the Dual Incorporation of Cu and Te. *J. Am. Chem. Soc.* **2020**, *142*, 15172–15186. [CrossRef] [PubMed]
136. Lucovsky, G.; White, R.M. Effects of resonance bonding on the properties of crystalline and amorphous semiconductors. *Phys. Rev. B* **1973**, *8*, 660–667. [CrossRef]

Article

The Effect of Specimen Size and Test Procedure on the Creep Behavior of ME21 Magnesium Alloy

Bettina Camin * and Maximilian Gille

Faculty III-Process Sciences, Institute of Materials Sciences and Technology, Technische Universität Berlin, Sekr. BH18, Ernst-Reuter-Platz 1, 10587 Berlin, Germany; gille@tu-berlin.de
* Correspondence: camin@physik.tu-berlin.de; Tel.: +49-30-314-22096

Abstract: Lightweight constructions and materials offer the opportunity to reduce CO_2 emissions in the transport sector. As components in vehicles are often exposed to higher temperatures above 40% of the melting temperature, there is a risk of creep. The creep behavior usually is investigated based on standard procedures. However, lightweight constructions frequently have dimensions not adequately represented by standardized specimen geometries. Therefore, comparative creep experiments on non-standardized miniature and standardized specimens are performed. Due to a modified test procedure specified by a miniature creep device, only the very first primary creep stage shows a minor influence, but subsequently, no effect on the creep process is detected. The creep behavior of hot extruded and heat treated ME21 magnesium alloy is investigated. It is observed that the creep parameters determined by the miniature and standard creep tests are different. As the deviations are systematic, qualitatively, evidence of the creep behavior is achieved. The creep parameters obtained, and particularly the creep strain and the strain rate, show a higher creep resistance of the miniature specimen. An initial higher number of twinned grains and possible multiaxiality in the gauge volume of the miniature specimen can be responsible.

Keywords: creep testing; ME21; magnesium alloy; size effects; miniature specimen

check for **updates**

Citation: Camin, B.; Gille, M. The Effect of Specimen Size and Test Procedure on the Creep Behavior of ME21 Magnesium Alloy. *Crystals* **2021**, *11*, 918. https://doi.org/10.3390/cryst11080918

Academic Editors: Umberto Prisco and Bolv Xiao

Received: 3 July 2021
Accepted: 5 August 2021
Published: 7 August 2021

Publisher's Note: MDPI stays neutral with regard to jurisdictional claims in published maps and institutional affiliations.

1. Introduction

Since the signing of the Kyoto Protocol and a steadily increasing understanding of the effects of CO_2 emissions on the earth's climate, industry is interested in optimizing its processes and designs regarding CO_2 reduction. In the transport sector, in particular, the goal is to lower the overall weight of vehicles, aircraft, and rail vehicles to decrease fuel consumption and thus environmental pollution. In this context, the terms green technology and life cycle costs focus on product development. Connecting factors are lightweight constructions and materials as well as energy-saving manufacturing processes.

In the latter, characterizing the mechanical and thermo-mechanical properties of alloys is critical in developing novel non- or low-contaminant construction and production technologies as well as during operation. Hence, the investigation of titanium [1,2], aluminum [3], and magnesium alloys [4] as lightweight materials under extreme conditions is of particular interest. Mainly, it is essential to depict the responses of structural components to external mechanical loadings [5,6]. Although the mechanical loads investigated in this work are relatively moderate nevertheless, small dimensions additionally exposed to high temperature can already be rated as highly loaded during service.

In lightweight constructions, modifications of the design by wall thickness reduction are further solutions in this context. As these components are subjected to mechanical loads at room temperature and higher temperatures, their creep resistance is of great importance. Typically, creep experiments are performed on standardized samples according to, e.g., DIN EN ISO 204 (European Standard) [7] or ASTM E139 (American Standard) [8]. However, complex or thin-walled lightweight component geometries partially do not allow standardized specimens to be taken. Consequently, there has been a high interest in carrying out investigations on miniaturized specimen geometries. Furthermore, the smaller dimensions of the specimens enable detailed investigations of representative material states from components, whereby the miniature specimens can be taken in different orientations from a small material volume [9–12]. The influence of the creep parameters by the specimen sizes is provided, and the test method is known. However, the confidence of these results compared to standard sample sizes is of great importance. Miniature specimens, or creep specimens with non-standard geometry, have been used for some time to investigate steel's creep behavior in power plant construction, e.g., [13–18]. Pipelines and connectors are components subjected to creep stresses, which cannot be investigated reproducibly with the geometry of standard creep specimens.

However, especially in the transport sector, replacing construction materials such as steel with lightweight materials effectively reduces weight [19,20]. In addition to aluminum and titanium lightweight materials, the magnesium alloys show a high application potential for the mentioned requirements as they have the lowest density ($1.74\ \mathrm{g/cm^{-3}}$) and high specific strength and stiffness compared to all other metallic construction materials. However, most magnesium alloy's poor forming properties and creep resistance due to their hexagonal closed packed (hcp) structure currently severely limit their industrial application [19,21,22]. The magnesium wrought alloy ME21 containing rare earth (RE) element cerium and the element manganese is characterized by good formability at higher temperatures, corrosion, and creep resistance compared to other Mg alloys [19,23–31]. Ce is added to increase the high-temperature strength and creep resistance due to its pronounced solid solution hardening and low diffusion tendency [25]. Furthermore, the intermetallic phase $Mg_{12}Ce$ formed at grain boundaries prevents grain boundary sliding. Its good formability enables the production of thin sheets with a thickness of 1 mm. However, studies on the creep behavior of the magnesium alloy ME21 are rare. The creep properties of the ME21 alloy are of interest providing a more cost-effective and resource-saving alternative to the WE43 and WE54 alloys containing a higher amount of expensive rare earth elements currently used for components subjected to creep [20,32]. The compositions of the alloys ME21, WE43, and WE54 are given in Table 1 (c.f. Section 2.1).

Additionally, energy-saving manufacturing technologies are selected, enabling the final product to be manufactured with as few manufacturing steps as possible. For this purpose, the manufacturing process by hot extrusion for transport assemblies is established and used for many years, e.g., [33–36]. The production of larger components up to thin sheets of various materials is possible with this process.

As many assemblies in the transportation sector—including lightweight structures—are manufactured nowadays by hot extrusion, in this work, comparative creep experiments of the hot extruded magnesium alloy ME21 are carried out. Miniature and standard specimen geometries are investigated, the creep behavior itself, as well as the influence due to downsizing, are validated.

2. Materials and Methods

2.1. Material

The aluminum-free magnesium alloy ME21 is alloyed by 2 wt % Mn and 1 wt % Ce. The chemical composition is given in Table 1.

Table 1. Chemical composition of ME21 [37], WE43 [38], WE54 [39].

	Mn	Ce	Al	Nd	Pr	Si	Th	Y	Zn	Mg
ME21	2.1	0.7	$\leq$0.01	$\leq$ 0.015	$\leq$0.01	$\leq$0.015	0.14	0.04	$\leq$0.015	bal.
	Mn	Li	Ni	Nd	Zr	Si	Cu	Y	Zn	Mg
WE43	$\leq$0.15	$\leq$0.2	$\leq$0.005	2.4–4.4	0.4–1.0	$\leq$0.01	$\leq$0.03	3.7–4.3	$\leq$0.2	bal.
WE54	$\leq$0.15	$\leq$0.2	$\leq$0.005	2.0–4.0	0.4–1.0	$\leq$0.01	$\leq$0.03	4.75–5.5	$\leq$0.2	bal.

The hot extruded cylindrical bars made of ME21 manufactured by an 8 MN horizontal extrusion plant (SMS group GmbH, Düsseldorf, Germany) and provided by the Extrusion Research and Development Center of the Technische Universität Berlin (FZS). The detailed function of the plant and the extrusion process are described in [40,41]. Parameters of hot extrusion process such as the initial geometry of the billet, the temperatures of the billet ϑ_b, container ϑ_a and tool ϑ_w, the extrusion ratio V, and product speed v_p are shown in Table 2.

Table 2. Parameters of the hot extrusion process.

Technique	Billet Geometry	ϑ_a, ϑ_b, ϑ_w	V	v_p	Cooling
indirect	$\varnothing = 123$ mm $l = 150$ mm	450 °C	71	1.7 m min^{-1}	air

After the hot extrusion process, the extruded product in the as extruded condition (as ex) shows a fine-grained microstructure inappropriate for creep resistance. According to Brömmelhoff et al. [42], subsequent heat treatment at $T = 550$ °C for 24 hours followed by quenching in water leads to fine-dispersed Mn, bigger Mg$_{12}$Ce-precipitates, a coarse-grained and homogeneous microstructure. Optical microscopy (OM) analyses illustrate the microstructure (Figure 1). The grain size varies between 10 and 70 µm by an average of 38 µm, and an aspect ratio of 1 is measured. The visible Mg$_{12}$Ce-precipitates marked by a white arrow are homogeneously distributed throughout the microstructure (Figure 1a,c) and are mainly found at the grain boundaries (Figure 1b). Small Mn-precipitates are fine-dispersed in the matrix (Figure 1b, light grey arrows).

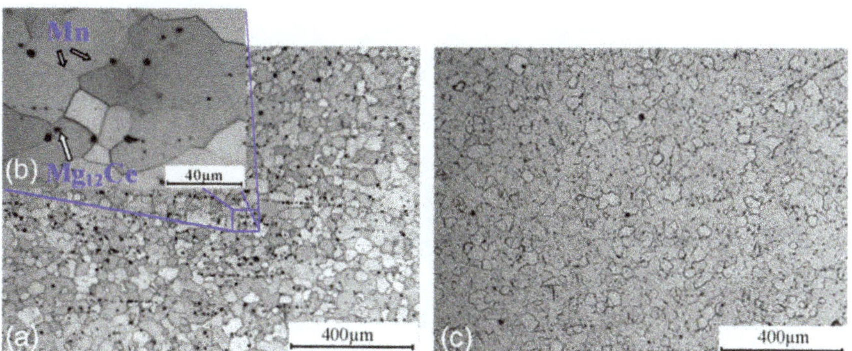

Figure 1. Microstructure of the heat treated ME21 material; (**a,b**) longitudinal section (=ED), (**a**) 50×, (**b**) 500×; (**c**) cross section, 200×.

2.2. Microscopy

The microstructure of the hot extruded and heat-treated initial material and of the miniaturized and standardized samples after the creep experiments is investigated by an optical microscope UnivaR Met (Reichert-Jung Optische Werke AG, Austria) equipped with a digital camera system Leica DFC 295 (Leica Microsystems GmbH, Germany) and by a scanning electron microscope JEOL JSM 640 (JEOL GmbH, Germany) equipped with a Noran detector (Thermo Fisher Scientific Inc., Germany) for EDS measurements. From the initial material sections in the extrusion direction (ED) and perpendicular to ED, the transverse direction (TD), are taken using a cutting-off machine WOCO 50 (Conrad, Germany). The creep-stressed specimens are investigated in ED only.

For metallographic preparation, the specimens are embedded in the two-component epoxy resin Technovit 4071 (Kulzer GmbH, Germany). First, the specimens are manually ground on a polishing and grinding machine, TegraPol-25 (Struers GmbH, Germany), followed by manual polishing of the specimens with diamond suspensions (6 to 1 μm). Subsequently, the samples were additionally chemically polished by immersing them in a CP2 solution for 2 s to remove near-surface deformation layers resulting from polishing and reopen pores closed by the preparation. The CP2 solution is made from 100 ml ethanol, 12 ml hydrochloric acid (25 percentage), and 8 ml nitric acid. For the OM examination of the microstructure, the sample surfaces are finally etched with a solution of 100 ml ethanol, 20 ml water, 6 ml glacial acetic acid, and 5 g picric acid for 3 s.

2.3. Creep Experiments

This work aims the investigation of the creep behavior of miniature specimens in comparison to specimens with standardized geometries—further referred to as standard specimens—according to DIN EN ISO 204 [7]. The dimensions of both creep specimens are shown in Figure 2.

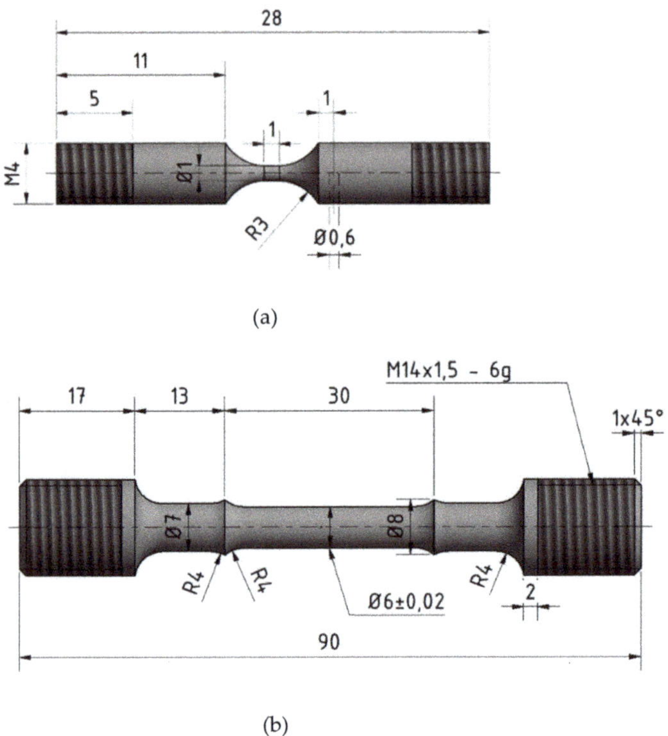

Figure 2. Geometries of the creep specimens, (**a**) miniature specimen, (**b**) standard specimen.

Two different creep devices were used for these studies. The creep tests using standard specimens are performed on a commercial creep testing machine DSM 6102 equipped with a three-stage vertical electric furnace type EO 4090 (Roell Amsler Prüfmaschinen GmbH & Co. KG, Germany), enabling temperatures up to 1100 °C. The force is applied via an electric motor, whereby the maximum test load of the system is 50 kN. The elongation of the sample is determined by an extensometer mechanically adapted directly to the specimen. This ensures only the change in length Δl of the specimen's gauge length l_g is determined. The creep experiments using miniature specimens are carried out applying a miniature creep device developed by TU Berlin for in situ synchrotron μ-tomography measurements (XCT) at the European Synchrotron Radiation Facility (ESRF) described in detail in [43–46]. In these works and furthermore, in [47–49], the applicability of this miniature creep device was shown. It should be mentioned here once again that the experimental procedure of creep experiments conducted by the miniature creep device (cf. [46]) is different from the procedure given by DIN EN ISO 204. Regarding the requirements of in situ XCT investigations at the beamlines of the ESRF, the following aspects are part of the experimental design of the miniature creep device:

- The mechanical load to the sample is applied *before* heating.
- The ratio of the initial gauge diameter $d_0 = 1$ mm and the initial gauge length $l_0 = 1$ mm of the sample is $d_0/l_0 = 1$ (Figure 2a).
- The elongation of the whole sample is measured during creep. As shown in [46], compared to the much larger diameter of the clamping, significant strain changes occur only in the small diameter of the gauge volume as soon as the thermal equilibrium is reached.

Comparative preliminary tests on standard samples (Figure 2b) in the DSM 6102 test rig are conducted to investigate the influence of the varied sequence of mechanical and thermal loading on the creep behavior. An uniaxial tensile mechanical load of $\sigma = 20$ MPa is applied in all preliminary creep tests. The two different test procedures are described as follows:

- In agreement with the standard creep test procedure specified by [7], the mechanical load is applied after the sample is heated up to the test temperature (sequence: heating $\rightarrow$ mechanical loading).
- In agreement with the creep test procedure specified by the miniature creep device, the mechanical load is applied at room temperature, and subsequent, the specimen is heated to the test temperature (sequence: mechanical loading $\rightarrow$ heating).

The analyses of the creep curves measured for both test procedures provide information about the applicability of the different sequences.

Another focus in this work is on comparative investigations of the creep behavior of standard and miniature creep specimens performed according to the test procedure given by the miniature creep device. The miniature specimens are tested in the miniature creep device, and the standard specimens are tested in the DSM 6102 test rig. All creep tests are carried out at a test temperature $T_{cr} = 523$ K due to the highest temperature expected in automobiles. Taking the melting temperature of magnesium $T_m = 923$ K into account, hence, a homologous temperature $T_{homolog} = T_{cr}/T_m = 0.58$ results. Mechanical loads $\sigma = 10, 15, 20, 25,$ and 30 MPa are selected. Due to the work of Stinton et al., phase stability can be expected [50]. For each mechanical load, two experiments are performed with the miniature specimens and one experiment with the standard specimens. Due to measured high strains at fracture ($\varepsilon_{fr} > 0.1$), the true creep curve $\varepsilon_{true} = f(t)$ and true strain rate curve $\dot{\varepsilon}_{true} = f(t)$ are calculated from the elongation of the sample Δl measured by extensometer in reference to the initial gauge length l_0 [51]:

$$\varepsilon_{true} = \ln\left(1 + \frac{\Delta l}{l_g}\right) \tag{1}$$

$$\dot{\varepsilon}_{true} = \frac{\varepsilon_{true}}{dt} \tag{2}$$

3. Results

3.1. Creep Tests

3.1.1. Preliminary Creep Tests

The comparison of the creep curves $\varepsilon_{true} = f(t)$ calculated from Equation (1) and strain rate curves $\dot{\varepsilon}_{true} = f(t)$ (Equation (2)) obtained in the preliminary creep tests are shown in Figure 3. It should be noted that the true strain is a dimensionless quantity. Overall, both creep curves basically show similar behavior. However, the strain develops differently within approximately first two hours of the creep experiments. While the creep curve of the experiment due to the standard sequence (heating $\rightarrow$ mechanical load) exhibits a typical behavior, the behavior of the varied creep test sequence (mechanical load → heating) is unusual in the early primary creep stage. After a minor positive initial strain, a negative strain seems to occur here, subsequently increasing again. During heating from room to test temperature under mechanical load competitive processes, time-independent and time-dependent plastic deformation occurs. The mechanisms behind, are not investigated in this work. Afterward, at the latest in the transition from the primary to the secondary stage, the creep curves of both test procedures more or less run parallel to each other. Finally, the creep time until fracture t_{fr} and the creep strain at fracture ε_{fr} are different (Figure 3a). The strain rate curves illustrate the different behavior in the early primary creep stage (Figure 3b). Apparently, even negative strain rates occur right at the beginning for the miniature specimen. After approximately 4 hours, the curves of both test procedures fit together, while the minimum creep rates $\dot{\varepsilon}_{min}$ are quite similar. The difference of the two measured values for the minimum creep rate is less than 4%, which is within typical experimental error limits. The changed sequence of the test procedure has no significant influence on the minimum creep rate determined for the given test parameters.

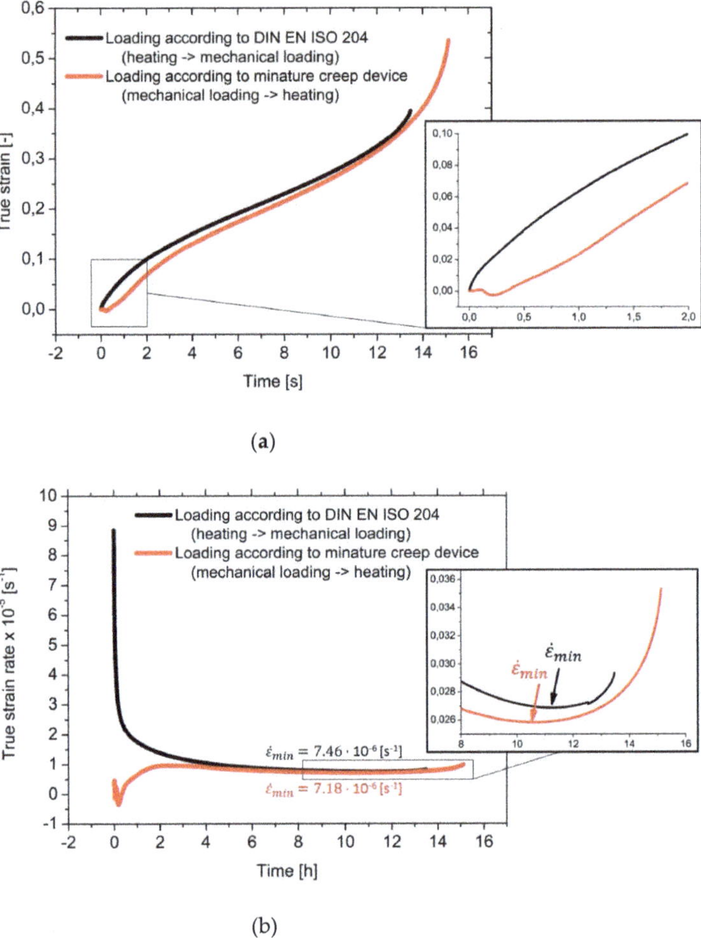

Figure 3. Comparison of different creep test procedures; (**a**) entire creep curves and detail of the very early primary creep stage, (**b**) entire curves of strain rate and detail of the region around minimum creep strain until fracture.

The obtained creep parameters of the preliminary experiments are shown in Table 3. Based on the procedure described in [46], the error limits are estimated at about approximately 45%.

Table 3. Creep parameters of preliminary creep tests.

Test Procedure	$\dot{\varepsilon}_{min}$ (s^{-1})	ε_{fr} $(-)$	t_{fr} (h)
Standard (Heating $\rightarrow$ mechanical loading)	$7.46 \times 10^{-6} \pm 3.36 \times 10^{-7}$	0.395 ± 0.178	13.5 ± 6.075
Variation due to miniature device (Mechanical loading $\rightarrow$ heating)	$7.18 \times 10^{-6} \pm 3.23 \times 10^{-7}$	0.537 ± 0.242	14.9 ± 6.705

In comparison from Table 3, it can be derived a slightly faster minimum creep strain rate $\dot{\varepsilon}_{min}$ leads to a shorter time to fracture t_{fr} as well as a smaller creep strain at fracture ε_{fr}, whereas the slower minimum creep rate $\dot{\varepsilon}_{min}$ in contrast lead to a while longer creep lifetime t_{fr} and higher creep strain at fracture ε_{fr}. Regarding the effect of a faster $\dot{\varepsilon}_{min}$ on ε_{fr}, this finding is not compliant to the literature, e.g., [51–53]: based on experience, a higher $\dot{\varepsilon}_{min}$ should lead to higher ε_{fr}. However, the minimum creep rate of the standard test procedure is reached at a later point in time compared to the varied test procedure due to miniature creep device. Regarding to a lower ε_{fr} this in turn is in accordance with the literature. It is known, the exact time to fracture t_{fr} and elongation at creep rupture ε_{fr} is strongly dependent on the local defect characteristics of the material, e.g., pores or cracks [53]. As these tests are performed only once without repetitions and the results agree pretty well, thus, a negligible influence of the test procedure on ε_{fr} and t_{fr} is assumed here. The varied creep test procedure of the miniature creep device is found to be suitable to standard specimen investigated in the DSM 6102 test rig.

3.1.2. Comparative Creep Tests

Comparative creep measurements (miniature and standard) are conducted applying the varied creep test procedure (mechanical load $\rightarrow$ heating). Due to the high values of the miniature specimens, the initial instantaneous strain in the very early primary creep stage is subtracted, so only the time-dependent true creep strain is plotted. Both the miniature creep specimens and the standard creep specimens show typical creep behavior, with a primary, secondary, and tertiary creep stage of the true creep curves $\varepsilon_{true} = f(t)$ (Figure 4). At $\sigma = 10$ MPa the creep experiments with miniature specimen are prematurely stopped at $t_{cr} = 255$ and 995 h due to expected very long t_{fr} (Figure 4a). Although the creep curves and consequently the respective creep parameters of the miniature specimens at $\sigma = 25$ MPa are very different the average values are calculated.

In general, with increasing stress on the specimens, an increasing slope of the secondary creep stage and a decreasing total creep time can be observed. Furthermore, for the same test parameters, longer times to fracture t_{fr} and a lower creep rate $\dot{\varepsilon}_{min}$ of the miniature specimens compared to the standard specimens are determined. The creep parameters derived from the creep curves are shown in Table 4. The ductility of the material expressed by the necking at fracture Z is determined by:

$$Z = \frac{A_0 - (\pi \cdot r_1 \cdot r_2)}{A_0} \cdot 100\%$$

(3)

where A_0 is the initial cross-section area and r_1, r_2 are the semi-axes of elliptical fracture surfaces. The average value of the necking at fracture of the standard specimens $Z_{stand} = 51\%$ is twice that of the miniature specimens $Z_{mini} = 25\%$. The error limits of the standard specimens are estimated by $\pm 45\%$ based on the procedure described in [46], whereas the error limits of the miniature specimens are calculated from the results of repeated creep experiments.

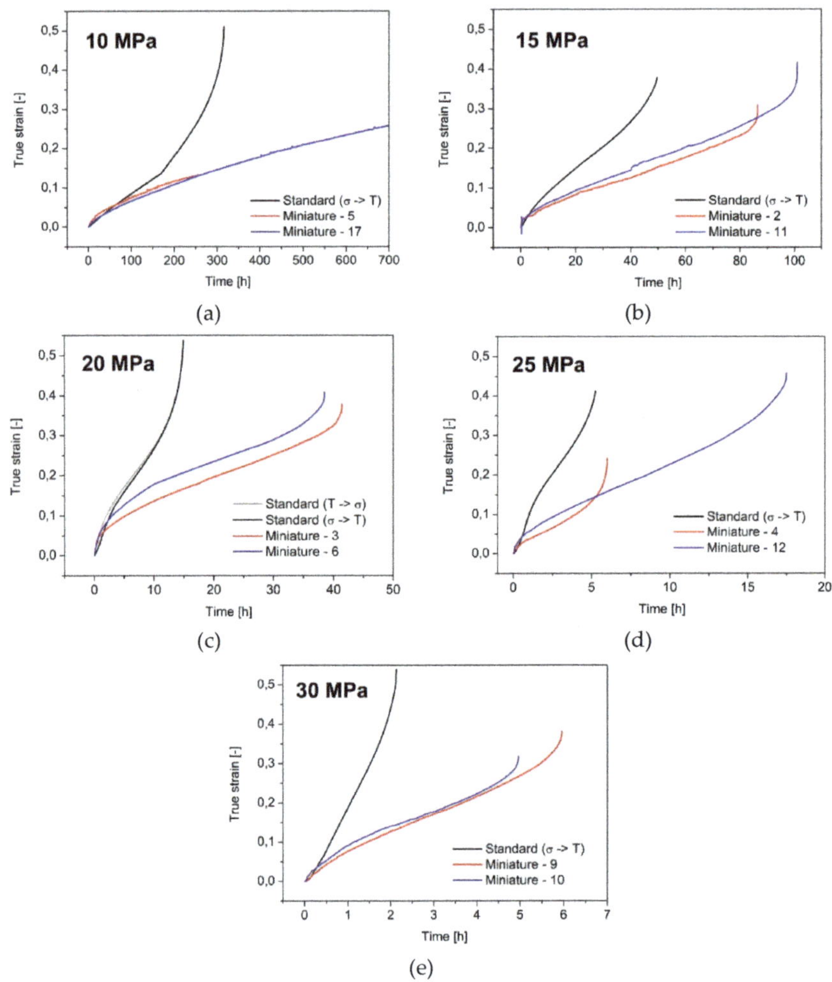

Figure 4. Creep curves, (**a**) 10 MPa, (**b**) 15 MPa, (**c**) 20 MPa, (**d**) 25 MPa, (**e**) 30 MPa.

3.2. Microscopy

OM of the standard and miniature specimen in the hot extruded and heat-treated condition after manufacturing is conducted in the region of the gauge length near the surface. Figure 5 shows the longitudinal sections of standard and miniature specimens not subjected to creep loading. Hence, the influence of the manufacturing process is shown. The region near the surface of the specimens manufactured by turning exhibits twinning due to deformation. As shown in Figure 5a, twinning of grains from the surface to a depth of 100 μm is detectable in the standard samples with a diameter of 6 mm. The grains of the miniature samples, on the other hand, are found to be twinned over the entire gauge volume diameter of 1 mm. The twinning density gradually decreases from the surface to the center of the sample (Figure 5b). Moreover, a lower twinning density is observed outside the gauge volume (Figure 5c). It should be noted that a strong scatter is observed within a miniature specimen batch regarding the degree of twinning. Some samples indicate a low twinning density, whereas other samples indicate a high twinning density.

Table 4. Creep parameters of all investigated samples.

Standard				
σ (MPa)	$\dot{\varepsilon}_{min}$ (s^{-1})	ε_{fr} (−)	t_{fr} (h)	Z (%)
10	$1.89 \cdot 10^{-7} \pm 8.505 \cdot 10^{-8}$	0.507 ± 0.228	464.16 ± 208.87	54 ± 2
15	$1.41 \cdot 10^{-6} \pm 6.345 \cdot 10^{-7}$	0.378 ± 0.17	50.03 ± 22.51	50 ± 1
20	$5.99 \cdot 10^{-6} \pm 2.7 \cdot 10^{-6}$	0.535 ± 0.241	15.14 ± 6.81	50 ± 2
25	$1.49 \cdot 10^{-5} \pm 6.705 \cdot 10^{-6}$	0.411 ± 0.185	5.47 ± 2.46	-
30	$5.90 \cdot 10^{-5} \pm 2.655 \cdot 10^{-5}$	0.507 ± 0.228	2.31 ± 1.04	51 ± 2
Miniature				
σ (MPa)	$\dot{\varepsilon}_{min}$ (s^{-1})	ε_{fr} (−)	t_{fr} (h)	Z (%)
10	$5.59 \cdot 10^{-8} \pm 1.3 \cdot 10^{-8}$	-	-	-
15	$5.09 \cdot 10^{-7} \pm 9.0 \cdot 10^{-9}$	0.487 ± 0.048	94.17 ± 7.35	27 ± 3
20	$1.47 \cdot 10^{-6} \pm 6.0 \cdot 10^{-8}$	0.558 ± 0.29	41.84 ± 29.47	26 ± 4
25	$4.56 \cdot 10^{-6} \pm 3.4 \cdot 10^{-7}$	0.457 ± 0.177	12.16 ± 5.77	-
30	$1.09 \cdot 10^{-5} \pm 8.5 \cdot 10^{-7}$	0.472 ± 0.019	5.86 ± 0.50	26 ± 5
35	$3.49 \cdot 10^{-5} \pm 1.57 \cdot 10^{-5}$	0.402 ± 0.181	1.38 ± 0.80	4 ± 0.5

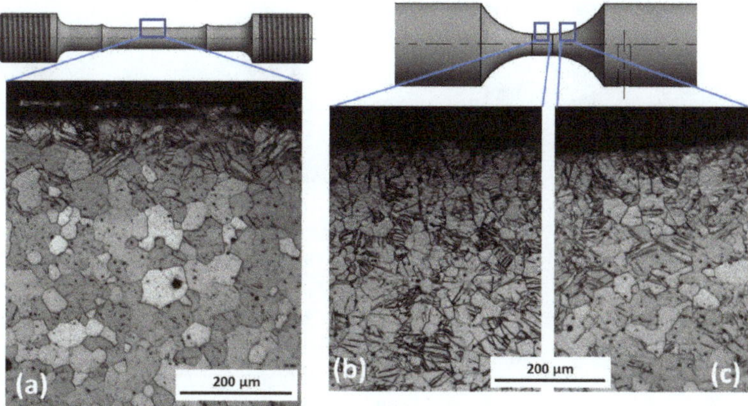

Figure 5. OM of the microstructure, longitudinal section, near-surface region, (**a**) standard specimen, (**b**) miniature specimen in the gauge length, (**c**) miniature specimen beyond the gauge length.

The microstructure of the standard and miniature samples after the creep experiments are also investigated. The observations described below occur in both the standard and the miniature specimens, although with different dimensions. First, the focus is on changes of the grain morphology: increasing mechanical stress is observed to be accompanied by increasing elongation of the grains. Starting from nearly globular grains (ratio = 1.0) at $\sigma = 10$ MPa, several of which at this load have twins, the aspect ratio is developing to 1.5 ($\sigma = 15$ MPa), 1.8 ($\sigma = 20$ MPa) to a maximum of 1.9 ($\sigma = 30$ MPa) illustrated exemplary on the standard specimen (Figure 6). The values of the aspect ratios are shown in Table 5.

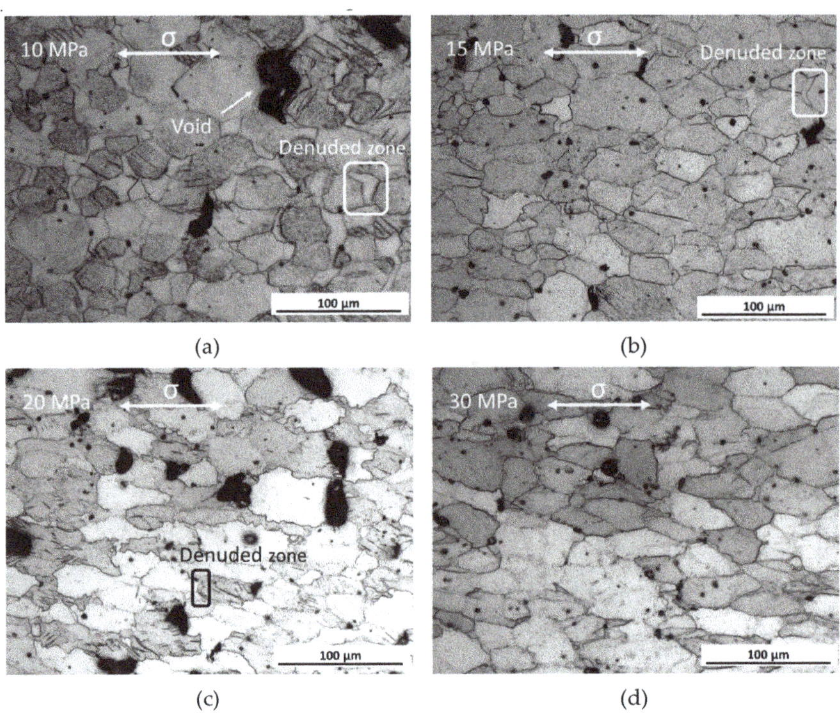

Figure 6. Longitudinal sections of crept standard samples, microstructure at different mechanical loads, (**a**) 10 MPa, (**b**) 15 MPa, (**c**) 20 MPa, (**d**) 30 MPa.

Table 5. Aspect ratios of elongated grains.

	10 MPa	15 MPa	20 MPa	25 MPa	30 MPa
Standard	1	1.5	1.8	1.8	1.9
Miniature	1	1.1	1.2	1.3	1.4

In addition, denuded zones—known from creep experiments on, for example, Zr containing Mg-alloys, e.g., [54–57]—are detected at the grain boundaries oriented perpendicular to the loading direction, the size of which is dependent on load and thus from the creep time. As higher the load, as smaller the denuded zones. While in the standard specimens at $\sigma = 10$ MPa, the width of the denuded zones $w_{dZ,\ 10MPA}$ is approx. 8 to 10 μm (Figure 6a), at $\sigma = 20$ MPa the width $w_{dZ,\ 20MPA}$ is approximately 1 to 2 μm (Figure 6c), whereas at $\sigma = 30$ MPa denuded zones are not observed (Figure 6d). In the miniature specimens denuded zones at $\sigma = 10$ MPa the width of the $w_{dZ,\ 10MPA}$ is approximately 6 μm. Here, the zones are detectable up to $\sigma = 20$ MPa.

Further investigations on the denuded zones are carried out by SE imaging and EDS measurements analyzing the structure and chemical composition. Due to large, denuded zones, the measurements were performed on a standard specimen loaded at $\sigma = 10$ MPa. Different grey values in the SE image already indicate an inhomogeneous distribution of the elements (Figure 7a). A pile-up of Mn-particles at grain boundaries parallel to the loading axis is detected (red arrow). An EDS mapping illustrates a depletion of the alloying element manganese in the denuded zone (Figure 7b) qualitatively. Quantitatively EDS single point measurements verify that the Mn concentration in the denuded zone is about 0.1 at%, whereas in the grains 0.8 at% to 0.9 at% are measured. Therefore, EDS line scans parallel to loading direction are conducted, showing in the adjacent grains (left and right) to the denuded zones that the Mn concentration and distribution are different. According to the line scan shown in

Figure 7c in the grain on the left from point (1), the Mn concentration is constant at an average of approx. 0.8 at%. Between the points (1) and (2)—the denuded zone—the Mn concentration drops to zero. To the right of point (2), there is a continuous increase in the Mn concentration, which initially fluctuates around an average of 2.5 at%, abruptly dropping back to the already known approx. 0.8 at%. The area around point (3) is enriched by Mn-precipitates marked with arrows in the SE image (Figure 7c, left side), explaining the concentration fluctuations around a mean value. The different Mn concentrations in the respective areas are assumed to be due to a directional diffusion of Mn.

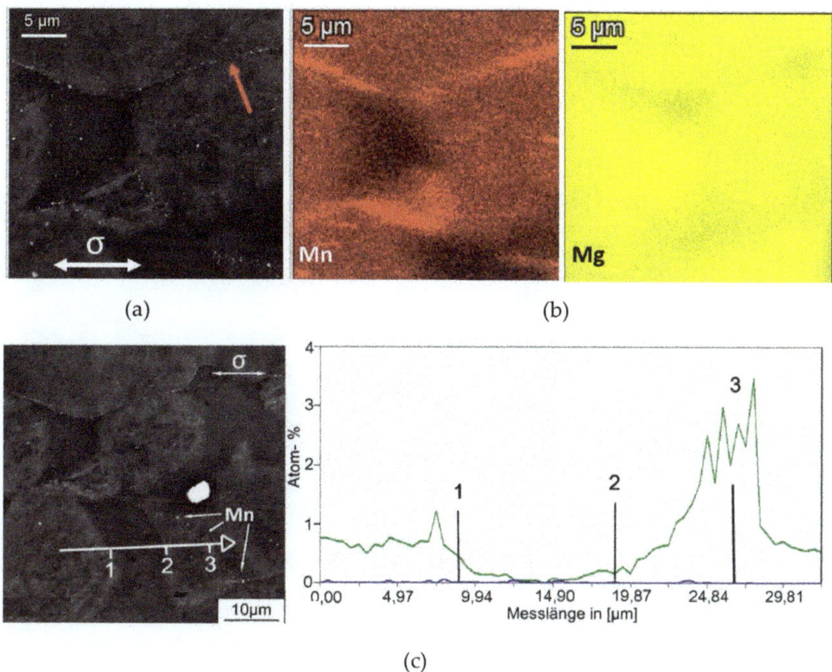

Figure 7. SEM, denuded zone, (**a**) SE image, (**b**) EDX-mapping of Mn + Mg, (**c**) EDS-line scan.

Additionally to the grain morphology and the denuded zones, the creep voids are characterized by shape and distribution. It has to be noticed, detailed analysis of the porosity fraction is not the subject of this work. The creep voids form preferentially at grain boundaries oriented perpendicular to the loading direction (Figure 6). Figure 8a–c show the longitudinal sections of a standard creep specimen at $\sigma = 20$ MPa. Next to the fracture surface a higher porosity is observed (Figure 8a). As already visually is apparent, with increasing distance from the fracture surface, a gradual decrease in void number and size occurs (Figure 8a–c). This behavior is detected in all measured specimens and thus appears to be independent of load. The shape of individual voids is predominantly elliptical, with the longitudinal axis oriented perpendicular to the loading direction. The same behavior is observed in the miniature creep samples, too (Figure 8d).

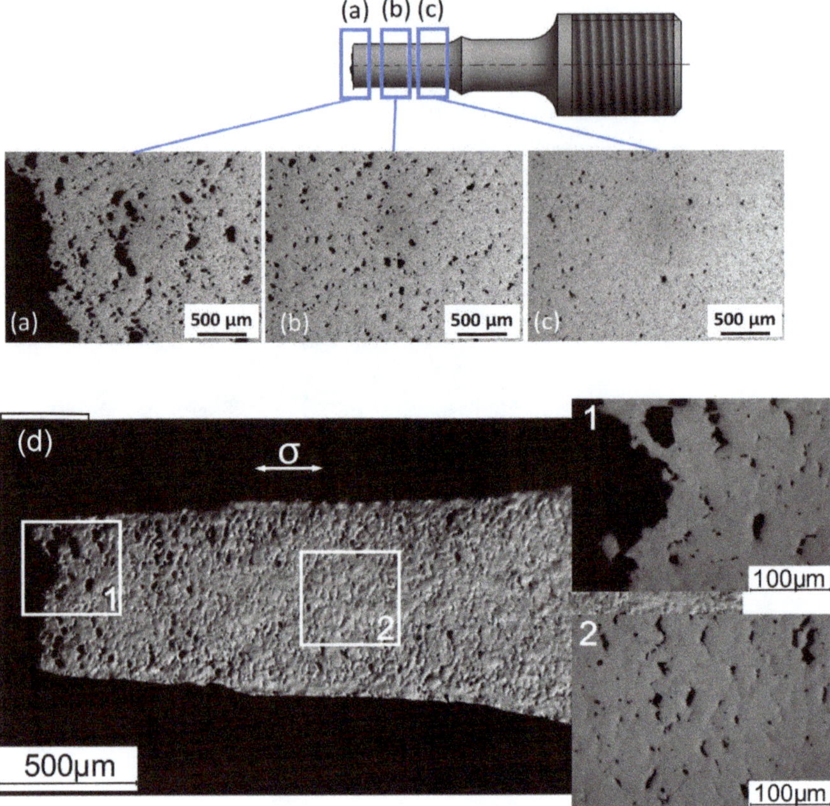

Figure 8. Creep damage, 20 MPa, longitudinal section; (**a–c**) standard specimen, (**a**) near fracture surface, (**b**) 5 mm distance to fracture surface, (**c**) 10 mm distance to fracture surface; (**d**) miniature specimen.

4. Discussion

4.1. Validation of Creep Experiments

As only a few experiments, the creep test using standard geometries merely once and with miniature specimens only one repetition test each, are performed the plausibility of the individual creep experiments is verified using the modified Monkman–Grant relationship [58]:

$$t_{fr} = \frac{C\prime \cdot \varepsilon_{fr}}{\dot{\varepsilon}_{min}^{m\prime}} \tag{4}$$

The ratio ε_{fr}/t_{fr} is plotted against $\dot{\varepsilon}_{min}$ in double logarithmic scale (Figure 9). Here, according to [58] the exponent $m\prime$ is proven and found to be 1. The error limits of the standard specimen tests are estimated from the error limits calculated by the test repetitions of the miniature creep experiments.

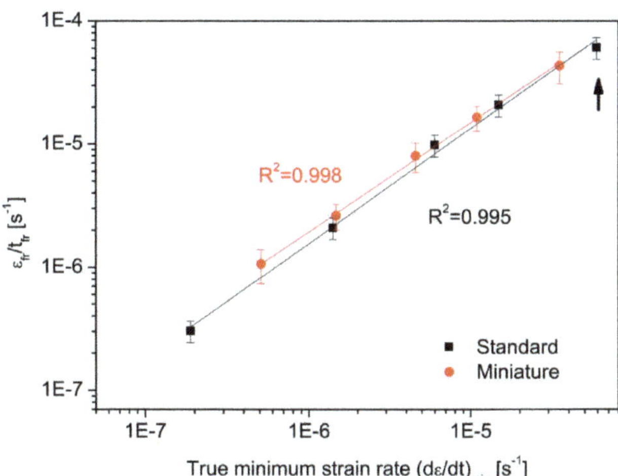

Figure 9. Illustration of the modified Monkman–Grant equation.

For both the standard and the miniature creep tests, a linear relationship can be determined for the individual measuring points using a regression coefficient R^2. The coefficients $R^2 = 0.995$ for the standard specimens and $R^2 = 0.998$ for the miniature specimens are observed. Thus, the validity of the individual tests is demonstrated, although, for the standard experiments, no repetitions are carried out. This is particularly relevant for validating the result of the creep test with standard geometry at $\sigma = 30$ MPa, marked by an arrow in Figure 9. As the value of $\dot{\varepsilon}_{min} = 5.9 \times 10^{-5}$ s^{-1} is slightly above the theoretical limit of creep tests of $\dot{\varepsilon}_{min} = 5.0 \times 10^{-5}$ s^{-1}, according to Maier identifying the transition to the power-law-breakdown [53]. However, due to the sufficiently accurate position on the regression line, the test can be used to characterize the creep behavior and determine the Norton stress exponent (cf. Section 4.3).

4.2. Test Procedure

As the sequence of the creep test procedure of the miniature creep device due to its design is not variable, the applicability to the standard specimen in a commercial test rig is validated. The force application is different in both test rigs, in the miniature device by spring and in the DSM 6102 test rig by an electric motor. The latter ensures constant force application to the specimen throughout the test. However, in the miniature creep device, the spring expands according to the elongation of the specimen. As a result, the loading force decreases proportionally to the specimen strain, while at the same time, the cross-section of the specimen is reduced. For low creep strains $\varepsilon < 0.4$, a change in mechanical loading of 5% is calculated. Consequently, the expansion of the spring can be neglected so that the assumption of force constancy can be justified (cf. [46]).

The change in the sequence of mechanical and thermal load application compared to the standard test procedure compliant to DIN EN ISO 204 in the varied test procedure according to the miniature device in the primary stage of the creep curve an untypical behavior occurs (cf. Figure 3). Plotting the curves of the creep strain and the creep rate in comparison to the specimen temperature, the atypical behavior in the primary creep stage is illustrated (Figure 10). It is found that the behavior of the strain and strain rate curve within the first quarter of an hour in all tested specimens independent from the mechanical load is the same. However, the characteristic of the curves depends on the amount of the mechanical load. It is assumed that in the very early primary creep stage, the strain and strain rate is essentially determined by the heating-up process than by time-dependent plastic deformation as described by Weertman [59,60], Cottrell [61], and the literature [53,62]. The investigation of the detailed microstructural processes at this stage is not the subject of this work. At the latest, when the test temperature and thermal equilibrium are reached, the strain and strain rate curves follow the typical creep curve shapes (cf. Figure 3), which is in good agreement with [10,12,46]. From this, it can be concluded, in the varied creep test procedure already in the later primary creep stage, the known hardening and softening mechanisms of the creep process correspond to those in the standard test procedure according to DIN EN ISO 204.

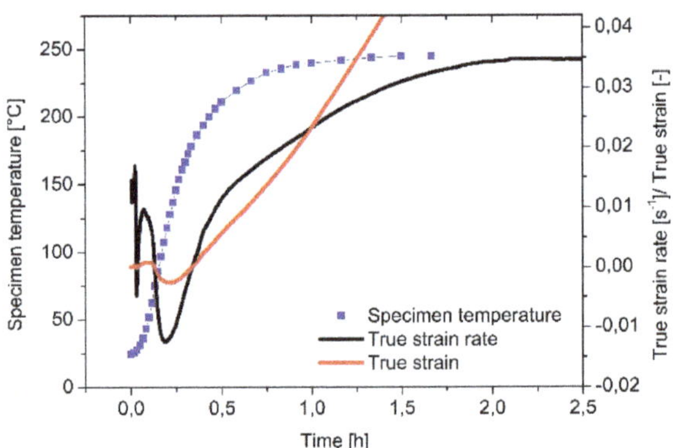

Figure 10. Comparison of the curves of true strain, true strain rate, and specimen temperature in the early primary creep stage.

4.3. Creep Behavior

The creep parameters (cf. Table 4) of the miniature and standard geometry experiments taken from the true strain and true strain rate curves show differences: for the miniature specimens ε_{fr} and $\dot{\varepsilon}_{min}$ are shifted to lower values, t_{fr} is higher by a factor of about 2 to 2.5, the proportion of the secondary creep stage of the total creep curve is percentual higher, and Z only half compared to the creep experiments with the standard specimen (Figure 11). Analyzing the creep parameters in detail, the true strain at fracture ε_{fr} for the standard specimens indicate no dependency from the mechanical stress. In contrast ε_{fr} of the miniature specimen within the error limits decreases with increasing stress (Figure 11a). The latter behavior is also known from the literature [52,53,62]. However, due to the small number of experiments—the standard experiments are carried out once, and the miniature experiments twice for each stress loading—the error limits are taken as an average from all miniature creep experiments are high. However, as it can be derived from Figure 4d in a single case, the difference can be much higher. As repeating measurements of standard specimens are missing here, qualitatively, the tendency to lower values of ε_{fr} as a function of σ for the miniature compared to the standard specimens can be determined, but a quantification is not possible. Furthermore, lower strain rate values $\dot{\varepsilon}_{min}$ for the miniature compared to the standard specimens are obtained. From the minimum strain rate values the Norton-exponent is determined verifying the creep mechanisms taking place (Figure 11b). A parallel shift of the Norton line to lower creep rates for the miniature specimens occurs. However, within the error limits, the value of the Norton-exponent is the same for both the miniature creep experiments n_{mini} =4.9 ± 0.2 and the standard creep experiments $n_{standard}$ =5.1 ± 0.2. Hence, a value of the Norton-exponents of approximately 5 indicates the creep mechanism of dislocation creep for both specimen geometries. The creep diagram in the double-logarithmic scale reveals a linear relationship of σ and t_{fr} holding the same power (−0.21) independent from the specimen geometry (Figure 11c). A parallel shift of the correlation line to higher values of the miniature specimens is observed here.

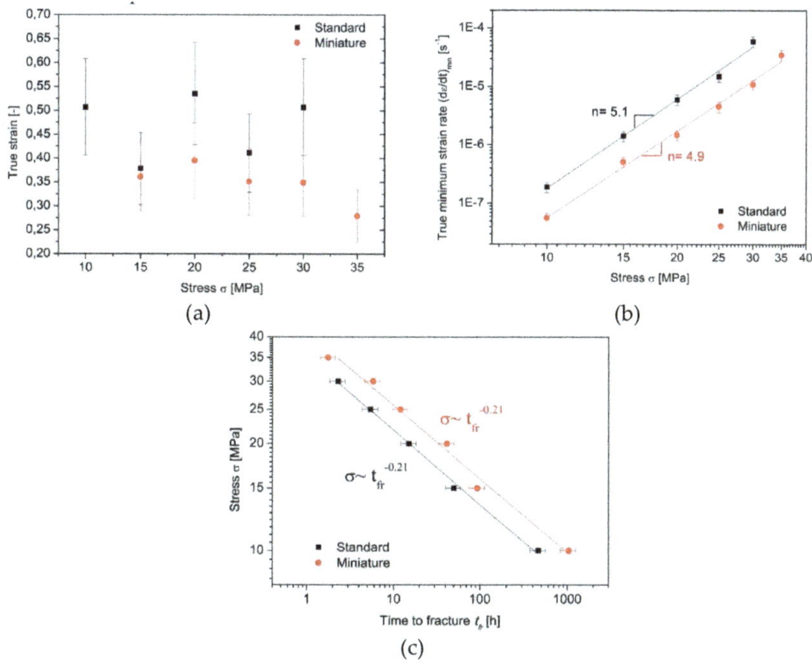

Figure 11. Creep parameter, (**a**) strain at fracture, (**b**) Norton-exponent, (**c**) creep diagram.

Lower ε_{fr} and $\dot{\varepsilon}_{min}$, higher t_{fr}, shifting of the Norton line and curve of creep rupture strength to higher values and lower Z indicate a higher creep resistance of the miniature specimens than standard specimens. The mechanisms behind, are discussed in the following section.

4.4. Creep Mechanisms

As shown in Section 4.3, the creep parameters determined by miniature experiments are not arbitrarily different from the determined creep parameters of the standard specimens but have systematic differences. This is taken from parallel shifts of the Norton line to lower creep rates and the creep rupture strength to higher values for the miniature specimens. While in the work of Kurumlu et al. [10,11] and Mälzer et al. [12], no size dependence of creep parameter of miniature and standard specimens is found, on the other hand nevertheless this is described by Olbricht et al. [9], Krompholz et al. [63], Schmieder [64], Camin [46], and Dymácek et al. [65]. Altogether, the results are inconsistent. In [46], the same miniature creep device as in this work was used. Consequently, the specimen dimensions and test procedure are the same; however, the heating in [46] was faster. As well as here, in [46], systematic deviation of the miniature creep experiments compared to standard experiments is determined from the values from the literature [66]. Nevertheless, different from here, the minimum creep rates of the miniature experiments show higher values. In [46], the creep behavior of metal matrix composites (MMC) materials was investigated in situ during creep by μ-tomography at the ESRF. An influence of the material can be excluded since [10,11] investigated the same class of material, obtaining contrary results. However, certainly miniaturized flat tensile specimens exhibiting a ratio of $l_0/d_0 \approx 6$ were used by [10,11], whereas [46] used miniaturized cylindrical specimens with a ratio of $l_0/d_0 = 1$. Cao et al. compared flat and mild notched cylindrical specimens under uniaxial load [67]. The notched cylindrical specimens with dimensions very similar to the specimens used in this work and [46] exhibited less strain than the flat samples [67], as observed here also. Cao et al. attributed this behavior to multiaxial stress states occurring in the notch. A multiaxial stress state in the miniature specimens used here may be one explanation for the higher creep resistance measured here. Furthermore, based on FEM results, Cao et al. concluded that unavoidable inaccuracies in manufacturing the miniature round notch specimens could be tolerated because only a minor influence ($\approx$5%) on the stress parameters characterizing the multiaxial stress states was found. This may apply to manufacturing tolerances, but not to changes in microstructure due to manufacturing, e.g., twins, as detected in this work and discussed below. Twinned grains and dislocations induced by cold working during manufacturing are beneficial for creep resistance interfering with dislocation movement.

The microstructural aspect has already been studied in other works. In ASTM E139 [8], it is assumed that specimen size effects expected to specimen miniaturization can be explained by the small number of grains in the specimen volume resulting in a higher impact of each individual grain orientation. This was investigated by Krompholz et al. using cylindrical specimens of different sizes. The creep parameters achieved, additionally to size effects, show unsystematic dependencies due to the test temperature and stress [63]. Krompholz et al. suggest that the diffusion processes, decisive for creep, are higher weighting in miniaturized specimens than in standard specimens, leading to differences in the creep parameters. Olbricht et al. in [9] criticized the latter interpretation of Krompholz et al. As it is elusive which diffusion process could be responsible for this.

Consequently, by combining the results from [9–11,46,63,67] and this work as interim results, it can be noted that not only the geometry of the miniature samples influences the systematic shift of the creep parameters measured here. As the effect of faster heating was not investigated in this work, the possible effect cannot be evaluated. Thus, comparative miniature creep experiments to standard geometries allow a qualitative characterization of the creep behavior due to the systematic deviations. In contrast, a direct comparison of quantitative parameters is not possible without knowledge and explanation of the reasons for the systematic deviation.

Hence, within this work's scope, comparative microstructural investigations of non-creep-stressed and creep-stressed miniature and standard specimens are carried out. After sample manufacturing, a significantly higher number of twins are detected in the miniature creep specimens (Ø 1 mm) than in the specimens with standard geometry (Ø 6 mm) due to higher stress by the centering mandrel and the machining, since the forces affecting a smaller cross-sectional area (cf. Figure 5). The observation of an increased number of twinned grains in the miniature specimens is therefore relevant, because in the work of Randle [68] and Alexandreanu et al. [69] in the context of "grain-boundary-engineering" an indirect influence of twins on material properties such as corrosion, ductility and creep strength are assumed. While Randle mentioned an influence by Σ3 oriented Coincidence Site Lattice (CSL) grain boundaries (twins) on creep properties in cubic materials, the studies of Alexandreanu et al. validating the Thaveeprungsriporn model [70] indicate that a reduced creep rate can be observed at increased Coincidence Site Lattice Boundary (CSLB) fractions in cubic crystal systems. These findings support the results of this work.

Furthermore, the microstructural investigations reveal a systematically increased necking at fracture Z by a factor of 2 for the creep-stressed standard compared to miniature specimens. The fracture necking attributed to the ductility of the material is strongly dependent on creep damage. Elliptical pores are formed at grain boundaries oriented perpendicular to the loading axis (cf. Figures 6 and 8). According to Cocks and Ashby [71], this is evidence of void growth by surface diffusion. As these voids act as small cracks, the damage in the miniature specimens due to their small size must be weighted more than in the standard specimen. This can explain the low fracture necking.

The stress-dependent Norton-exponent $n = 5$ indicating plastic deformation is typically associated with dislocation creep (cf. Section 4.3). Additionally, grain elongation oriented parallel to the load can be seen in the optical micrographs. This grain elongation can be explained by both plastic deformations due to dislocation movements and diffusion processes. According to Rösler et al. [52] and Kassner [72], increased concentration of vacancies can be expected at grain boundaries oriented perpendicular to the load. As a result, a mass flow occurs in the direction of the tensile load, causing a grain shape change, which is also detected here.

Furthermore, at lower loads $\sigma \leq 20$ MPa, denuded zones at grain boundaries oriented perpendicular to the load are observed (cf. Figure 7). In the past, many creep experiments on Mg-alloys containing the alloying element Zr were carried out investigating this phenomenon. However, the mechanisms behind are discussed widely and differently, e.g., in [54–57,73–75]. Squires et al. [54] observed a similar characteristic of zircon (Zr) depleted regions in Mg-0.5 wt % Zr alloys as detected here (cf. Figure 7). The authors attributed this to preferential magnesium bulk diffusion in the load direction. Studies by Burton and Reynolds [76], as well as Greenwood [74], Owen et al. [77], and Langdon [73] indicate that alloy element denuded zones are indicative for diffusion creep. Langdon describes the accompanied pile-up of particles at grain boundaries (cf. Figure 7a) and elongation of the grains (Figure 6) parallel to the loading axis, both observed here also, as attributed to diffusional creep. However, a Norton-exponent $n = 1$ related to diffusional creep is specified in [73], different from $n = 5$ related to dislocation creep found in this work. In [76] it is further discussed that grain boundaries are often not centrally located in the denuded zone, which would be expected assuming an equal flow of material from both adjacent grains toward the grain boundary. Burton and Reynolds attribute this to increased mobility of grain boundaries in diffusion zones responding to minor external disturbance by moving to the end of the denuded zone. Wolfenstine et al., however, refuted the relationship between denuded zones and diffusional creep [56]. Kloc, in turn, refers to the work of Wolfenstine et al., and in contradiction to this, it is argued that a unilateral occurrence of a denuded zone is entirely compatible with the theory of diffusion creep. However, it is justified to the asymmetry of the random grain boundaries [75]. Wadsworth et al. proposed a model for the formation of denuded zones and illustrated schematically the mechanisms behind. The mechanism postulated is grain boundary sliding rate-controlled by dislocation climb [57]. From their point of view, a simultaneous action of grain boundary sliding, and grain boundary migration are responsible for dissolving small particles, dragging of undissolved particles growing bigger as small particles dissolve, and diffusion of solute atoms. The SEM and EDS investigations presented in Figure 7 are supported by the proposed model of Wadsworth et al.: Due to uniaxial mechanical loading denuded zones are observed only at boundaries transverse to the tensile direction (Figure 12a, yellow arrow). The grain boundary is located between the denuded zone and a Mn-containing precipitate zone. The denuded zone introducing a fresh amorphous-like region is depleted of the alloying element Mn (Figure 12b,c, between red arrows). Undissolved Mn-precipitates are dragged by the migrating boundary (Figure 12b,c, yellow arrow and dashed line). Some larger particles brake away from the migrating boundary remaining in the denuded zone (Figure 12b,c, magenta arrow). Furthermore, on lateral grain boundaries precipitates occur due to diffusion of solute atoms, mentioned by [73] also (Figure 12b,c, green arrows).

The results of the microstructural investigations thus indicate that the plastic deformation of the coarse-grained ME21 creep specimens at σ up to maximum 20 MPa can be explained by dominating dislocation creep accompanied by a minor process of grain boundary sliding and migrating, whereas, at $\sigma > 20$ MPa, the plastic deformation is due to dislocation creep only.

Apart from a different number of twinned grains and different levels of fracture necking, on the observed scale, no microstructural differences can be found in the miniature and standard specimens. Consequently, in this work, the microstructural creep mechanisms activated in the material are independent of the specimen size. However, a decreased creep rate and increased creep rupture time indicate different kinetics of the mechanisms, which are not investigated in this work.

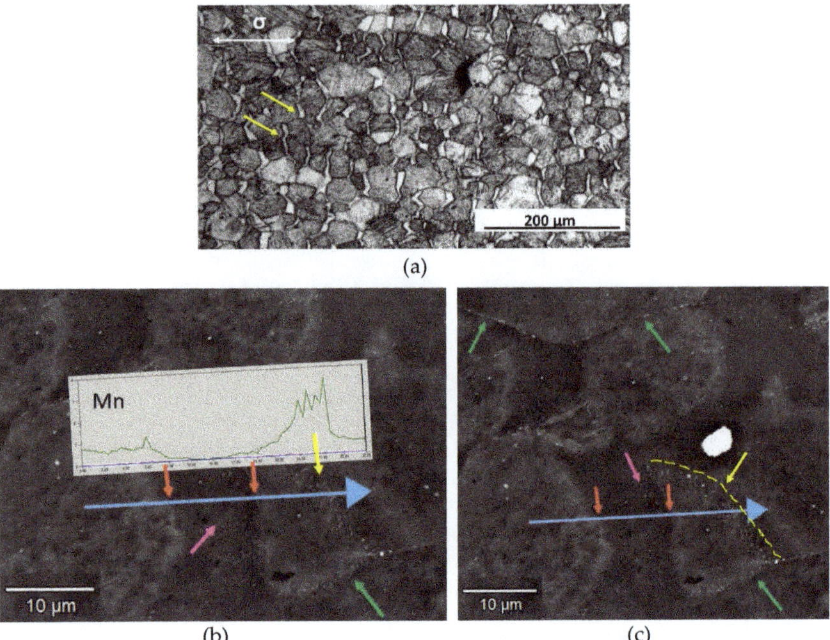

(a)

(b)   (c)

Figure 12. Microstructure after creep, $\sigma = 10$ MPa: (**a**) OM, yellow arrows: denuded zones; (**b**,**c**) EDS measurements, blue arrow indicates the line scan, Mn depleted zones between red arrows, yellow dashed line: migration grain boundary, yellow arrow: undissolved Mn-precipitates, magenta arrow: large Mn-particles, green arrows: Mn-precipitates on grain boundaries.

5. Conclusion

Comparative uniaxial creep measurements on miniature and standard specimens of different geometries on hot extruded and heat treated ME21 magnesium alloy are performed according to the test procedure of a miniature creep device initially developed for in situ XCT experiments. The sequence of loading of the test procedure (mechanical load → heating) different from the standard creep test (heating → mechanical loading) lead to a different creep behavior in the very early primary creep stage. However, the higher initial creep strain of the miniature samples is associated with the miniature creep device and does not affect the entire creep behavior. Overall, in all experiments—miniature and standard—a typical primary, secondary, and tertiary creep behavior are observed. Hence, the applicated test procedure provides reliable results.

The main creep mechanism is identified as dislocation creep, at lower stresses accompanied by grain boundary sliding and migrating of minor influence, occurring identically in both specimen geometries and, thus, is independent of specimen size.

However, the analyses of the creep parameters detect a higher creep resistance for the miniature creep specimens. A higher number of twinned grains originated by the manufacturing process indicates a higher work hardening interfering with the dislocation movement, being beneficial for creep resistance. Therefore, it can be assumed, a higher fraction of twinned grains within the gauge volume gains more importance in small specimens. In addition, in a notch-like gauge volume of the miniature specimens, an emerging multiaxial stress state is possible. Therefore, the higher creep resistance of the miniature specimens measured here is probably dependent on both the twinned grains and geometry in the gauge volume. In a good approximation, the creep behavior of a material can be determined by miniature creep specimens; however, a direct comparison of quantitative creep parameters is not possible without the knowledge and explanation of the reasons for systematic deviations.

Author Contributions: Conceptualization, B.C.; methodology, B.C.; validation, M.G.; formal analysis, B.C. and M.G.; investigation, B.C.; writing—original draft preparation, B.C. and M.G.; writing—review and editing, B.C. and M.G.; visualization, B.C. and M.G. All authors have read and agreed to the published version of the manuscript.

Funding: The publication of this article was funded by the Open Access publication fund of the Technische Universität Berlin.

Acknowledgments: We acknowledge support from the Open Access publication fund of the Technische Universität Berlin and the Extrusion Research and Development Center of the Technische Universität Berlin (FZS) for providing the material.

Conflicts of Interest: The authors declare no conflict of interest.

References

1. MacLeod, S.; Errandonea, D.; Cox, G.A.; Cynn, H.; Daisenberger, D.; Finnegan, S.; McMahon, M.; Munro, K.; Popescu, C.; Storm, C. The phase diagram of Ti-6Al-4V at high-pressures and high-temperatures. *J. Phys. Condens. Matter* **2021**. [CrossRef]
2. Smith, D.; Joris, O.P.J.; Sankaran, A.; Weekes, H.E.; Bull, D.J.; Prior, T.J.; Dye, D.; Errandonea, D.; Proctor, J.E. On the high-pressure phase stability and elastic properties of β-titanium alloys. *J. Phys. Condens. Matter* **2017**, *29*, 155401. [CrossRef]
3. Dorward, R.C.; Pritchett, T.R. Advanced aluminium alloys for aircraft and aerospace applications. *Mater. Des.* **1988**, *9*, 63–69. [CrossRef]
4. Wang, G.G.; Bos, J. A study on joining magnesium alloy high pressure die casting components with thread forming fasteners. *J. Magnes. Alloy.* **2018**, *6*, 114–120. [CrossRef]
5. Straumal, B.B.; Korneva, A.; Kilmametov, A.R.; Lityńska-Dobrzyńska, L.; Gornakova, A.S.; Chulist, R.; Karpov, M.I.; Zięba, P. Structural and Mechanical Properties of Ti−Co Alloys Treated by High Pressure Torsion. *Materials* **2019**, *12*, 426. [CrossRef] [PubMed]
6. Errandonea, D.; Burakovsky, L.; Preston, D.L.; MacLeod, S.G.; Santamaría-Perez, D.; Chen, S.; Cynn, H.; Simak, S.I.; McMahon, M.I.; Proctor, J.E.; et al. Experimental and theoretical confirmation of an orthorhombic phase transition in niobium at high pressure and temperature. *Commun. Mater.* **2020**, *1*. [CrossRef]
7. DIN EN ISO 204-2019-04—Beuth.de. Available online: https://www.beuth.de/de/norm/din-en-iso-204/289976402 (accessed on 8 June 2021).
8. E28 Committee. *Test Methods for Conducting Creep, Creep-Rupture, and Stress-Rupture Tests of Metallic Materials*; ASTM International: West Conshohocken, PA, USA, 2011.
9. Olbricht, J.; Bismarck, M.; Skrotzki, B. Characterization of the creep properties of heat resistant 9–12% chromium steels by miniature specimen testing. *Mater. Sci. Eng. A* **2013**, *585*, 335–342. [CrossRef]
10. Kurumlu, D. Mechanische und Mikrostrukturelle Untersuchungen an Einer Kurzfaserverstärkten Aluminiumlegierung. Ph.D. Thesis, Ruhr-Universität Bochum, Universitätsbibliothek, Bochum, Germany, 2010.
11. Kurumlu, D.; Payton, E.J.; Young, M.L.; Schöbel, M.; Requena, G.; Eggeler, G. High-temperature strength and damage evolution in short fiber reinforced aluminum alloys studied by miniature creep testing and synchrotron microtomography. *Acta Mater.* **2012**, *60*, 67–78. [CrossRef]
12. Mälzer, G.; Hayes, R.W.; Mack, T.; Eggeler, G. Miniature Specimen Assessment of Creep of the Single-Crystal Superalloy LEK 94 in the 1000 °C Temperature Range. *Met. Mat. Trans. A* **2007**, *38*, 314–327. [CrossRef]
13. Hiyoshi, N.; Itoh, T.; Sakane, M.; Tsurui, T.; Tsurui, M.; Hisaka, C. Development of miniature cruciform specimen and testing machine for multiaxial creep investigation. *Theor. Appl. Fract. Mech.* **2020**, *108*, 102582. [CrossRef]
14. Hyde, T.H.; Hyde, C.J.; Sun, W. Theoretical basis and practical aspects of small specimen creep testing. *J. Strain Anal. Eng. Des.* **2013**, *48*, 112–125. [CrossRef]
15. Morris, A.; Cacciapuoti, B.; Sun, W. The role of small specimen creep testing within a life assessment framework for high temperature power plant. *Int. Mater. Rev.* **2018**, *63*, 102–137. [CrossRef]
16. Yu, B.; Han, W.; Tong, Z.; Geng, D.; Wang, C.; Zhao, Y.; Yang, W. Application of Small Specimen Test Technique to Evaluate Creep Behavior of Austenitic Stainless Steel. *Materials* **2019**, *12*, 2541. [CrossRef] [PubMed]
17. Chvostová, E.; Džugan, J. Creep test with use of miniaturized specimens. *IOP Conf. Ser. Mater. Sci. Eng.* **2017**, *179*, 12032. [CrossRef]
18. Luan, L.; Riesch-Oppermann, H.; Heilmaier, M. Tensile creep of miniaturized specimens. *J. Mater. Res.* **2017**, *32*, 4563–4572. [CrossRef]
19. Kammer, C.; Aluminium-Zentrale, D. *Düsseldorf Aluminium-Zentrale*; Aluminium: Düsseldorf, Germany, 2000; ISBN 3870172649.
20. Kielbus, A. The influence of ageing on structure and mechanical properties of WE54 alloy. *J. Achiev. Mater. Manuf. Eng.* **2007**, *1*, 27–30.

21. von Buch, F. Magnesium-Eigenschaften, Anwendungen, Potentiale: Vortragstexte eines Fortbildungsseminars der Deutschen Gesellschaft für Materialkunde e.V. in Zusammenarbeit mit dem Institut für Werkstofforschung des GKSS-Forschungszentrum Geesthacht GmbH. In *Eigenschaften von Magnesiumlegierungen und deren Beeinflussung*; Kainer, K.U., Ed.; Wiley-VCH: Weinheim, Germany, 2000; ISBN 3527299793.

22. Pekguleryuz, M.O.; Kaya, A.A. Creep Resistant Magnesium Alloys for Powertrain Applications. *Adv. Eng. Mater.* **2003**, *5*, 866–878. [CrossRef]

23. Leontis, T.E. The properties of sand cast magnesium-rare earth alloys. *JOM* **1949**, *1*, 968–983. [CrossRef]

24. Neite, G.; Kubota, K.; Higashi, K.; Hehmann, F. Magnesium-Based Alloys. In *Materials Science and Technology*; Cahn, R.W., Haasen, P., Kramer, E.J., Eds.; Wiley-VCH Verlag GmbH & Co. KGaA: Weinheim, Germany, 2006; ISBN 9783527603978.

25. Polmear, I.J. *Light Alloys: Metallurgy of the Light Metals*, 2nd ed.; Arnold: London, UK, 1989; ISBN 0340491752.

26. Kennedy, A.J. The Physical Metallurgy of Magnesium and Its Alloys. *J. R. Aeronaut. Soc.* **1959**, *63*, 737–738. [CrossRef]

27. Schemme, K. Magnesiumwerkstoffe fur die Neunziger Jahre. *Aluminium* **1991**, *67*, 167–178.

28. Kurz, G.; Petersen, T.; Bohlen, J.; Letzig, D. Variation of Rare Earth Elements in the Magnesium Alloy ME21 for the Sheet Production. In *Magnesium Technology 2017*; Solanki, K.N., Orlov, D., Singh, A., Neelameggham, N.R., Eds.; Springer International Publishing: Berlin/Heidelberg, Germany, 2017; pp. 353–363. ISBN 978-3-319-52391-0.

29. Mo, N.; Tan, Q.; Bermingham, M.; Huang, Y.; Dieringa, H.; Hort, N.; Zhang, M.-X. Current development of creep-resistant magnesium cast alloys: A review. *Mater. Des.* **2018**, *155*, 422–442. [CrossRef]

30. Neh, K.; Ullmann, M.; Oswald, M.; Berge, F.; Kawalla, R. Twin Roll Casting and Strip Rolling of Several Magnesium Alloys. *Mater. Today Proc.* **2015**, *2*, S45–S52. [CrossRef]

31. Wang, J.G.; Hsiung, L.M.; Nieh, T.G.; Mabuchi, M. Creep of a heat treated Mg–4Y–3RE alloy. *Mater. Sci. Eng. A* **2001**, *315*, 81–88. [CrossRef]

32. Kang, Y.H.; Wang, X.X.; Zhang, N.; Yan, H.; Chen, R.S. Effect of initial temper on the creep behavior of precipitation–hardened WE43 alloy. *Mater. Sci. Eng. A* **2017**, *689*, 419–426. [CrossRef]

33. Automotive. Available online: https://www.hydro.com/en/aluminium/industries/automotive/ (accessed on 22 June 2021).

34. How is Aluminum Extrusion used in Auto Industry? | AEC. Available online: https://www.aec.org/page/extrusion-applications-auto-industry (accessed on 22 June 2021).

35. Aluminium Insider. Auto Industry Looks to Aluminum Extrusions for Increased Performance and Cost-Effective Solutions—Aluminium Insider. Available online: https://aluminiuminsider.com/auto-industry-looks-to-aluminum-extrusions-for-increased-performance-and-cost-effective-solutions/ (accessed on 22 June 2021).

36. Isogai, M.; Murakami, S. Development and application of aluminum extrusion for automotive parts, mainly bumper reinforcement. *J. Mater. Process. Technol.* **1993**, *38*, 635–654. [CrossRef]

37. Huppmann, M.; Gall, S.; Müller, S.; Reimers, W. Changes of the texture and the mechanical properties of the extruded Mg alloy ME21 as a function of the process parameters. *Mater. Sci. Eng. A* **2010**, *528*, 342–354. [CrossRef]

38. Magnesium WE43-T6, Cast. Available online: http://www.matweb.com/search/datasheet.aspx?matguid=4b8a8c13cf354fc5893a40cf8eca022c&ckck=1 (accessed on 23 July 2021).

39. Magnesium WE54-T6, Cast. Available online: http://www.matweb.com/search/datasheet.aspx?matguid=f41ac8bc1ddd4c1d85247e646b3ec2be (accessed on 23 July 2021).

40. Gall, S. Grundlegende Untersuchungen zum Strangpressen von Magnesiumblechen und deren Weiterverarbeitung: Mikrostruktur und mechanische Eigenschaften. *DepositOnce* **2013**. [CrossRef]

41. Müller, S. *Weiterentwicklung des Strangpressens von AZ Magnesiumlegierungen im Hinblick auf eine Optimierung der Mikrostruktur, des Gefüges und der Mechanischen Eigenschaften*, 1st ed.; Cuvillier Verlag: Göttingen, Germany, 2007; ISBN 9783736923706.

42. Brömmelhoff, K.; Huppmann, M.; Reimers, W. The effect of heat treatments on the microstructure, texture and mechanical properties of the extruded magnesium alloy ME21. *Int. J. Mater. Res.* **2011**, *102*, 1133–1141. [CrossRef]

43. Pyzalla, A.; Camin, B.; Buslaps, T.; Di Michiel, M.; Kaminski, H.; Kottar, A.; Pernack, A.; Reimers, W. Simultaneous tomography and diffraction analysis of creep damage. *Science* **2005**, *308*, 92–95. [CrossRef]

44. Huppmann, M.; Camin, B.; Pyzalla, A.R.; Reimers, W. In-situ observation of creep damage evolution in Al–Al2O3 MMCs by synchrotron X-ray microtomography. *Int. J. Mater. Res.* **2010**, *101*, 372–379. [CrossRef]

45. Camin, B.; Hansen, L. In Situ 3D-µ-Tomography on Particle-Reinforced Light Metal Matrix Composite Materials under Creep Conditions. *Metals* **2020**, *10*, 1034. [CrossRef]

46. Camin, B. In-situ Untersuchung des Schädigungsverhaltens mehrphasiger Werkstoffe unter thermischer und mechanischer Beanspruchung. *DepositOnce* **2015**. [CrossRef]

47. Borbély, A.; Dzieciol, K.; Sket, F.; Isaac, A.; Di Michiel, M.; Buslaps, T.; Kaysser-Pyzalla, A.R. Characterization of creep and creep damage by in-situ microtomography. *JOM* **2011**, *63*, 78–84. [CrossRef]

48. Isaac, A.; Sket, F.; Reimers, W.; Camin, B.; Sauthoff, G.; Pyzalla, A.R. In situ 3D quantification of the evolution of creep cavity size, shape, and spatial orientation using synchrotron X-ray tomography. *Mater. Sci. Eng. A* **2008**, *478*, 108–118. [CrossRef]

49. Isaac, A.; Sket, F.; Borbély, A.; Sauthoff, G.; Pyzalla, A.R. Study of Cavity Evolution During Creep by Synchrotron Microtomography Using a Volume Correlation Method. *Pract. Metallogr.* **2008**, *45*, 242–245. [CrossRef]

50. Stinton, G.W.; MacLeod, S.G.; Cynn, H.; Errandonea, D.; Evans, W.J.; Proctor, J.E.; Meng, Y.; McMahon, M.I. Equation of state and high-pressure/high-temperature phase diagram of magnesium. *Phys. Rev. B* **2014**, *90*. [CrossRef]

51. Abe, F. Development of creep-resistant steels and alloys for use in power plants. *Struct. Alloy. Power Plants* **2014**, 250–293. [CrossRef]
52. Rösler, J.; Harders, H.; Bäker, M. *Mechanisches Verhalten der Werkstoffe*, 6th ed.; Springer Vieweg: Wiesbaden, Germany, 2019; ISBN 9783658268015.
53. Maier, H.J.; Niendorf, T.; Bürgel, R. *Handbuch Hochtemperatur-Werkstofftechnik: Grundlagen, Werkstoffbeanspruchungen, Hochtemperaturlegierungen und -beschichtungen*, 6th ed.; Springer Vieweg: Wiesbaden, Germany, 2019; ISBN 3658253134.
54. Squires, R.L.; Weiner, R.T.; Phillips, M. Grain-boundary denuded zones in a magnesium-wt% zirconium alloy. *J. Nucl. Mater.* **1963**, *8*, 77–80. [CrossRef]
55. Poirier, J.-P. *Creep of Crystals*; Cambridge University Press (CUP): London, UK, 1985.
56. Wolfenstine, J.; Ruano, O.A.; Wadsworth, J.; Sherby, O.D. Refutation of the relationship between denuded zones and diffusional creep. *Scr. Metall. Mater.* **1993**, *29*, 515–520. [CrossRef]
57. Wadsworth, J.; Ruano, O.A.; Sherby, O.D. Denuded zones, diffusional creep, and grain boundary sliding. *Metall. Mat. Trans. A* **2002**, *33*, 219–229. [CrossRef]
58. Dunand, D.C.; Han, B.Q.; Jansen, A.M. Monkman-grant analysis of creep fracture in dispersion-strengthened and particulate-reinforced aluminum. *Metall. Mat. Trans. A* **1999**, *30*, 829–838. [CrossRef]
59. Weertman, J. Theory of Steady-State Creep Based on Dislocation Climb. *J. Appl. Phys.* **1955**, *26*, 1213–1217. [CrossRef]
60. Weertman, J. Dislocation Climb Theory of Steady-State Creep. *Trans. ASM* **1968**, *161*, 681–694. [CrossRef]
61. Cottrell, A.H. The time laws of creep. *J. Mech. Phys. Solids* **1952**, *1*, 53–63. [CrossRef]
62. Gottstein, G. *Materialwissenschaft und Werkstofftechnik: Physikalische Grundlagen*, 4th ed.; Springer Vieweg: Berlin, Germany, 2014; ISBN 9783642366031.
63. Krompholz, K.; Kalkhof, D. Size effect studies of the creep behaviour of a pressure vessel steel at temperatures from 700 to 900 °C. *J. Nucl. Mater.* **2002**, *305*, 112–123. [CrossRef]
64. Meeting on current work on behavior of materials at elevated temperatures. In *Size Effect in Creep and Rupture Tests on Unnotched and Notched Specimens of Cr–Mo–V Steel: Reports of Current Work on Behavior of Materials at Elevated Temperatures*; Schmieder, A.K. (Ed.) American Society of Mechanical Engineers: New York, NY, USA, 1974.
65. Dymáček, P.; Jarý, M.; Dobeš, F.; Kloc, L. Tensile and Creep Testing of Sanicro 25 Using Miniature Specimens. *Materials* **2018**, *11*, 142. [CrossRef]
66. Requena, G.; Telfser, D.; Hörist, C.; Degischer, H.P. Creep behaviour of AA 6061 metal matrix composite alloy and AA 6061. *Mater. Sci. Technol.* **2002**, *18*, 515–521. [CrossRef]
67. Cao, L.; Bürger, D.; Wollgramm, P.; Neuking, K.; Eggeler, G. Testing of Ni-base superalloy single crystals with circular notched miniature tensile creep (CNMTC) specimens. *Mater. Sci. Eng. A* **2018**, *712*, 223–231. [CrossRef]
68. Randle, V. Twinning-related grain boundary engineering. *Acta Mater.* **2004**, *52*, 4067–4081. [CrossRef]
69. Alexandreanu, B.; Sencer, B.H.; Thaveeprungsriporn, V.; Was, G.S. The effect of grain boundary character distribution on the high temperature deformation behavior of Ni–16Cr–9Fe alloys. *Acta Mater.* **2003**, *51*, 3831–3848. [CrossRef]
70. Thaveeprungsriporn, V.; Was, G.S. The role of coincidence-site-lattice boundaries in creep of Ni-16Cr-9Fe at 360 °C. *Met. Mat. Trans. A* **1997**, *28*, 2101–2112. [CrossRef]
71. Cocks, A.; Ashby, M.F. On creep fracture by void growth. *Prog. Mater. Sci.* **1982**, *27*, 189–244. [CrossRef]
72. Kassner, M.E. *Fundamentals of Creep in Metals and Alloys*, 3rd ed.; Elsevier Science: Amsterdam, The Netherlands, 2015; ISBN 9780080994277.
73. Langdon, T.G. Identifiying creep mechanisms at low stresses. *Mater. Sci. Eng. A* **2000**, *283*, 266–273. [CrossRef]
74. Greenwood, G.W. Denuded zones and diffusional creep. *Scr. Metall. Mater.* **1994**, *30*, 1527–1530. [CrossRef]
75. Kloc, L. On the symmetry of denuded zones in diffusional creep. *Scr. Mater.* **1996**, *35*, 539–541. [CrossRef]
76. Burton, B.; Reynolds, G.L. In defense of diffusional creep. *Mater. Sci. Eng. A* **1995**, *191*, 135–141. [CrossRef]
77. Owen, D.M.; Langdon, T.G. Low stress creep behavior: An examination of Nabarro—Herring and Harper—Dorn creep. *Mater. Sci. Eng. A* **1996**, *216*, 20–29. [CrossRef]

crystals

MDPI

Article

Thermal Convection in the Core of Ganymede Inferred from Liquid Eutectic Fe-FeS Electrical Resistivity at High Pressures

Joshua A. H. Littleton *, Richard A. Secco and Wenjun Yong

Department of Earth Sciences, University of Western Ontario, London, ON N6A5B7, Canada;
secco@uwo.ca (R.A.S.); wyong4@uwo.ca (W.Y.)
* Correspondence: jlittlet@uwo.ca

Abstract: The core of Ganymede is suggested to be mainly Fe but with a significant proportion of S. Effects of S as a core constituent are freezing-point depression, allowing for a molten core at relatively low core temperatures, and modification of transport properties that can influence the dynamo and thermal evolution. The electrical resistivity of solid and liquid Fe-FeS (~24–30 wt.% S) was measured up to 5 GPa and thermal conductivity was calculated using the Wiedemann–Franz law. These first well-constrained experimental data on near eutectic Fe-FeS compositions showed intermediate values of electrical and thermal conductivities compared to the end-members. Eutectic temperatures were delineated from the solid to liquid transition, inferred from sharp changes in electrical resistivity, at each pressure. Combined with thermal models, our calculated estimates of the adiabatic heat flow of a molten Fe-FeS eutectic composition core model of Ganymede showed that thermal convection is permissible.

Keywords: electrical resistivity; iron sulfides; high pressure; high temperature; Ganymede; thermal convection

Citation: Littleton, J.A.H.; Secco, R.A.; Yong, W. Thermal Convection in the Core of Ganymede Inferred from Liquid Eutectic Fe-FeS Electrical Resistivity at High Pressures. *Crystals* **2021**, *11*, 875. https://doi.org/10.3390/cryst11080875

Academic Editors: Simone Anzellini and Daniel Errandonea

Received: 24 June 2021
Accepted: 26 July 2021
Published: 28 July 2021

Publisher's Note: MDPI stays neutral with regard to jurisdictional claims in published maps and institutional affiliations.

1. Introduction

The dipolar magnetic field of Ganymede may be produced by internal convection of a liquid iron (Fe) outer core, similar to Earth's geodynamo [1]. Convective motions in liquid cores may be derived from two broad sources: (i) thermal and (ii) compositional [2]. The former requires heat transfer out of the core exceeding the heat transferred by conduction of the core. The latter is related to the gradual cooling of the core and consequent inner core formation, and density contrasts between precipitated core chemical species and residual liquid (e.g., Fe snow or FeS floatation) [3–5]. While Ganymede is believed to have a predominantly Fe core, it has been suggested that the core is composed of more sulfur (S) compared to other terrestrial Fe cores [3,6–10]. A significant effect due to the presence of S as a core element is freezing-point depression. For instance, the eutectic temperature (T) in the Fe-FeS system at 1 atm is ~1260 K, approximately 600 K lower than the melting T of Fe [11]. Impurities and more abundant elemental constituents may also affect transport properties such as electrical resistivity (ρ) and thermal conductivity (κ) [12], which are two critical parameters in magnetic field generation via planetary body dynamos.

Experimental investigations of ρ and κ of Fe (e.g., [13–19]) and FeS [16,20,21] at core conditions have shown varying degrees of agreement and consistency. However, typical estimates of the S content in the core of Ganymede, based on internal structure and magnetic field generation models, are adjacent to the Fe-FeS eutectic [3–5,9,10,22]. If core S composition is eutectic or eutectic-adjacent (i.e., within a few weight percent), this may allow the core of Ganymede to be molten and permit thermally driven convection at relatively low core T (<1400 K) that can power a dynamo-produced magnetic field. In this work, we measured ρ of eutectic-adjacent Fe-FeS in both solid and liquid states at pressures (P) up to 5 GPa. The measured results were used to delineate the P-dependent eutectic T and to calculate κ and adiabatic heat flow (Q_a) to determine if thermal convection is

59

permissible in a molten, eutectic-adjacent S composition core of Ganymede constrained to a low core T.

2. Materials and Methods

Fe and FeS powders were purchased from ESPI Metals (99.95% purity) and Alfa Aesar (99.98% purity), respectively, and were mixed together to attain an S content close to the eutectic composition (Table 1) [23]. All experiments were conducted in a 1000-ton cubic anvil press, as described by Secco [24]. A three-sectioned cubic P cell design and four-wire electrical resistance technique using Type S (platinum (Pt) and rhodium (Rh) alloy) thermocouples for all experiments were the same as those used and described by Littleton et al. [21]. Experimental specifications remained largely the same, with two minor alterations: (i) the highest T reached was ~1430 K since the liquidus T's are considerably lower for the Fe-FeS system investigated than for FeS; and (ii) P- and T-dependent contributions to the measured voltage from tungsten (W) disks placed between the thermocouple junctions and powder mixture sample were accounted for [25] since these contributions were proportionally larger and non-negligible compared to measurements of FeS under similar conditions.

Table 1. Values of targeted eutectic Fe-FeS sample compositions and post-experiment analysis results of sample compositions for each pressure in this study. Post-experiment sample compositions are noted to be either Fe-rich or FeS-rich relative to the target pressure-dependent eutectic composition.

Pressure (GPa)	Target Eutectic Composition (wt.% S)	Sample Composition (Post-Experiment) (wt.% S)	Relative Location Adjacent to Eutectic
2	28.0	29.56 ± 0.05	FeS-rich
3	26.5	26.17 ± 0.05	Fe-rich
4	25.1	25.68 ± 0.07	FeS-rich
5	23.8	23.76 ± 0.06	Fe-rich

3. Results and Discussion

Figure 1a,b show measured values of ρ and calculated values of κ of Fe-FeS, respectively, up to 5 GPa and ~1430 K from this study. Also shown are ρ and κ values of FeS [21], Fe-FeS (20 wt.% S) [16], and Fe [15] from previous studies for comparison. Fe-FeS was observed to have intermediate values of ρ compared to the end-members. This result was expected since the addition of electrically conductive Fe to FeS should decrease ρ, or equivalently the addition of semiconducting FeS to Fe should increase ρ. The Wiedemann–Franz law (WFL), $\kappa = L{\cdot}T/\rho$, using a Lorenz number (L) equal to the theoretical Sommerfeld value ($2.445{\cdot}10^{-8}$ W$\cdot\Omega\cdot$K^{-2}), was used to calculate the electronic component of κ. Similarly, as expected, the results showed that Fe-FeS is more thermally conductive than FeS but less so than Fe.

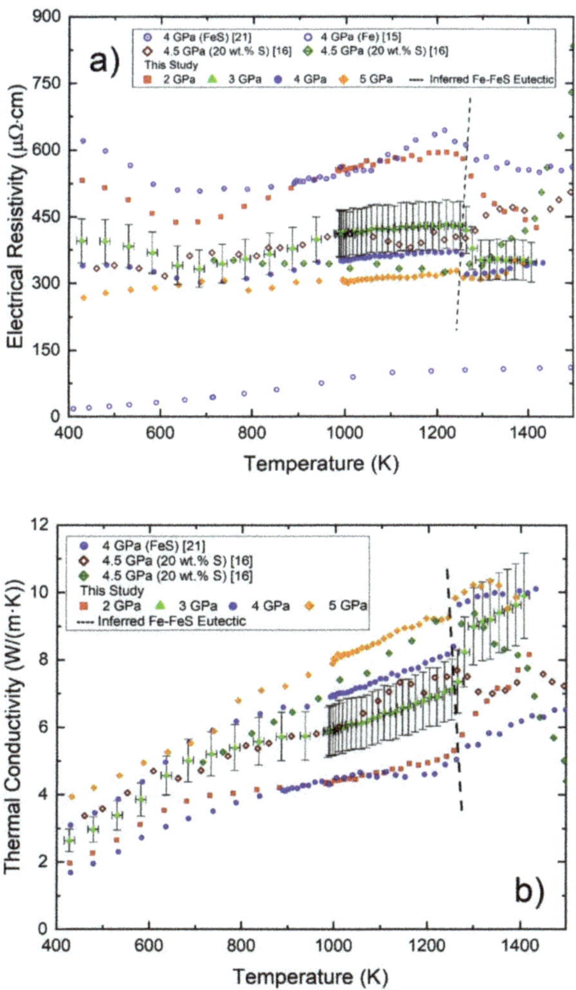

Figure 1. (a) Measured electrical resistivity of Fe-FeS at pressures of 2–5 GPa as a function of temperature. Sulfur contents (wt. %) of the samples at each pressure are: 29.56 ± 0.05 (2 GPa); 26.17 ± 0.05 (3 GPa); 25.68 ± 0.07 (4 GPa); 23.76 ± 0.06 (5 GPa). Results of this study are compared to three previous studies and include end-member compositions. The dashed line indicates the transition from solid to molten states and was used to delineate eutectic temperatures. (b) Electronic components of thermal conductivity as a function of temperature of Fe-FeS calculated from the electrical resistivity measurements using the Wiedemann–Franz law and the Sommerfeld value of the Lorenz number. Results of this study are compared to two previous studies. Data of the pure Fe end-member are not shown for scaling purposes ($>12 \ \text{W} \cdot \text{m}^{-1} \cdot \text{K}^{-1}$). The dashed line indicates the transition from solid to partially molten states.

The results of this study showed the T-coefficient of ρ, $(\partial \rho / \partial T)_P$, from room T up to a few hundred degrees increased as a function of increasing P. The T-coefficient was most negative at 2 GPa, which gradually became shallower with each P increment until 5 GPa, at which a shallow positive T-coefficient was observed. The negative T-coefficient for 2–4 GPa is consistent with the trends observed for FeS IV, the hexagonal phase of FeS, at similar conditions [21]; however, the positive T-coefficient at 5 GPa is a behavior more comparable to a metallic electrical conductor (e.g., [15–17,26–28]). One interpretation is that Fe-FeS behavior changes to become more conductor-like with increasing P at low T. Alternatively, as discussed later, this observation may be the result of the sample containing the largest proportion of Fe to account for the P-dependency of the eutectic composition [23]. Below the eutectic T at our experimental P, the binary system exhibits a two-phase solid-state regime (Fe + FeS) [11,29]. We doubt this observation is related to a solid-state phase transition since the reported P conditions at which other solid-state phases (e.g., Fe_3S, Fe_3S_2) have been observed well exceed ours (>10 GPa) [29–34]. Two sets of low T measurements of Fe 20 wt.% S at 4.5 GPa reported by Pommier [16] are more similar to our observations at 5 GPa. Those results show, however, conflicting solid-state trends as one exhibits a positive T-coefficient while the other is negative. With continued heating and increasing T, the T-coefficient of ρ at all P is positive and approaches a nearly linear trend at T leading up to the transition from solid to liquid states. These T-dependencies are also consistent with the trends observed for FeS V [16,21], the Ni-As-type phase of FeS, and Fe [14–17,19].

The T-dependent trends leading up to the eutectic T observed in this study are in good agreement with those measured by Pommier [16] who also showed, in full context of that study, that ρ and κ of Fe-FeS had intermediate values of the end-members. With the expectation that 20 wt.% S is more electrically conductive than our higher S contents, as discussed later, the measured values of ρ and κ are also in good agreement within reported error ranges up to the eutectic T. At T above the eutectic, the measured values differ considerably, with high T measurements at 4.5 GPa approaching values as large as 3000 μΩ·cm while values of ρ in this study remain nearly a magnitude less. The corresponding values of κ calculated via the WFL are expectedly low by comparison to this work. Similar observations and comparisons to Pommier [16] were made in the FeS investigations by Littleton et al. [21] in which they suggested that a possible explanation for the rapidly increasing values of ρ was apparent incomplete liquid confinement and reduced thermocouple/electrode chemical integrity. For the 4.5 GPa experiment by Pommier [16] that utilized molybdenum (Mo) electrodes and Type-C (tungsten (W)-rhenium (Re) alloys) thermocouples, a post-experiment cross-section SEM image of the same sample clearly showed and annotated liquid migration and complete dissolution of the electrodes. We echo the same interpretation as Littleton et al. [21] for these investigations on Fe-FeS.

A representative post-experiment cross-section is shown in Figure 2a,b. Figure 2a shows an image of the cross-section of the 4 GPa pressure cell centered on the sample and Figure 2b shows a back-scattered electron image of the same sample. Tabulated electron microprobe results of 15 locations correspond to labeled sites on the Figure 2b image. The bulk of the sample retained an Fe-S composition and the bulk of the W disks and arms of the Type-S TC wires retained high chemical purity. After normalizing the Fe and S content values, the microprobe analyses were used to determine an average S-content (wt.% S) of the samples: 29.56 ± 0.05 (2 GPa); 26.17 ± 0.05 (3 GPa); 25.68 ± 0.07 (4 GPa); 23.76 ± 0.06 (5 GPa). For comparison, estimates of the P-dependent eutectic composition of the Fe-FeS system, using the equation reported by Buono and Walker [23], are (wt. % S): 28.0 (2 GPa); 26.5 (3 GPa); 25.1 (4 GPa); 23.8 (5 GPa), suggesting our sample compositions are eutectic or eutectic-adjacent.

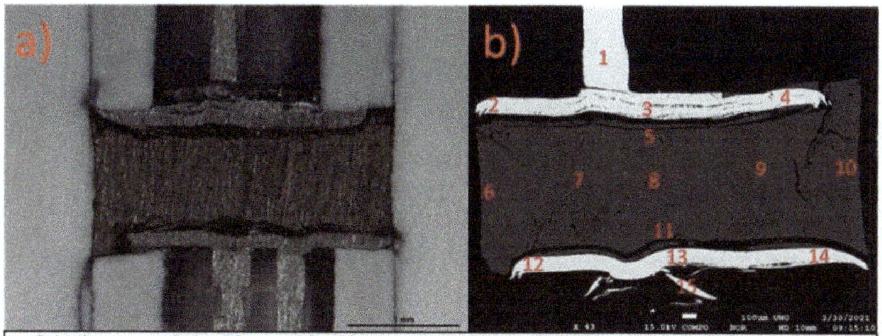

Point #	Fe (wt. %)	S (wt. %)	W (wt. %)	Pt (wt.%)	Rh (wt. %)	Total (wt. %)
1	0.00	0.07	0.00	88.56	9.20	97.84
2	0.00	0.04	97.01	0.31	0.00	97.35
3	0.00	0.03	97.40	0.28	0.00	97.71
4	0.00	0.03	96.38	0.31	0.00	96.73
5	69.49	26.33	0.02	0.00	0.00	95.84
6	73.19	22.88	0.06	0.00	0.01	96.15
7	65.38	31.16	0.06	0.01	0.00	96.60
8	72.53	23.55	0.05	0.01	0.00	96.13
9	64.62	32.25	0.06	0.00	0.00	96.92
10	70.64	25.57	0.00	0.00	0.01	96.22
11	71.58	25.18	0.08	0.00	0.01	96.86
12	0.02	0.03	97.18	0.32	0.00	97.54
13	0.00	0.04	97.62	0.29	0.00	97.95
14	0.01	0.03	96.76	0.32	0.00	97.12
15	0.00	0.09	0.00	97.49	0.40	97.97

Figure 2. (**a**) Cross-sectional view of the post-experiment 4 GPa pressure cell; (**b**) backscattered electron image of the sample from (**a**) at a different depth due to additional grinding and polishing required for electron microprobe analysis. Results of the microprobe are tabulated.

In this study, the abrupt decrease of ρ following the near linear trends is indicative of T exceeding the eutectic T and a state change of the sample from solid to partially liquid. The decrease of ρ at the solidus is consistent with other Fe alloys (e.g., [28,35]) and FeS [21]. Compared to FeS, the decrease is significantly sharper. Although fast heating rate and high measurement frequency were also used here, the difference in the sharpness of this transition is due to a sample composition in the proximity of the eutectic. An exception, however, are the 2 GPa results that show a broader and more gradual decrease similar to FeS. This could be attributed to the composition of the sample being the most S-rich and furthest from the eutectic, resulting in a larger partial melting region. In other words, the more gradual decrease of ρ reflects the gradual production of liquid with increasing T past the eutectic. While previous works have investigated a broader range of S contents [16,20], our results are the first well-constrained experimental data on near eutectic compositions in the Fe-FeS system.

A line was drawn on Figure 1a to estimate the eutectic T at each P where the resistivity trend began to decrease. Immediately left of and right of the line represent the last solid and initial liquid state measurement of the sample, respectively. For the 5 GPa experiment, the results do not show an observable partial melting region. This may indicate that the S content is either very close to or at the eutectic composition. Thus, the measurement immediately to the right of the line represents the initial measurement of a completely liquid state sample. Figure 3 compares the eutectic T estimates of this study to prior works at high P. Our results are in good agreement with several prior studies at similar P conditions and indicate that the eutectic T up to 5 GPa does not deviate far from the eutectic at 1 atm. Moreover, our results indicate a negative P-dependency of the eutectic T, a trend also reported by Fei et al. [29] and Morard et al. [39] at higher P. However, we note that our P-dependent trend is shallower by comparison. The difference in the trends could be related to the methodology used for determining the eutectic T. Fei et al. [29] determined eutectic melting on the basis of quenched textures and chemical mapping, while Morard et al. [39] used in situ X-ray diffraction. Moreover, Buono and Walker [23] asserted that the presence of hydrogen, from the breakdown of trapped water within the sample material and/or sample enclosure, may be responsible for the significant eutectic T depression in the Fe-FeS system observed by Fei et al. [29] and Morard et al. [39].

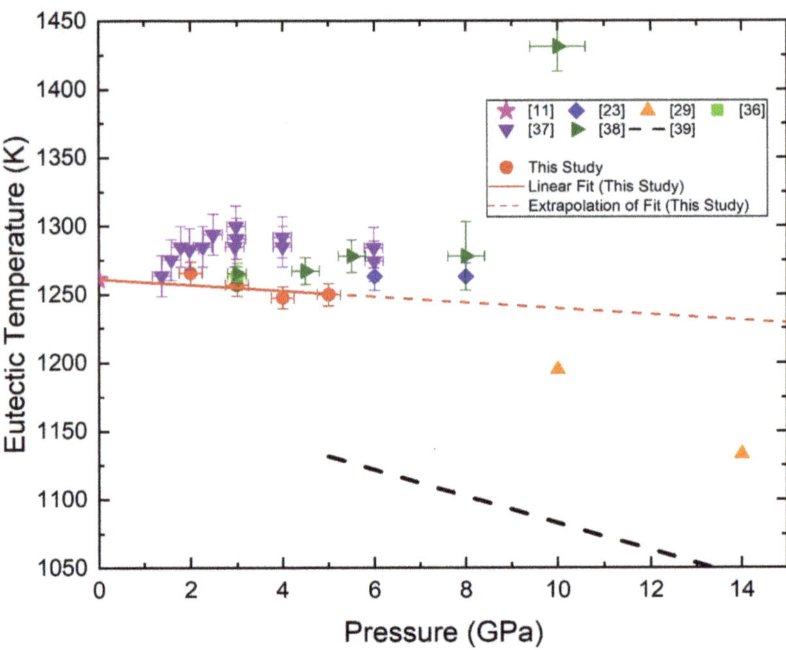

Figure 3. Experimentally determined eutectic temperatures of the Fe-FeS system as a function of pressure. The results of this work are compared to several previous works [11,23,29,36–39].

Estimates of P within the core of Ganymede range from ~5 GPa at the core–mantle boundary (CMB) to ~11 GPa at its center and T at the CMB span ~1250–2100 K [3,5,7,9,10]. We adopted the same procedure to calculate adiabatic heat flow (Q_a) at the top of Ganymede's core as Littleton et al. [21], although for the current study two linear fits to interpolate and extrapolate values of ρ at 5 GPa between 1250 and 1450 K were used, as shown in the Appendix (Figure A1). A positive linear fit (with $(\partial\rho/\partial T)_P > 0$ $\mu\Omega\cdot$cm/K) was used to account for all measurements at T above the estimated eutectic T, while a horizontal linear fit (with $(\partial\rho/\partial T)_P = 0$ $\mu\Omega\cdot$cm/K and with $\rho = 315$ $\mu\Omega\cdot$cm) was used to account for all measurements excluding the highest two temperatures. These were excluded because they may indicate the onset of deteriorating thermocouple integrity and W contamination. Q_a on the core side of the CMB was calculated using Equation (1) below:

$$Q_a = -4\pi r^2 \, \kappa (\partial T/\partial r)_a \qquad (1)$$

where $(\partial T/\partial r)_a$ is the adiabatic thermal gradient adopted from Breuer et al. [40]. Figure 4 shows Q_a at the top of the core with radius varying between 700 and 1200 km and with a CMB T from 1250 to 1450 K and P of 5 GPa alongside estimates of the heat flow through the CMB for comparison. The specified T range was chosen to allow for an entirely molten core. Our estimates of Q_a using a horizontal linear model ranged from ~8 GW for a CMB T of 1250 K and core radius of 700 km up to ~32 GW for a CMB T of 1450 K and core radius of 1200 km. These heat flow estimates are similar in magnitude to those reported by Littleton et al. [21] for a molten FeS core, which ranged from ~11 GW up to ~37 GW. However, it is important to note that the estimates of this study are for a significantly cooler molten core allowed by a eutectic-adjacent S composition. For instance, if the horizontal linear fit were extrapolated to a core T between 1600 and 1700 K used by Littleton et al. [21], the lower-bound and upper-bound Q_a in the core would be ~13 GW and ~44 GW, respectively. The heat flow in the core using the positive linear fit is more constrained than the horizontal linear fit, with estimates of Q_a ranging from ~8 GW up to ~28 GW. This result is due to the competing effects of increasing T and ρ, which are directly proportional and inversely proportional to κ, respectively, via the WFL.

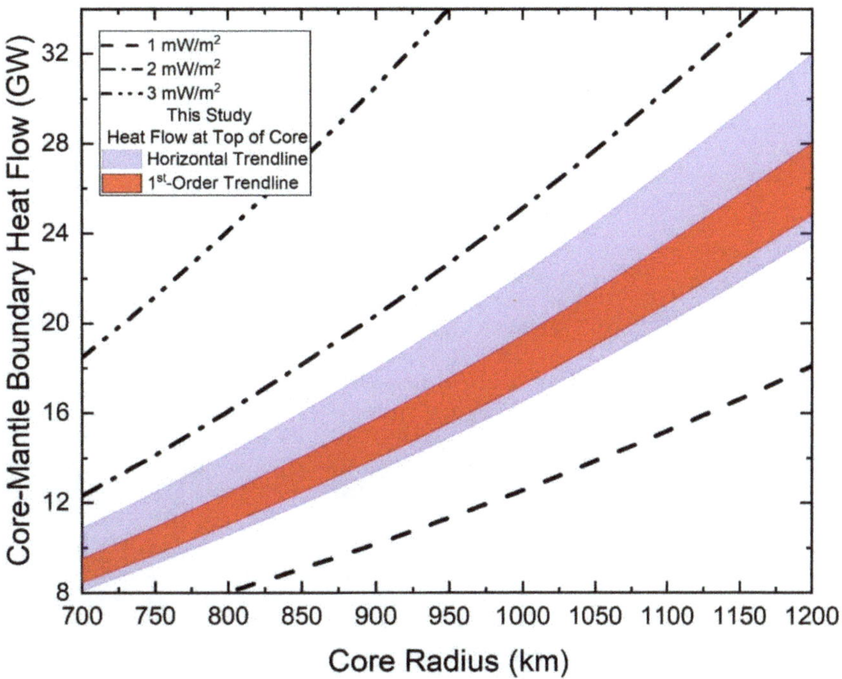

Figure 4. Calculated adiabatic heat flow at the core–mantle boundary (CMB) of Ganymede of a fully molten eutectic-adjacent core at 5 GPa. The differently dashed lines each represent different estimates of the heat flow through the CMB [3–5,10]. The shaded regions represent all values of calculated heat flow values on the core-side of the CMB. The lighter blue and darker red shaded regions are calculated values using a horizontal and linear fitting trendline, as shown in the Supporting Information section. The CMB temperature ranges from 1250 K at the bottom to 1450 K at the top of the shaded regions of this study. Propagated uncertainty for the calculations ranges from ~1 to 1.5% of the reported values.

Estimates of the heat flux out of the core and through the CMB of Ganymede range from ~1 to 6 mW/m² [3–5,10]. Both the linear and horizontal models showed that thermal convection of a molten Fe-FeS core at relatively low core T is permissible provided the heat flux on the mantle-side of the CMB exceeds ~ 1.5 mW/m². Littleton et al. [21] showed that thermal convection can carry up to one-third of the heat load to the CMB for a heat transfer of 3 mW/m² through the CMB in a Ganymede core of liquid FeS. The results of this study for a Ganymede core of near eutectic Fe-FeS show a similar heat load proportion that can be carried by thermal convection for our core model, especially when the difference in core T estimates are taken into account. Thus, our results indicate that thermal convection is permissible for the majority of the 1250–1450 K T-range and may be a source of energy to power an internal core dynamo to produce the magnetic field of Ganymede. We note that with respect to P and T, these estimates of convective heat load represent a lower bound. Based on the trends observed in this study and other investigations on the Fe-FeS system [16], increasing P will result in decreased values of ρ and increased values of κ and Q_a. Similarly, based on the results of this study, the net effect of increasing T will also result in increased values of κ and Q_a. Acting singly or together, both effects produce a more thermally conductive core and could potentially diminish the effectiveness of, or completely shut down, heat transfer by thermal convection. For instance, extrapolated core heat flow values (~13–44 GW) of our horizontal linear fit applied to a 1600–1700 K liquid core model [21] suggest that the heat load carried by thermal convection to the CMB can be reduced to one-fifth for a heat transfer of 3 mW/m^2 through the CMB. With respect to S content, these estimates provide information for a middle-ground core composition between the lower (FeS) and upper (Fe) bounds. Core thermal conductivity is expected to increase as composition becomes more Fe-rich and decrease as composition becomes more FeS-rich in other core composition models.

We note the absence of consensus regarding the P-dependence of the eutectic T in this system. A linear or non-linear increase (e.g., [38,41]) or decrease (e.g., [29,39] and this study) of the eutectic T can significantly change the lowest-bound T that allows for an entirely molten Fe-FeS core of Ganymede and, consequently, constrains the effectiveness of thermally driven convection as a heat transport mechanism. A higher eutectic T than reported here would result in a core that is more thermally conductive since the lower-bound of Q_a increases and thus decreases reliance on thermally driven convection to power an internal dynamo. Conversely, a lower eutectic T would result in a core that is less thermally conductive since the lower-bound of Q_a decreases and thus increases reliance on thermally driven convection. A CMB pressure of 5 GPa is the lowest expected value. If pressure increases at the CMB, the eutectic composition shifts towards the Fe-end of the binary system and ρ is expected to decrease while κ and Q_a are expected to increase. The uncertain behavior of the eutectic T implies the lower-bound T for either Fe or FeS crystallization regimes is uncertain. Beyond this, it is difficult to precisely describe the extent of the effect of the uncertain eutectic T on both crystallization regimes (bottom-up or top-down) and chemical- or buoyancy-driven convection. This is due to the non-linearity of the liquidus boundaries with increasing P, which marks the onset of crystallization. An increased or decreased eutectic T may widen or shrink the T range between the liquidus and solidus but may have little to no significant effect on the liquidus T for some compositions.

4. Conclusions

The presence of S as an element in the Fe-rich core of Ganymede can allow for an entirely molten core at relatively low core T due to freezing-point depression, while also affecting core transport properties influencing magnetic field generation via an internal dynamo. This study provided measurements of the ρ of Fe-FeS with eutectic-adjacent S-contents in solid and liquid states at P from 2–5 GPa, where the transition from solid to liquid states was inferred from measurements of ρ. The phase transition was used for delineation of the eutectic T of the Fe-FeS system, which showed a small negative P-dependency. Our results are the first well-constrained experimental data on near eutectic compositions in this binary system. The electronic component of κ was calculated via the WFL using the measured values of ρ, and was subsequently used to estimate Q_a on the core-side of Ganymede's CMB. The results showed that both ρ and κ had intermediate values between the end-members of the system, and that thermal convection may be permissible in the core to transport heat and act as a dynamo energy source.

Author Contributions: Conceptualization, R.A.S.; methodology, J.A.H.L., R.A.S. and W.Y.; validation, R.A.S.; formal analysis, J.A.H.L.; investigation, J.A.H.L. and W.Y.; resources, R.A.S.; data curation, J.A.H.L.; writing—original draft preparation, J.A.H.L.; writing—review and editing, R.A.S. and W.Y.; supervision, R.A.S.; project administration, R.A.S.; funding acquisition, R.A.S. All authors have read and agreed to the published version of the manuscript.

Funding: This work was supported by Canada Foundation for Innovation (11860) and Natural Sciences and Engineering Research Council of Canada (RGPIN-2018-05021).

Institutional Review Board Statement: Not applicable.

Informed Consent Statement: Not applicable.

Data Availability Statement: The data presented in this study are openly available in Mendeley Data at doi:10.17632/4zphh54798.1 [42].

Acknowledgments: We thank Jonathan Jacobs for help with machining of experimental components and two anonymous reviewers for providing insightful and constructive criticisms that helped improve the quality of this manuscript.

Conflicts of Interest: The authors declare no conflict of interest. The funders had no role in the design of the study; in the collection, analyses, or interpretation of data; in the writing of the manuscript, or in the decision to publish the results.

Appendix A

The following figure (Figure A1) shows the values and function used to interpolate the value of electrical resistivity at temperatures between 1250 K and 1450 K. The interpolated and extrapolated values were used to calculate the adiabatic conductive heat flow at the top of the core of Ganymede.

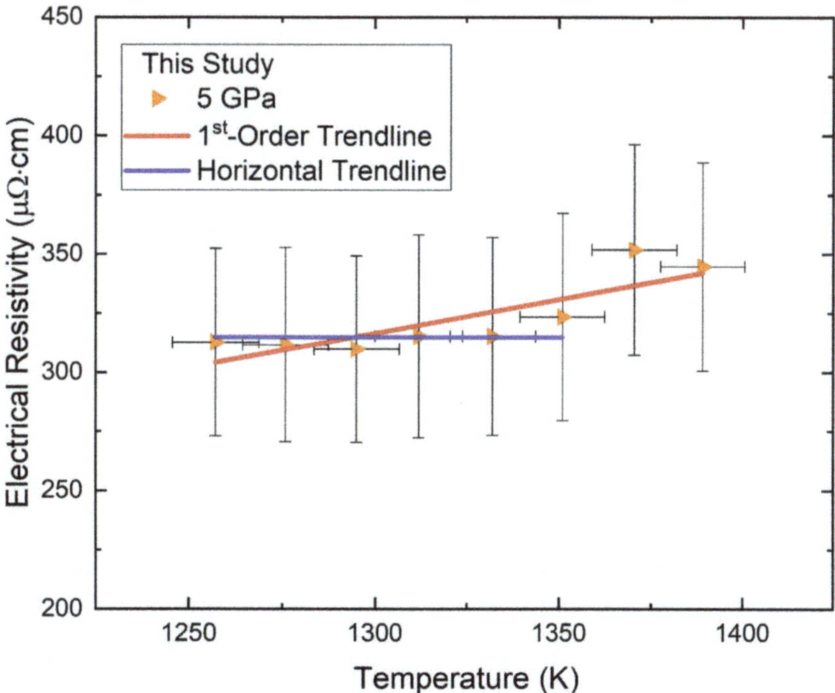

Figure A1. A first-order polynomial (red) was fitted to eight measurements and a horizontal line (315 μΩ·cm; blue) was fitted to six measurements of the 5 GPa experiment. Both fits were used to interpolate and extrapolate the values of electrical resistivity for temperatures from 1250 K to 1450 K. The values obtained from these fits were used to calculate the adiabatic conductive heat flow for the core of Ganymede described in the main text.

References

1. Connerney, J.E.P. Planetary Magnetism. In *Treatise on Geophysics*, 1st ed.; Spohn, T., Schubert, G., Eds.; Elsevier Ltd.: Amsterdamn, The Netherlands, 2007; Volume 10, pp. 243–280.
2. Olson, P. Mantle control of the geodynamo: Consequences of top-down regulation. *Geochem. Geophys. Geosyst.* **2016**, *17*, 1935–1956. [CrossRef]
3. Hauck II, S.A.; Aurnou, J.M.; Dombard, A.J. Sulfur's impact on core evolution and magnetic field generation on Ganymede. *J. Geophys. Res. Planet.* **2006**, *111*, E09008. [CrossRef]
4. Rückriemen, T.; Breuer, D.; Spohn, T. The Fe snow regime in Ganymede's core: A deep-seated dynamo below a stable snow zone. *J. Geophys. Res. Planet.* **2015**, *120*, 1095–1118. [CrossRef]
5. Rückriemen, T.; Breuer, D.; Spohn, T. Top-down freezing in a Fe-FeS core and Ganymede's present-day magnetic field. *Icarus* **2018**, *307*, 172–196. [CrossRef]
6. Schubert, G.; Zhang, K.; Kivelson, M.G.; Anderson, J.D. The magnetic field and internal structure of Ganymede. *Nature* **1996**, *384*, 544–545. [CrossRef]
7. Sohl, F.; Spohn, T.; Breuer, D.; Nagel, K. Implications from Galileo observations on the interior structure and chemistry of the Galilean satellites. *Icarus* **2002**, *157*, 101–119. [CrossRef]
8. Scott, H.P.; Williams, Q.; Ryerson, F.J. Experimental constraints on the chemical evolution of large icy satellites. *Earth Planet. Sci. Lett.* **2002**, *203*, 399–412. [CrossRef]
9. Bland, M.T.; Showman, A.P.; Tobie, B. The production of Ganymede's magnetic field. *Icarus* **2008**, *198*, 384–399. [CrossRef]
10. Kimura, J.; Nakagawa, T.; Kurita, K. Size and compositional constraints of Ganymede's metallic core for driving an active dynamo. *Icarus* **2009**, *202*, 216–224. [CrossRef]
11. Kubaschewski, I. Iron-Sulphur. In *IRON—Binary Phase Diagrams*; Springer: Berlin/Heidelberg, Germany; Düsseldorf, Germany, 1982; pp. 125–128.
12. Gomi, H.; Yoshino, T. Impurity resistivity of fcc and hcp Fe-based alloys: Thermal stratification at the top of the core of super-Earths. *Front. Earth Sci.* **2018**, *6*, 217. [CrossRef]
13. Konôpková, Z.; McWilliams, R.S.; Gómez-Pérez, N.; Goncharov, A.F. Direct measurement of thermal conductivity of solid iron at planetary core conditions. *Nature* **2016**, *534*, 99–101. [CrossRef]
14. Ohta, K.; Kuwayama, Y.; Hirose, K.; Shimizu, K.; Ohishi, Y. Experimental determination of the electrical resistivity of iron at Earth's core conditions. *Nature* **2016**, *534*, 95–98. [CrossRef]
15. Silber, R.E.; Secco, R.A.; Yong, W.; Littleton, J.A.H. Electrical resistivity of liquid Fe to 12 GPa: Implications for heat flow in cores of terrestrial bodies. *Sci. Rep.* **2018**, *8*, 10758. [CrossRef]
16. Pommier, A. Influence of sulfur on the electrical resistivity of a crystallizing core in small terrestrial bodies. *Earth Planet. Sci. Lett.* **2018**, *496*, 37–46. [CrossRef]
17. Yong, W.; Secco, R.A.; Littleton, J.A.H.; Silber, R.E. The iron invariance: Implications for thermal convection in Earth's core. *Geophys. Res. Lett.* **2019**, *46*, 11065–11070. [CrossRef]
18. Hsieh, W.-P.; Goncharov, A.F.; Labrosse, S.; Holtgrewe, N.; Lobanov, S.S.; Chuvashova, I.; Deschamps, F.; Lin, J.-F. Low thermal conductivity of iron-silicon alloys at Earth's core conditions with implications for the geodynamo. *Nat. Commun.* **2020**, *11*, 3332. [CrossRef]
19. Ezenwa, I.C.; Yoshino, T. Martian core heat flux: Electrical resistivity and thermal conductivity of liquid Fe at Martian core P-T conditions. *Icarus* **2021**, *360*, 114367. [CrossRef]
20. Manthilake, G.; Chantel, J.; Monteux, J.; Andrault, D.; Bouhifd, M.A.; Casanova, N.B.; Boulard, E.; Guignot, N.; King, A.; Itie, J.P. Thermal conductivity of FeS and its implications for Mercury's long-sustaining magnetic field. *J. Geophys. Res. Planet.* **2019**, *124*, 2359–2368. [CrossRef]
21. Littleton, J.A.H.; Secco, R.A.; Yong, W. Electrical resistivity of FeS at high pressures and temperatures: Implications of thermal transport in the core of Ganymede. *J. Geophys. Res. Planet.* **2021**, *126*, e2020JE006793. [CrossRef]
22. Kuskov, O.L.; Kronrod, V.A. Core sizes and internal structure of Earth's and Jupiter's satellites. *Icarus* **2001**, *151*, 204–227. [CrossRef]
23. Buono, A.S.; Walker, D. H, not O or pressures, causes eutectic T depression in the Fe-FeS system to 8 GPa. *Meteorit. Planet. Sci.* **2015**, *50*, 547–554. [CrossRef]
24. Secco, R.A. High p,T physical property studies of Earth's interior: Thermoelectric power of solid and liquid Fe up to 6.4 GPa. *Can. J. Phys.* **1995**, *73*, 287–294. [CrossRef]
25. Littleton, J.A.H.; Secco, R.A.; Yong, W.; Berrada, M. Electrical resistivity and thermal conductivity of W and Re up to 5 GPa and 2300 K. *J. Appl. Phys.* **2019**, *125*, 135901. [CrossRef]
26. Pommier, A. Experimental investigations of the effect of nickel on the electrical resistivity of Fe-Ni and Fe-Ni-S alloys under pressure. *Am. Mineral.* **2020**, *105*, 1069–1077. [CrossRef]
27. Silber, R.E.; Secco, R.A.; Yong, W.; Littleton, J.A.H. Heat Flow in Earth's Core From Invariant Electrical Resistivity of Fe-Si on the Melting Boundary to 9 GPa: Do Light Elements Matter? *J. Geophys. Res. Solid Earth* **2019**, *124*, 5521–5543. [CrossRef]
28. Berrada, M.; Secco, R.A.; Yong, W.; Littleton, J.A.H. Electrical resistivity measurements of Fe-Si with implications for the early lunar dynamo. *J. Geophys. Res. Planet.* **2020**, *125*, e2020JE006380. [CrossRef]

29. Fei, Y.; Bertka, C.M.; Finger, L.W. High-Pressure Iron-Sulfur Compound, Fe$_3$S, and Melting Relations in the Fe-FeS System. *Science* **1997**, *275*, 1621–1623. [CrossRef]
30. Fei, Y.; Li, J.; Bertka, C.M.; Prewitt, C.T. Structure type and bulk modulus of Fe$_3$S, a new iron-sulfur compound. *Am. Mineral.* **2000**, *85*, 1830–1833. [CrossRef]
31. Li, J.; Fei, Y.; Mao, H.K.; Hirose, K.; Shieh, S.R. Sulfur in the Earth's inner core. *Earth Planet. Sci. Lett.* **2001**, *193*, 509–514. [CrossRef]
32. Stewart, A.J.; Schmidt, M.W.; van Westrenen, W.; Liebske, C. Mars: A new core-crystallization regime. *Science* **2007**, *316*, 1323–1325. [CrossRef]
33. Morard, G.; Andrault, D.; Guignot, N.; Sanloup, C.; Mezouar, M.; Petitgirard, S.; Fiquet, G. In situ determination of Fe-Fe$_3$S phase diagram and liquid structural properties up to 65 GPa. *Earth Planet. Sci. Lett.* **2008**, *272*, 620–626. [CrossRef]
34. Kamada, S.; Terasaki, H.; Ohtani, E.; Sakai, T.; Kikegawa, T.; Ohishi, Y.; Hirao, N.; Sata, N.; Kondo, T. Phase relationships of the Fe-FeS system in conditions up to the Earth's outer core. *Earth Planet. Sci. Lett.* **2010**, *293*, 94–100. [CrossRef]
35. Pommier, A.; Leinenweber, K.; Tran, T. Mercury's thermal evolution controlled by an insulating liquid outermost core? *Earth Planet. Sci. Lett.* **2019**, *517*, 125–134. [CrossRef]
36. Brett, R.; Bell, P.M. Melting relations in the Fe-rich portion of the system Fe-FeS at 30 kb pressure. *Earth Planet. Sci. Lett.* **1969**, *6*, 479–482. [CrossRef]
37. Ryzhenko, B.; Kennedy, G.C. The effect of pressure on the eutectic in the system Fe-FeS. *Am. J. Sci.* **1973**, *273*, 803–810. [CrossRef]
38. Usselman, T.M. Experimental approach to the state of the core: Part, I. The liquidus relations of the Fe-rich portion of the Fe-Ni-S system from 30 to 100 kb. *Am. J. Sci.* **1975**, *275*, 278–290. [CrossRef]
39. Morard, G.; Sanloup, C.; Fiquet, G.; Mezouar, M.; Rey, N.; Poloni, R.; Beck, P. Structure of eutectic Fe-FeS melts to pressures up to 17 GPa: Implications for planetary cores. *Earth Planet. Sci. Lett.* **2007**, *262*, 128–139. [CrossRef]
40. Breuer, D.; Rückriemen, T.; Spohn, T. Iron snow, crystal floats, and inner-core growth: Modes of core solidification and implications for dynamos in terrestrial planets and moons. *Prog. Earth Planet. Sci.* **2015**, *2*, 39. [CrossRef]
41. Boehler, R. Fe-FeS eutectic temperatures to 620 kbar. *Phys. Earth Planet. Inter.* **1996**, *96*, 181–186. [CrossRef]
42. Littleton, J.A.H.; Secco, R.A.; Yong, W. *Thermal Convection in The Core of Ganymede Inferred from Liquid Eutectic Fe-FeS Electrical Resistivity at High Pressures*; [dataset]; Mendeley Data; Elsevier: Amsterdam, The Netherlands, 2020; Version 1. [CrossRef]

crystals

Article

Effect of High Pressure and Temperature on the Evolution of Si Phase and Eutectic Spacing in Al-20Si Alloys

Rong Zhang, Chunming Zou, Zunjie Wei * and Hongwei Wang *

School of Materials Science and Engineering, Harbin Institute of Technology, Harbin 150001, China;
zhang_r1992@163.com (R.Z.); zouchunming1977@163.com (C.Z.)
* Correspondence: weizj@hit.edu.cn (Z.W.); wanghw@hit.edu.cn (H.W.)

Abstract: The microstructure of the Si phase in Al-20Si alloys solidified under high pressure was investigated. The results demonstrate that the morphology of Si phase transformed (bulk→short rod→long needle) with the increase of superheat temperature under high pressure. At a pressure of 3 GPa and a superheat temperature of 100 K, a microstructure with a uniform distribution of fine Si phases on the α-Al matrix was obtained in the Al-20Si alloy. In addition, a mathematical model was developed to analyze the spacing variation of the lamellar Al-Si eutectics under the effect of pressure. The lamellar Al-Si eutectics appeared at 2 GPa and superheat temperatures of 70–150 K, and at 3 GPa and superheat temperatures of 140–200 K. With the increase of pressure from 2 GPa to 3 GPa, the average spacing of lamellar Al-Si eutectics decreased from 1.2–1.6 μm to 0.9–1.1 μm. In binary alloys, the effect of pressure on the eutectic spacing is related to the volume change of the solute phase from liquid to solid. When the volume change of the solute phase from liquid to solid is negative, the lamellar eutectic spacing decreases with increasing pressure. When it is positive, the eutectic spacing increases with increasing pressure.

Keywords: high pressure; eutectic spacing; Al-Si alloy; superheat

Citation: Zhang, R.; Zou, C.; Wei, Z.; Wang, H. Effect of High Pressure and Temperature on the Evolution of Si Phase and Eutectic Spacing in Al-20Si Alloys. *Crystals* **2021**, *11*, 705. https://doi.org/10.3390/cryst11060705

Academic Editor: Marek Szafrański, Simone Anzellini and Daniel Errandonea

Received: 21 May 2021
Accepted: 17 June 2021
Published: 20 June 2021

Publisher's Note: MDPI stays neutral with regard to jurisdictional claims in published maps and institutional affiliations.

1. Introduction

The study of the influence of compression in the properties of metallic alloys, including morphology and microstructure, is extremely relevant for multiple technological applications [1–3]. At present, solidification phenomena are mainly described by nucleation [4–6], constitutional undercooling [7], interface stability [8–13], eutectic growth [14], dendritic growth [15,16], etc. These theories mainly consider the effects of temperature and concentration but ignore that of pressure on the solidification process. The high pressure at the GPa-level causes the microstructure of the material to be greatly refined [17–19], while at the same time, the solubility of the solute in the melt increases [20–22]. Therefore, the solidification process must take into account the effect of GPa-level pressure.

Similarly, the morphology and spacing of the eutectic can be affected by the GPa-level pressure. Temperature, pressure and composition are important variables in the study of microstructure evolution. While considering the effect of GPa-level pressure on the microstructure, the effect of temperature on the microstructure cannot be ignored. Under the combined effect of pressure and temperature, higher pressure enables the Al-20Si alloy to maintain stable interfacial morphology at higher temperatures [23]. The increase in pressure from 1 atm to 3 GPa leads to an increase in the tensile strength of the Al-20Si alloy and the refinement of the eutectic [24]. The pressure has a significant effect on the eutectics. It is necessary to study the eutectic spacing of binary alloys under the effect of high pressure.

In this paper, the Si phase in Al-20Si alloy solidified under high pressure was investigated. Then, based on the classical eutectic growth model, the mathematical model of eutectic spacing under the effect of pressure was developed. Finally, the eutectic spacing was investigated by this mathematical model.

71

2. Materials and Methods

Al-20Si (wt%) powders (Tian Jiu Technology, purity 99.9 %, 9–11 μm, Changsha, China) were used as the raw materials. The powders were cold-pressed under 300 MPa to obtain Φ20 × 18 mm cylinders. Figure 1 shows the HTDS-032F high-pressure equipment. The pressure was calibrated by the I-II phase transition of Bi (at 2.55 GPa) in our previous works [25]. The temperature was calibrated by a thermocouple (B-type). The details of the cell assembly can be found in [26]. The samples were pressurized to the target pressure (2 GPa and 3 GPa), then a current was applied to the graphite heater to heat the samples to the target temperature for 5 min, and finally, the current was disconnected and cooled to room temperature at a rate of 20 K/S. The specimens were etched with 0.5 vol% hydrofluoric acid solution for 10 s or 20 min (deep-etched). The microstructure was analyzed using a scanning electron microscope (Merlin Compact, ZEISS, Germany) operated at 20 kV.

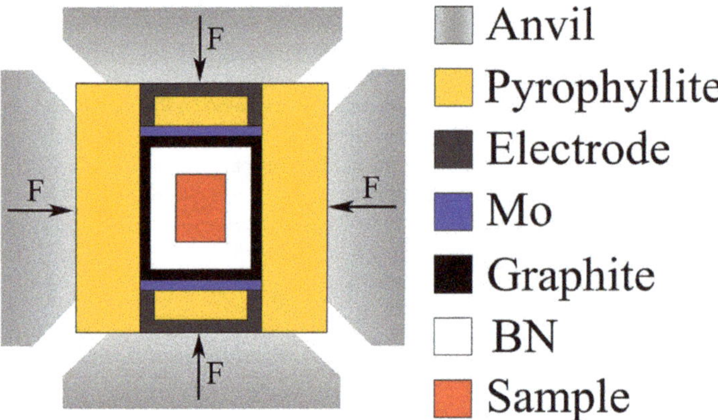

Figure 1. High-pressure experiment.

3. Results

3.1. Microstructure of Al-Si Alloys under High Pressure

Figures 2 and 3 show the microstructure morphologies of the lamellar Al-Si eutectics and anomalous Si phases under high pressure, respectively. The dark phases are the α-Al phases, while the bright phases are the Si phases. In our previous study [23], the microstructure morphologies of the α-Al phases in the high-pressure solidified Al-20Si alloy underwent a process of change from dendritic to cellular, spherical and planar as the melt superheat temperature decreased. It is clearly observed in Figure 2 that the Al-Si eutectics have lamellar microstructure morphologies when the α-Al phases grow in dendritic and cellular forms. As can be clearly observed in Figure 3, anomalous Si phases appear when the α-Al phases grow in planar and spherical forms, and the Si phases have bulk morphologies. Under high-pressure solidification conditions, changes in the superheat temperature of the melt had an effect not only on the microstructures of the α-Al phases but also on that of the Si phases. With an increase in the superheat temperature of the melt, the microstructures of the Si phases in the high-pressure solidified Al-20Si alloy changed from bulk to short rods and then to long needles.

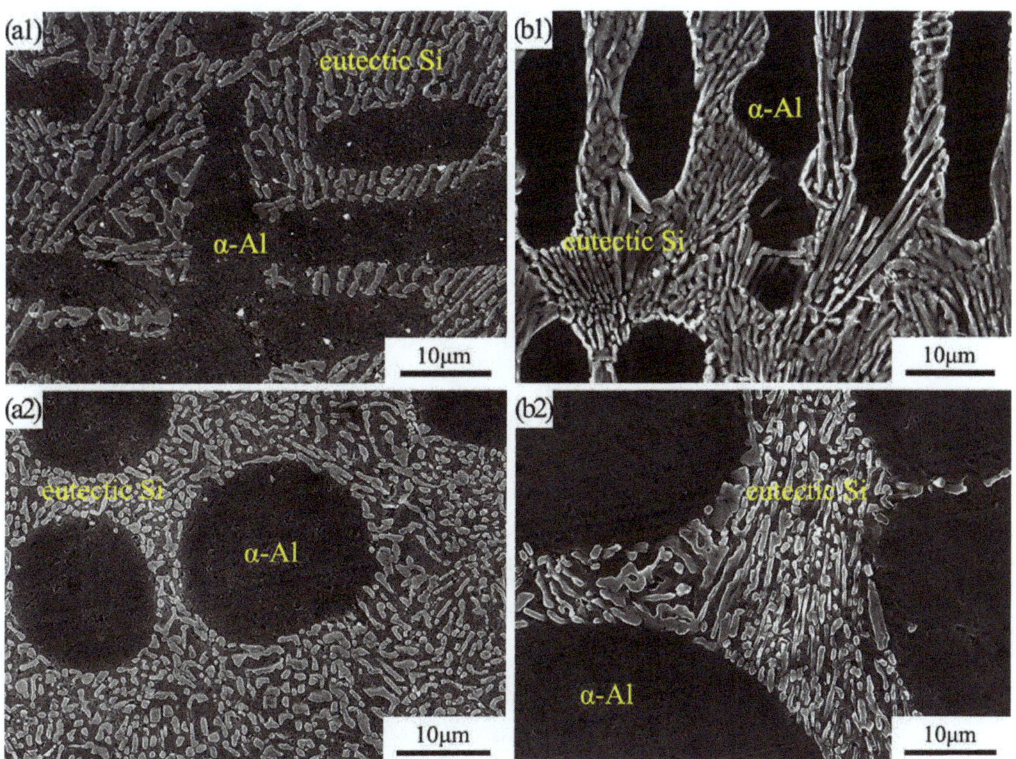

Figure 2. Microstructures of lamellar Al-Si eutectics at different pressures and superheat temperatures: (**a1,a2**) 2 GPa; (**b1,b2**) 3 GPa; (**a1**) 150 K; (**a2**) 70 K; (**b1**) 200 K; (**b2**) 140 K.

It should be noted that what appears on Figure 2 as needles and rods in fact is a section view of plates (lamella). The "aspect ratio" was defined as the ratio between the longest and shortest lines joining two points of the Si phase contour and passing through the centroid. The aspect ratios of the Si phases at different pressures and superheat temperatures are shown in Figure 4. The aspect ratios of the Si phases increased with increasing superheat temperature, corresponding to the transformation of the Si phases (bulk→short rod→long needle).

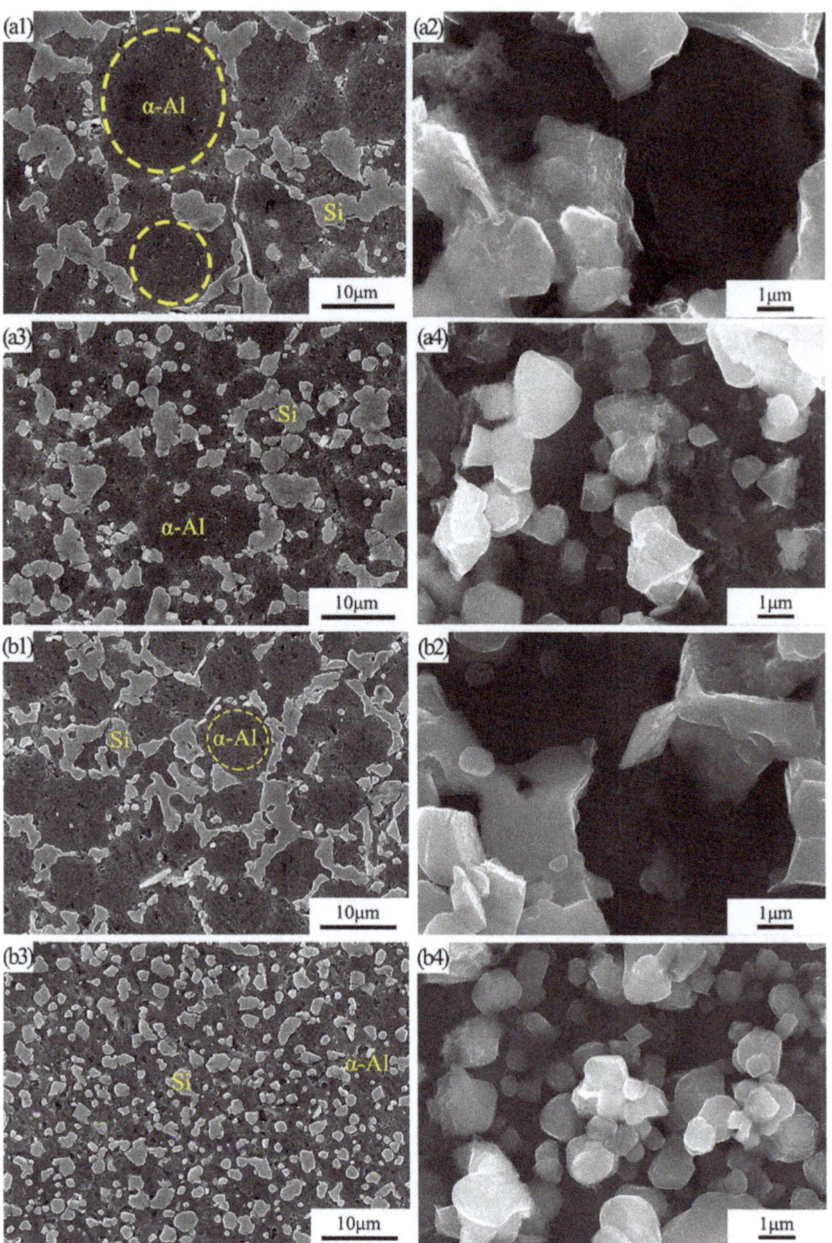

Figure 3. Microstructures of anomalous Si phases at different pressures and superheat temperatures: (**a1–a4**) 2 GPa; (**b1–b4**) 3 GPa; (**a1,a2**) 60 K; (**a3,a4**) 50 K; (**b1,b2**) 110 K; (**b3,b4**) 100 K.

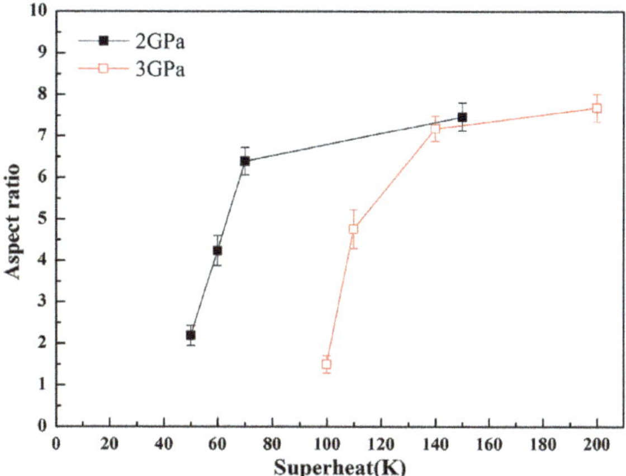

Figure 4. Aspect ratios of Si phases.

The spacing of lamellar Al-Si eutectics was measured by the random intercepts, as shown in Figure 5. Only the eutectic spacings when the Al-Si eutectics were lamellar were counted and averaged in the figure. The average eutectic spacings of the lamellar Al-Si eutectics decreased with increasing pressure.

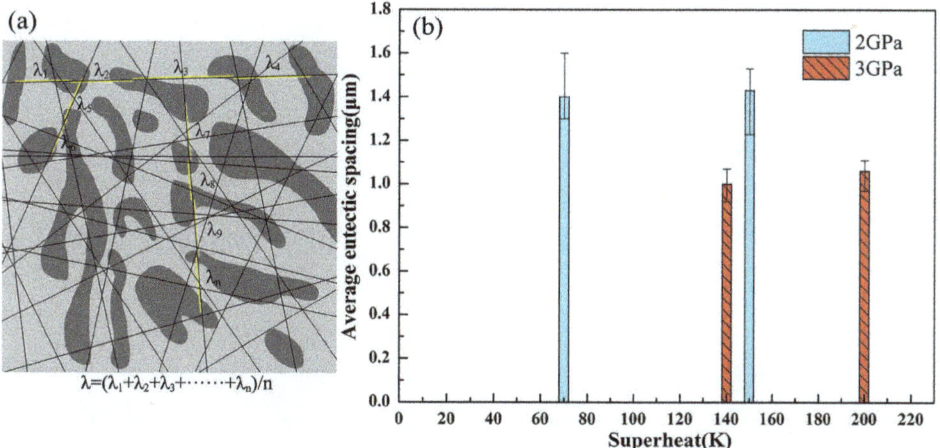

Figure 5. Average eutectic spacings of lamellar Al-Si eutectics: (**a**) random intercepts; (**b**) statistics.

It is worth mentioning that the effect of pressure on eutectic spacing should include the condition of temperature, and the two variables of pressure and temperature could not be considered independently. Based on the evolution of microstructures of Al-20Si alloy in Figures 2 and 3, it could be inferred that when the superheat temperature was 100 K, it was lamellar eutectic at 2 GPa, while it was fine Si phases at 3 GPa. It was not possible to compare the average spacings between the two.

The lamellar Al-Si eutectics appeared at 2 GPa and superheat temperatures of 70–150 K, and at 3 GPa and superheat temperatures of 140–200 K. When the Al-Si eutectics were lamellar, the spacings of the two could be compared. In the above superheat temperature range, the effect of temperature on the average eutectic spacing was small. With the increase of pressure from 2 GPa to 3 GPa, the average spacing of lamellar Al-Si eutectics decreased from 1.2–1.6 μm to 0.9–1.1 μm. The following discussions of eutectic spacing were based on the foundation that the Al-Si eutectics were lamellar.

3.2. Morphological Evolution of the Si Phase

The increase in pressure not only increases the eutectic temperature of the Al-Si alloy, but also increases the Si content corresponding to the eutectic point [27,28]. Al-20Si alloy was actually in the hypoeutectic zone at pressures of 2 GPa and 3 GPa. The α-Al phases precipitated first, and the growth and distribution of the Si phases were influenced by the solute diffusion. The solute diffusion coefficient under high pressures can be calculated by Ref. [29]:

$$D_P = D \exp\left(-\frac{PV_0}{RT}\right) \tag{1}$$

where P (Pa) is the pressure, V_0 (m^3/mol) is the original volume of the liquid phase, $R = 8.314$ (J/(mol·K)) is the gas constant, T (K) is the temperature, D (m^2/s) is the solute diffusion coefficient in liquid. It can be seen from Equation (1) that the increase in pressure caused an exponential decrease in the solute diffusion coefficient.

When the superheat temperatures were lower, the α-Al phase grew in planar and spherical forms, while the temperature of the solid-liquid interface front increased due to the release of latent heat. Due to the lower melt temperature and exponential decrease of solute diffusion coefficient, Si atomic clusters were enriched at the solid-liquid interface front and did not have time to diffuse. When the compositional supercooling at the solid-liquid interface front reached the nucleation supercooling of the Si phase, the Si phases nucleated and grew into bulk crystals. The Si phase in the α-Al matrix was precipitated in bulk form, resulting in the disappearing of the typical lamellar eutectic and turning into an anomalous Si phase.

When the superheat temperatures were higher, the α-Al phases grew in cellular and dendritic forms. Due to the higher temperature of the melt, the Si atoms were enriched at the front of the solid-liquid interface had a certain amount of time to diffuse laterally, leading to the formation of lamellar Al-Si eutectics. The increase in pressure led to an exponential decrease in the solute diffusion coefficient, resulting in the diffusion distance of Si atoms under a pressure of 3 GPa being smaller than that of Si atoms under a pressure of 2 GPa, and thus the eutectic spacing became progressively smaller with the increase in pressure. Analyzed from the aspect of solute diffusion coefficient, the pressure had an effect on the solute diffusion coefficient, which in turn had an effect on the eutectic spacing. In addition, the increase in melt temperature led to a longer solidification time, so the morphology of the Si phase changed from bulk→short rod→long needle, making the aspect ratio of the Si phase increase. In fact, the lamellae of Al-Si eutectics became thinner and longer.

3.3. Effect of Pressure on Eutectic Spacing

For the lamellar eutectic growth problem, Jackson and Hunt developed a J-H mathematical model to calculate and analyze the eutectic growth process [30]. Kurz extended and simplified the theoretical expressions to predict the eutectic spacing for both the high Péclet number and the low Péclet number [13]:

$$\lambda^2 V = \frac{K_r}{K_c} \tag{2}$$

$$K_c = \frac{|m_\alpha||m_\beta|}{|m_\alpha| + |m_\beta|} \cdot \frac{1 - K}{2\pi D} \cdot \frac{2\pi/P_c}{\left[1 + (2\pi/P_c)^2\right]^{1/2} - 1 + 2k} \tag{3}$$

$$K_r = \frac{2(1 - f)|m_\beta|\Gamma_\alpha \sin\theta_\alpha + 2f|m_\alpha|\Gamma_\beta \sin\theta_\beta}{f(1 - f)(|m_\alpha| + |m_\beta|)} \tag{4}$$

where λ (m) is the lamellar eutectic spacing, m_α is the liquidus slope of α phase, m_β is the liquidus slope of β phase, V (m/s) is the growth rate, P_c is the solute Péclet number, f is the volume fraction of α phase, θ (°) is wetting angle, k is the solute distribution coefficient, Γ is the Gibbs-Thomson coefficient.

According to Kurz's theoretical derivation [13], inserting Equations (3) and (4) into Equation (2) and separating the terms related to P_c, the Equation (2) is obtained as:

$$\lambda^2 V = D\Theta \frac{\left(P_c^2 + 4\pi^2\right)^{1/2} - P_c + 2kP_c}{2\pi(1 - K)} \tag{5}$$

where Θ is a constant which is a combination of all fixed parameters in Equations (2)–(4).

The solute Péclet number can be expressed as [13]:

$$P_c = \frac{\lambda V}{2D} \tag{6}$$

As the pressure increased, the solute diffusion coefficient decreased exponentially, so the solute Péclet number increased, $P_c \gg 2\pi$ under high pressure. Equation (5) is simplified as follows:

$$\lambda = \Theta \frac{k}{2\pi(1 - K)} \tag{7}$$

The effect of pressure on the solute distribution coefficient can be calculated by Refs. [31–33]:

$$k_P = K\left(1 + \frac{\Delta V_m^B \Delta P}{RT_m}\right) \tag{8}$$

where subscript m represents melting, ΔVBm (m³/mol) is the volume change of component B, ΔP (Pa) is the change in pressure.

By inserting Equation (8) into Equation (7), Equation (7) is obtained as:

$$\lambda = \Theta \frac{k\left(1 + \frac{\Delta V_m^B \Delta P}{RT_m}\right)}{2\pi\left[1 - K\left(1 + \frac{\Delta V_m^B \Delta P}{RT_m}\right)\right]} \tag{9}$$

Equation (9) reflected the relationship between pressure and eutectic spacing. According to Equation (9), the result of calculation for Al-20Si alloy is shown in Figure 6. The calculated results were consistent with the variation pattern of eutectic spacing of the Al-20Si alloy observed in Figure 2. The increase in solidification pressure decreased the eutectic spacing. It can be seen from Figure 6 that the calculated results agree with the results for atmospheric pressure and 1 GPa [28].

It can be seen from Equation (9) that the eutectic spacing under the effect of high pressure depended on the change of the molar volume of the solute phase in the binary alloy during the liquid-solid phase transition. For the Al-20Si alloy, the molar volume of the Si phase becomes larger during the transition from liquid to solid ($\Delta VSim$ = –1.122 cm³/mol [34]). The k_P and λ decreased with increasing pressure. When the volume of the solute phase in the alloy decreased during the transition from liquid to solid, the variation of k_P and λ with pressure was opposite to the above trend. Therefore, once the trend of the molar volume of the solute phase in the alloy during the transition from liquid to solid was determined, the trend of the effect of pressure on the average eutectic spacing could be predicted. This equation was applicable to the condition that the eutectic phase grew in lamellar form.

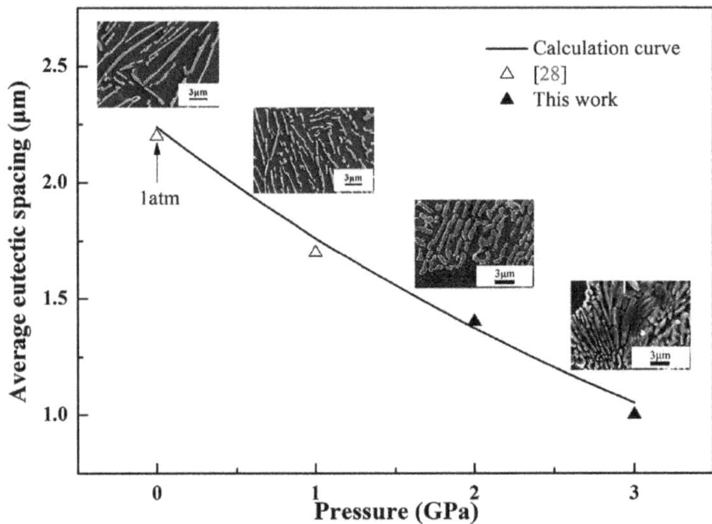

Figure 6. Relationship between pressure and eutectic spacing in Al-20Si alloys solidified under different pressures.

4. Conclusions

The microstructure of the Si phase of the Al-20Si alloy under high pressure was investigated. Then, a mathematical model was developed to analyze the trend of eutectic spacing of Al-Si alloy under the effect of pressure. The results are obtained as follows:

(1) The morphology of Si phase transformed (bulk→short rod→long needle) with the increase of superheat temperature under high pressure. At a pressure of 3 GPa and a superheating temperature of 100 K, a microstructure with a uniform distribution of fine Si phases on the α-Al matrix was obtained in the Al-20Si alloy.

(2) The lamellar Al-Si eutectics appeared at 2 GPa and superheat temperatures of 70–150 K, and at 3 GPa and superheat temperatures of 140–200 K. With the increase of pressure from 2 GPa to 3 GPa, the average spacing of lamellar Al-Si eutectics decreased from 1.2–1.6 μm to 0.9–1.1 μm.

(3) In binary alloys, the effect of pressure on the eutectic spacing is related to the volume change of the solute phase from liquid to solid. When the volume change of the solute phase from liquid to solid is negative, the lamellar eutectic spacing decreases with increasing pressure. When it is positive, the eutectic spacing increases with increasing pressure.

Author Contributions: Conceptualization, R.Z. and C.Z.; methodology, R.Z.; formal analysis, R.Z., C.Z. and Z.W.; investigation, R.Z. and H.W.; resources, Z.W.; data curation, R.Z. and Z.W.; writing—original draft preparation, R.Z. and C.Z.; writing—review and editing, R.Z., Z.W. and H.W. All authors have read and agreed to the published version of the manuscript.

Funding: This work was supported by the National Natural Science Foundation of China [Nos. 51774105] and the "Head Goose" team project [XNAUEA5640208420].

Data Availability Statement: All data and models during the study appear in the submitted article.

Conflicts of Interest: The authors declare no conflict of interest.

References

1. Mostafa, A.; Adaileh, W.; Awad, A.; Kilani, A. Mechanical Properties of Commercial Purity Aluminum Modified by Zirconium Micro-Additives. *Crystals* **2021**, *11*, 270. [CrossRef]
2. MacLeod, S.; Errandonea, D.; Cox, G.A.; Cynn, H.; Daisenberger, D.; Finnegan, S.; McMahon, M.; Munro, K.; Popescu, C.; Storm, C. The phase diagram of Ti-6Al-4V at high-pressures and high-temperatures. *J. Phys. Condens. Matter* **2021**, *33*, 154001. [CrossRef] [PubMed]
3. Smith, D.; Joris, O.P.J.; Sankaran, A.; Weekes, H.E.; Bull, D.J.; Prior, T.J.; Dye, D.; Errandonea, D.; Proctor, J.E. On the high-pressure phase stability and elastic properties of β-titanium alloys. *J. Phys. Condens. Matter* **2017**, *29*, 155401. [CrossRef] [PubMed]
4. Thompson, C.V.; Spaepen, F. Homogeneous crystal nucleation in binary metallic melts. *Acta Metall.* **1983**, *31*, 2021–2027. [CrossRef]
5. Ishihara, K.; Maeda, M.; Shingu, P. The nucleation of metastable phases from undercooled liquids. *Acta Metall.* **1985**, *33*, 2113–2117. [CrossRef]
6. Cantor, B.; Doherty, R. Heterogeneous nucleation in solidifying alloys. *Acta Metall.* **1979**, *27*, 33–46. [CrossRef]
7. Tiller, W.; Jackson, K.; Rutter, J.; Chalmers, B. The redistribution of solute atoms during the solidification of metals. *Acta Metall.* **1953**, *1*, 428–437. [CrossRef]
8. Mullins, W.W.; Sekerka, R.F. Morphological stability of a particle growing by diffusion or heat flow. *J. Appl. Phys.* **1963**, *34*, 323–329. [CrossRef]
9. Mullins, W.W.; Sekerka, R.F. Stability of a planar interface during solidification of a dilute binary alloy. *J. Appl. Phys.* **1964**, *35*, 444–451. [CrossRef]
10. Sekerka, R.F. A Stability function for explicit evaluation of the Mullins-Sekerka Interface Stability Criterion. *J. Appl. Phys.* **1965**, *36*, 264–268. [CrossRef]
11. Sekerka, R.F. Morphological stability. *J. Cryst. Growth* **1968**, *3–4*, 71–81. [CrossRef]
12. Trivedi, R.; Kurz, W. Morphological stability of a planar interface under rapid solidification conditions. *Acta Metall.* **1986**, *34*, 1663–1670. [CrossRef]
13. Kurz, W.; Fisher, D.J. *Fundamentals of Solidification*, 4th ed.; Trans Tech Publications LTD: Zurich, Switzerland, 1986.
14. Trivedi, R.; Magnin, P.; Kurz, W. Theory of eutectic growth under rapid solidification conditions. *Acta Metall.* **1987**, *35*, 971–980. [CrossRef]
15. Trivedi, R. The role of interfacial free energy and interface Kinetics during the growth of precipitate plates and needles. *Metall. Mater. Trans. B* **1970**, *1*, 921–927.
16. Lipton, J.; Kurz, W.; Trivedi, R. Rapid dendrite growth in undercooled alloys. *Acta Metall.* **1987**, *35*, 957–964. [CrossRef]
17. Wei, Z.J.; Wang, Z.L.; Wang, H.W.; Cao, L. Evolution of microstructures and phases of Al-Mg alloy under 4GPa high pressure. *J. Mater. Sci.* **2007**, *42*, 7123–7128. [CrossRef]
18. Xu, R. The effect of high pressure on solidification microstructure of Al-Ni-Y alloy. *Mater. Lett.* **2005**, *59*, 2818–2820. [CrossRef]
19. Jie, J.; Zou, C.; Brosh, E.; Wang, H.; Wei, Z.; Li, T. Microstructure and mechanical properties of an Al-Mg alloy solidified under high pressures. *J. Alloys Compd.* **2013**, *578*, 394–404. [CrossRef]
20. Jie, J.; Wang, H.; Zou, C.; Wei, Z.; Li, T. Precipitation in Al-Mg solid solution prepared by solidification under high pressure. *Mater. Charact.* **2014**, *87*, 19–26. [CrossRef]
21. Zhang, R.; Zou, C.M.; Wei, Z.J.; Wang, H.W.; Liu, C. Interconnected SiC-Si network reinforced Al-20Si composites fabricated by high pressure solidification. *Ceram. Int.* **2021**, *47*, 3597–3602. [CrossRef]
22. Liu, X.; Ma, P.; Jia, Y.D.; Wei, Z.J.; Suo, C.J.; Ji, P.C.; Shi, X.R.; Yu, Z.S.; Prashanth, K.G. Solidification of Al-xCu alloy under high pressures. *J. Mater. Res. Technol.* **2020**, *9*, 2983–2991. [CrossRef]
23. Zhang, R.; Zou, C.M.; Wei, Z.J.; Wang, H.W.; Ran, Z.; Fang, N. Effects of high pressure and superheat temperature on microstructure evolution of Al-20Si alloy. *J. Mater. Res. Technol.* **2020**, *9*, 11622–11628. [CrossRef]
24. Ma, P.; Wei, Z.; Jia, Y.; Zou, C.; Scudino, S.; Prashanth, K.; Yu, Z.; Yang, S.; Li, C.; Eckert, J. Effect of high pressure solidification on tensile properties and strengthening mechanisms of Al-20Si. *J. Alloys Compd.* **2016**, *688*, 88–93. [CrossRef]
25. Wang, X.; Dong, D.; Zhu, D.; Wang, H.; Wei, Z. The Microstructure Evolution and Mass Transfer in Mushy Zone during High-Pressure Solidifying Hypoeutectic Al-Ni Alloy. *Appl. Sci.* **2020**, *10*, 7206. [CrossRef]
26. Zhang, R.; Zou, C.M.; Wei, Z.J.; Wang, H.W. In situ formation of SiC in Al-40Si alloy during high-pressure solidification. *Ceram. Int.* **2021**. [CrossRef]
27. Batashef, A.E. *Crystallization of Metals and Alloys under Pressure*, 1st ed.; Moscow Metallurgy: Moscow, Russia, 1977.
28. Ma, P.; Zou, C.; Wang, H.; Scudino, S.; Fu, B.; Wei, Z.; Kühn, U.; Eckert, J. Effects of high pressure and SiC content on microstructure and precipitation Kinetics of Al-20Si alloy. *J. Alloys Compd.* **2014**, *586*, 639–644. [CrossRef]
29. Yu, X.F.; Zhang, G.Z.; Wang, X.Y.; Gao, Y.Y.; Jia, G.L.; Hao, Z.Y. Non-equilibrium microstructure of hyper-eutectic Al-Si alloy solidified under superhigh pressure. *J. Mater. Sci.* **1999**, *34*, 4149–4152. [CrossRef]
30. Jackson, K.; Hunt, J. Lamellar and rod eutectic growth. In *Dynamics of Curved Fronts*; Pelcé, P., Ed.; Academic Press: San Diego, CA, USA, 1988.
31. Koutsoyiannis, D. Clausius-Clapeyron equation and saturation vapour pressure: Simple theory reconciled with practice. *Eur. J. Phys.* **2012**, *33*, 295–305. [CrossRef]
32. Hu, H.Q. *Fundamentals of Metal Solidification*; China Machine Press: Beijing, China, 1999; p. 36. (In Chinese)

33. Huang, X.; Han, Z.; Liu, B. Study on the effect of pressure on the equilibrium and stability of the solid-liquid interface in solidification of binary alloys. *Sci. China Technol. Sci.* **2011**, *54*, 479–483. [CrossRef]
34. Hallstedt, B. Molar volumes of Al, Li, Mg and Si. *Calphad* **2007**, *31*, 292–302. [CrossRef]

Article
Ab Initio Phase Diagram of Copper

Samuel R. Baty [1,†], Leonid Burakovsky [1,*,†] and Daniel Errandonea [2,†]

[1] Los Alamos National Laboratory, Los Alamos, NM 87545, USA; srbaty@lanl.gov
[2] MALTA Consolider Team, Departamento de Física Aplicada-ICMUV, Universidad de Valencia, Edificio de Investigación, C/Dr. Moliner 50, 46100 Valencia, Spain; Daniel.Errandonea@uv.es
*   Correspondence: burakov@lanl.gov; Tel.: +1-505-667-5222
†   These authors contributed equally to this work.

Abstract: Copper has been considered as a common pressure calibrant and equation of state (EOS) and shock wave (SW) standard, because of the abundance of its highly accurate EOS and SW data, and the assumption that Cu is a simple one-phase material that does not exhibit high pressure (P) or high temperature (T) polymorphism. However, in 2014, Bolesta and Fomin detected another solid phase in molecular dynamics simulations of the shock compression of Cu, and in 2017 published the phase diagram of Cu having two solid phases, the ambient face-centered cubic (fcc) and the high-PT body-centered cubic (bcc) ones. Very recently, bcc-Cu has been detected in SW experiments, and a more sophisticated phase diagram of Cu with the two solid phases was published by Smirnov. In this work, using a suite of *ab initio* quantum molecular dynamics (QMD) simulations based on the Z methodology, which combines both direct Z method for the simulation of melting curves and inverse Z method for the calculation of solid–solid phase boundaries, we refine the phase diagram of Smirnov. We calculate the melting curves of both fcc-Cu and bcc-Cu and obtain an equation for the fcc-bcc solid–solid phase transition boundary. We also obtain the thermal EOS of Cu, which is in agreement with experimental data and QMD simulations. We argue that, despite being a polymorphic rather than a simple one-phase material, copper remains a reliable pressure calibrant and EOS and SW standard.

Keywords: quantum molecular dynamics; melting curve; solid–solid phase transition boundary; equation of state; multi-phase materials

Citation: Baty, S.R.; Burakovsky, L.; Errandonea, D. *Ab Initio* Phase Diagram of Copper. *Crystals* **2021**, *11*, 537. https://doi.org/10.3390/cryst11050537

Academic Editor: Francesco Montalenti

Received: 13 April 2021
Accepted: 5 May 2021
Published: 12 May 2021

Publisher's Note: MDPI stays neutral with regard to jurisdictional claims in published maps and institutional affiliations.

1. Introduction

Copper is one of the most studied d-block transition metals. It is a common pressure calibrant because of the availability of its accurate shock compression data [1–3]. Copper has also been used to calibrate two of the important standards for static X-ray diffraction experiments, ruby and gold [4]. Copper is useful because it provides more accurate pressure (P) determination than other standards, including Pt, Mo, and W, due to its larger compressibility and the presumed lack of phase changes. Indeed, previous shock studies suggest that Cu remains in face-centered cubic (fcc) structure from ambient conditions until melting [5]. These studies reached peak conditions by utilizing a single shock. Using ramp compression, in which peak conditions are reached through a series of shocks, the ambient fcc phase is observed to be stable to TPa pressures [6]. However, a metastable body-centered cubic (bcc) structure is observed in the pseudomorphic Cu films grown on the {100} surfaces of Pd, Pt, Ag, and Fe, or as small precipitates in a bcc-Fe matrix [7], and is suggested to possibly be stable at higher temperatures [8]. A recent computational study by Neogi and Mitra [9] based on density functional theory (DFT) suggests the possible existence of a bcc or body-centered tetragonal (bct) structure at a temperature (T) of 1520 K and $P > 100$ GPa. In 2019, Sims et al. [10] completed a series of laser shock experiments with in situ X-ray diffraction using the Dynamic Compression Sector at Argonne National Laboratory in order to study the phase diagram and melting curve of Cu. At $P \sim 240$ GPa they observed the appearance of bcc-Cu with a density of 14.02 g/cc. The bcc phase consistently coexists with melt, suggesting a high thermodynamic, or possibly kinetic barrier to transformation. The experimental Ts were higher than those in the ramp compression studies, which implies that bcc-Cu is thermodynamically stable at high T only. In a more recent X-ray diffraction study of shock-compressed copper by Sharma et al. [11] the fcc-bcc transformation is observed at $\sim$180 GPa; specifically, the line profile of Cu at 181.5 GPa shows the appearance of a new peak indexed as the {110} bcc peak that partially overlaps with the {111} fcc peak indicating a mixed fcc-bcc phase. At 211.5 GPa the fcc peaks completely disappear and two additional {200} and {211} bcc peaks appear instead, which implies a wide P interval of $\sim$30 GPa of fcc-bcc coexistence in the shock-wave experiment.

The physical properties of bcc-Cu have been studied theoretically using both classical and *ab initio* approaches. At ambient P and $T = 0$, bcc-Cu is mechanically unstable $(C' \equiv (C_{11} - C_{12})/2 < 0)$; it becomes mechanically stable at $P \gtrsim 7.5$ GPa at $T = 0$ [12], or above $\sim$600 K at $P = 0$ [13]. The equations of state of both fcc-Cu and bcc-Cu are predicted to be very close to each other [12], in terms of the very similar values of the corresponding atomic volumes, bulk moduli, and their pressure derivatives. At ambient P, bcc-Cu is higher in energy than fcc-Cu by $\sim$2.9 mRy/atom, or $\sim$40 meV/atom [14], and their energy difference increases with increasing P [14] and/or volumetric strain [13]. As shown in [14], fcc-Cu remains the most thermodynamically stable solid structure of copper up to at least 10 TPa, which is confirmed by the very recent ramp compression experiments to 2.3 TPa [6]. However, as the example of niobium clearly demonstrates [15], an energy excess as high as $\sim$450 meV/atom can be overcome with the proper entropy gain by another solid structure at high T (in the case of Nb, it is the high-T orthorhombic Pnma vs. the ambient bcc), so that bcc-Cu can in principle be expected to become thermodynamically competitive with fcc-Cu at high T. In fact, as recent shock compression experiments demonstrate, bcc-Cu may be the physical solid phase of Cu at high-PT.

To the best of our knowledge, bcc-Cu was consistently studied for the first time by Bolesta and Fomin using classical molecular dynamics (CMD) code LAMMPS [16]. They simulated the shock-wave loading of fcc-Cu and observed a transition to bcc at P above $\sim$80 GPa and the corresponding T above $\sim$2000 K. Then, they calculated the phase diagram of Cu by considering both structures, and found out that bcc-Cu becomes thermodynamically stable at high-PT conditions; specifically, the fcc-bcc-liquid triple point is at $(P, T) = (80 \text{ GPa}, 3490 \text{ K})$, and the entropy difference between the two solid structures at the triple point is $\Delta s \equiv s_{\text{bcc}} - s_{\text{fcc}} = 0.12\ k_{\text{B}}$ [16]. Furthermore, Bolesta and Fomin were the first to detect the appearance of bcc-Cu above 100 GPa and 2000 K in molecular dynamic simulations of the shock compression of copper [17].

In a very recent paper, Smirnov [18] presents the phase diagrams of copper, silver and platinum calculated using a first principles based approach. He claims that all three substances are complex materials with at least two different solid phases on their phase diagrams. Specifically, all three phase diagrams contain both fcc (which is the ambient phase of each of the three substances) and bcc phases. The purpose of this work is to calculate the phase diagram of copper and to confirm that bcc does become the physical phase of copper at high-PT conditions.

To clarify the issues related to the phase diagram of copper, in the present work we carried out a systematic DFT-based study. Specifically, we calculated equations of state of both fcc and bcc, their melting curves using *ab initio* quantum molecular dynamics (QMD) simulations implemented with VASP (Vienna Ab initio Simulation Package), and estimated the P-T location of the fcc-bcc solid–solid phase transition boundary. Our theoretical results appear to be in excellent agreement with all the available relevant experimental data as well as the theoretical calculations of Bolesta and Fomin [16] and Smirnov [18].

2. Equations of State

For our theoretical study of the phase diagram of Cu, we used the following electron core-valence representation: $[^{12}\text{Mg}]\ 3p^6\ 3d^{10}\ 4s^1$, i.e., we assigned the 17 outermost electrons of Cu to the valence. The valence electrons were represented with a plane-wave basis set with a cutoff energy of 460 eV, while the core electrons were represented by projector augmented-wave (PAW) pseudopotentials. We used the generalized gradient approximation (GGA) with the Perdew–Burke–Ernzerhof (PBE) exchange-correlation functional.

We first calculated the $T = 0$ equations of state (EOS) of both fcc-Cu and bcc-Cu. We used unit cells with volumes that correspond to a P range of ~ -10–1000 GPa, and very dense k-point meshes of $60 \times 60 \times 60$ for fcc-Cu and $75 \times 75 \times 75$ for bcc-Cu. With such a dense k-point mesh, full energy convergence to $\lesssim 0.1$ meV/atom is achieved in the whole P range for both fcc-Cu and bcc-Cu. The third-order Birch–Murnaghan forms of the two $T = 0$ EOSs are (in the following ρ stands for density, in g cm^{-3}, B and B' for bulk modulus, in GPa, and its pressure derivative, and the subscript 0 means $(T = 0, P = 0)$)

$$P(\rho) = \frac{3}{2} B_0 \left(\eta^{7/3} - \eta^{5/3} \right) \left[1 + \frac{3}{4} (B_0' - 4) \left(\eta^{2/3} - 1 \right) \right], \tag{1}$$

where $\eta = \rho / \rho_0$, and

$$\text{fcc-Cu:} \quad \rho_0 = 9.02, \quad B_0 = 133.0, \quad B_0' = 5.1,$$

$$\text{bcc-Cu:} \quad \rho_0 = 8.98, \quad B_0 = 131.7, \quad B_0' = 5.1.$$

In each of the two cases, this analytic form is expected to be reliable to ~ 1000 GPa. So, indeed, the two EOSs are very similar to each other. Our value of ρ_0 coincides with the experimental one [19], and those of B_0 and B_0' for fcc-Cu are in excellent agreement with $B_0 = 133$ and $B_0' = 5.2 \pm 0.2$ from experiment [20].

We also note that the finite-T counterparts of the above two EOSs can be written approximately as

$$P(\rho, T) = P(\rho) + \alpha T, \quad \alpha_{\text{fcc}} = 6.13 \cdot 10^{-3}, \quad \alpha_{\text{bcc}} = 8.26 \cdot 10^{-3}. \tag{2}$$

These α values were chosen to match the two ambient melting points in the ρ-T coordinates which are, respectively, (8.361, 1357.6) from experiment [19,21], and (8.112, 1252) from extrapolating our QMD data on the melting curve of bcc-Cu discussed below to $P = 0$.

The reliability of these two "thermal EOSs" is demonstrated by comparing the values of the melting P (P_m) that they give when the corresponding ρs and melting Ts (T_m) are used to those that come directly from QMD melting simulations, see Tables 1 and 2. Let us now show another example. The "thermal EOS" (1), (2) for bcc-Cu on the Hugoniot with $P = 240$ GPa and the corresponding T of 6721.1 K, which comes from $T = T(P)$ along the Hugoniot discussed below, gives $\rho = 14.02$ g cm^{-3}, in exact agreement with [10].

The similarity of the two sets of the EOS parameters (ρ_0, B_0, B_0') and the corresponding two values of α implies that, if the fcc-bcc phase transition does occur in Cu, at the transition (P, T) the two volumes are expected to be close to each other, so that the corresponding volume change is small, and therefore the fcc-bcc phase transition boundary is rather flat, in view of the Clausius–Clapeyron formula. Our *ab initio* phase diagram of copper discussed below demonstrates exactly that.

Table 1. The six *ab initio* melting points of fcc-Cu, $(P_m, T_m \pm \Delta T_m)$, obtained from the Z method implemented with VASP.

Lattice Constant (Å)	Density (g/cm^3)	P_m (GPa)	P_m from (1), (2)	T_m (K)	ΔT_m (K)
3.71	8.2657	−1.20	−1.453	1280	125.0
3.51	9.7607	26.5	26.45	2220	125.0
3.31	11.639	87.4	87.35	3620	250.0
3.21	12.761	140	140.5	4570	250.0
3.11	14.032	218	218.5	5780	312.5
3.01	15.478	333	332.9	7230	375.0

Table 2. The six *ab initio* melting points of bcc-Cu, $(P_m, T_m \pm \Delta T_m)$, obtained from the Z method implemented with VASP.

Lattice Constant (Å)	Density (g/cm^3)	P_m (GPa)	P_m from (1), (2)	T_m (K)	ΔT_m (K)
2.95	8.2205	1.42	1.363	1290	125.0
2.80	9.6138	27.8	27.78	2070	125.0
2.65	11.341	85.9	85.97	3640	250.0
2.49	13.670	220	219.8	6880	250.0
2.42	14.891	320	320.2	9110	312.5
2.35	16.262	462	461.8	12,130	375.0

3. Melting Curves

Our QMD melting simulations were carried out using the Z method implemented with VASP, which is described in detail in Refs. [22–24]. We used supercells of ∼500 atoms; specifically, a 500-atom ($5 \times 5 \times 5$) one for fcc-Cu and a 512-atom one ($8 \times 8 \times 8$ 109.5°-rhombohedral) for bcc-Cu. The simulations were carried out with a single Γ-point (with such a large supercell, full energy convergence, to $\lesssim$1 meV/atom, was achieved in every case considered) having 17 outermost electrons of Cu in the valence, so that our system had ∼8500 valence electrons; to the best of our knowledge, QMD simulations of a similar magnitude (∼9000 valence electrons) have been previously undertaken only once [22]. We simulated six melting points each of fcc-Cu and bcc-Cu. To this end, an average of five computer runs per point were performed (for a total of ∼60 computer runs for both structures), with a time step of 1 fs, of a total length of 15,000–25,000 time steps per run.

Figures 1–4 offer two examples of our Z method melting simulations. They correspond to the first of the six T_ms in each of the two cases, and show the time evolution of T and P, respectively, during the corresponding computer runs. Consider, for example, Figures 1 and 2. During the $T_0 = 2500$ K run, the system remains a superheated solid: both the average T and P stay virtually the same during the 20 ps of running time. The $T_0 = 2750$ K run is the melting run [23], during which a melting occurs: it starts after ∼11 ps of running time, and the melting process takes about 2 ps. It results in the decrease of average T from ∼1400 to 1280 K, and the corresponding increase of average P from ∼−2 to −1.2 GPa. This is so because the total energy, $E \sim k_B T + PV$, is conserved, and V is fixed. For the same reason, Figures 1 and 2, and Figures 3 and 4 are "mirror images" of each other; see [23] for more detail. In the run with higher $T_0 = 3000$ K, the melting starts after only 2 ps of running time, and the melting process takes only 1 ps. For a sufficiently high initial T, the system melts virtually immediately.

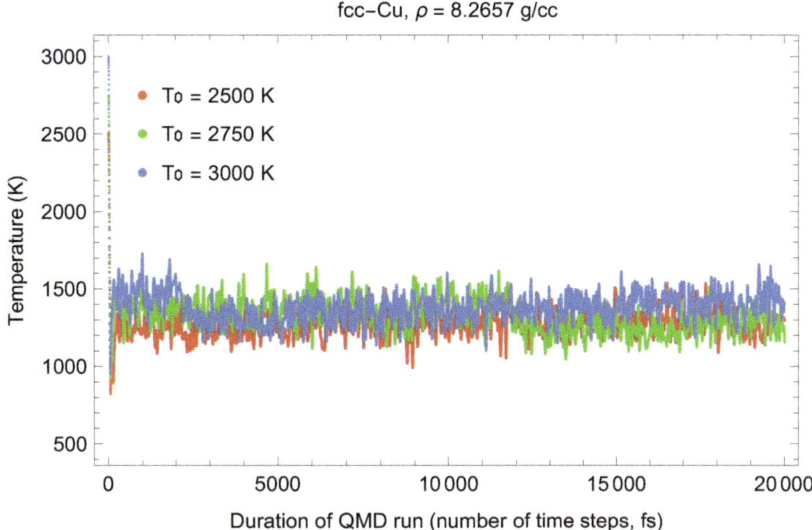

Figure 1. The melting point of fcc-Cu at a density of 8.2627 g/cc: melting T from *ab initio* Z method implemented with VASP.

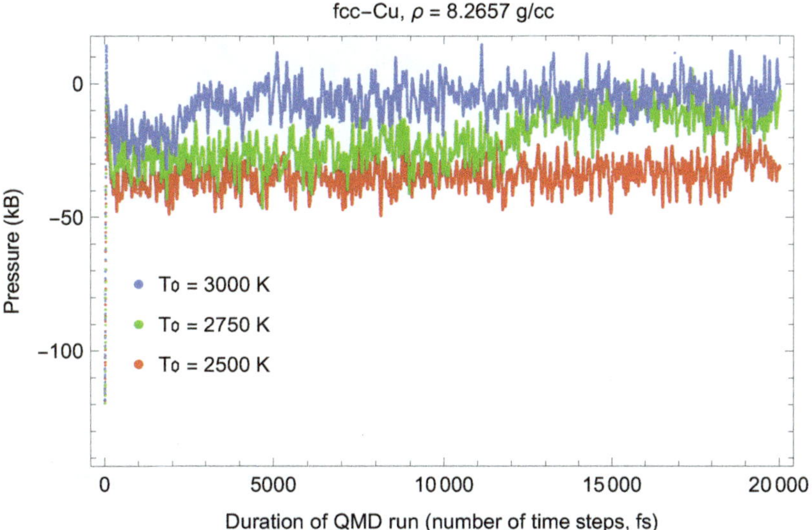

Figure 2. The same as in Figure 1 for melting P.

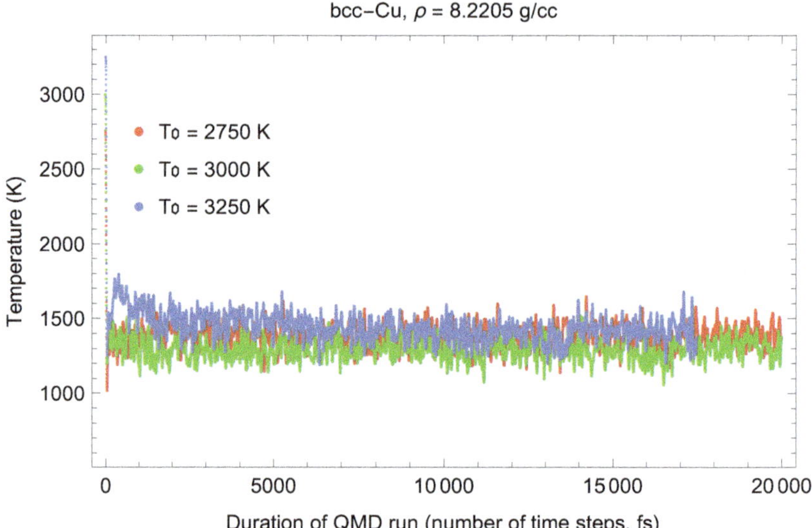

Figure 3. The same as in Figure 1 for bcc-Cu at a density of 8.2205 g/cc.

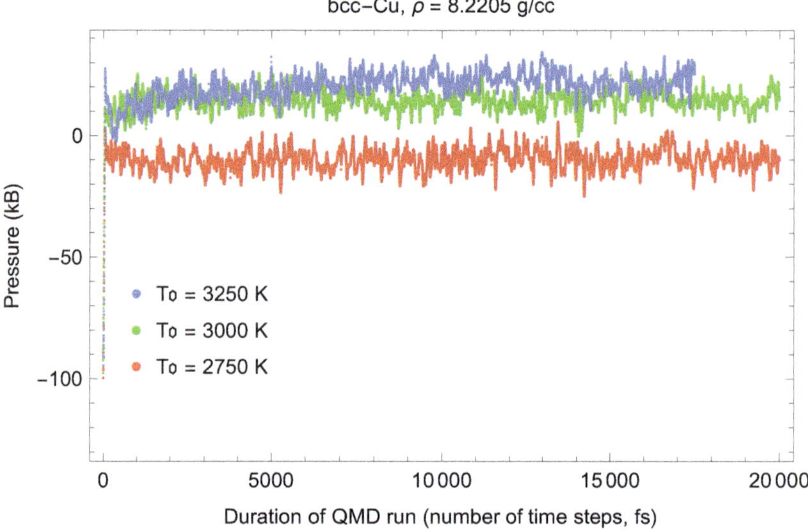

Figure 4. The same as in Figure 3 for melting P.

The results of our melting simulations are summarized in Tables 1 and 2. The errors in melting T (T_m) are half of the increment of the initial T for a series of computer runs at the corresponding density [23]. We chose these increments to be 250 K for the 1st and 2nd, 500 K for the 3rd and 4th, 625 K for the 5th, and 750 K for the 6th T_m in each of the two cases. The corresponding T_m errors are listed in the tables as ΔT_m. The errors in melting P (P_m) are negligibly small, of the order 1–2 GPa in each case. Tables 1 and 2 also include the values of P_m that come from each of the thermal EOSs (1), (2) at the corresponding T_ms.

The best fits to the corresponding six datapoints are the corresponding melting curves:

$$T_m^{\text{fcc}-\text{Cu}}(P) = 1358 \left(1 + \frac{P}{19.7}\right)^{0.58}, \tag{3}$$

and

$$T_m^{\text{bcc}-\text{Cu}}(P) = 1252 \left(1 + \frac{P}{34.3}\right)^{0.85}. \tag{4}$$

These fits are very accurate: the corresponding values of χ^2 per d.o.f. are ≈ 1 for bcc-Cu, and ≈ 1.5 for fcc-Cu if $T_m(\rho_m)$ is fixed at 1358 K, in agreement with the experiment, or otherwise ≈ 1, if it is considered as a free parameter; in the latter case, its value is ≈ 1357.9 K.

The first three terms of the power-series expansion of (3) are $1358 + 39.98\,P - 0.4262\,P^2$. This is in good agreement with $T_m(P) = 1355(5) + 44.5(31)\,P - 0.61(21)\,P^2$ from the low-P experimental study of ref. [25]. Our value of dT_m/dP at $P = 0$, 40 K/GPa, is in the middle of the range of lower-P experiments: 44.5 [25], 43 ± 2 [26], 42 [27], 42 [28], 41.8 [29], 41 [30], 36.5 ± 2.7 [31], 36.4 [32]. The vast majority of the theoretical melting curves of fcc-Cu have similar values for the initial slope; e.g., 50 [33], 39 [34,35], 38 [36], 37.75 [37], 36.7 [38,39]. Ref. [39] does not offer an explicit value of dT_m/dP at $P = 0$; however, the Cu melting curve of [39] is virtually identical to that of [38].

At higher P, the melting curve of bcc-Cu is higher than that of fcc-Cu, therefore, bcc-Cu is thermodynamically more stable and represents the physical solid phase of copper. The melting curves cross each other at $(P$ in GPa, T in K$)$ $(P, T) = (79.2, 3461.2)$, which is the fcc-bcc-liquid triple point. We note that our triple-point P-T coordinates are in excellent agreement with those from Ref. [16]: $(80, 3490)$.

4. fcc-bcc Solid-Solid Phase Transition Boundary

Here we discuss the fcc-bcc solid–solid phase transition boundary in copper. We start our discussion with the derivation of a formula for the initial slope of a solid–solid phase transition boundary in general case.

4.1. Theoretical Estimate of the Initial Slope

Here, we derive a formula for the initial slope of a solid–solid phase transition boundary at the solid$_1$-solid$_2$-liquid triple point.

Let V_1, V_2 and V_ℓ be the volumes of, respectively, solid$_1$, solid$_2$ and liquid at the solid$_1$-solid$_2$-liquid triple point, and S_1, S_2 and S_ℓ the corresponding entropies. According to the Clausius–Clapeyron formula, the slopes of the two melting curves at the triple point (where they cross each other) are

$$T_1' \equiv \frac{dT_1}{dP} = \frac{V_\ell - V_1}{S_\ell - S_1}, \quad T_2' \equiv \frac{dT_2}{dP} = \frac{V_\ell - V_2}{S_\ell - S_2}. \tag{5}$$

Then, the slope of the solid–solid phase transition boundary, also from the Clausius–Clapeyron formula and the above relations, is

$$T_{12}' = \frac{V_2 - V_1}{S_2 - S_1} = \frac{(V_2 - V_1)\,T_1'\,T_2'}{(V_2 - V_1)\,T_1' + \delta\,V_1\,(T_2' - T_1')}. \tag{6}$$

where δ stands for volume change at melt of solid$_1$: $V_\ell \equiv (1 + \delta)V_1$. The values of V_1 and V_2 come from the corresponding thermal EOSs, and those of T_1' and T_2' from the corresponding melting curve equations, at $(P, T) = (79.161, 3461.18)$ of the fcc-bcc-liquid triple point. Specifically, $V_1 = 5.557$ cm^3/mol, $V_2 = 5.684$ cm^3/mol (i.e., $V_2 > V_1$, thus the phase boundary has a positive slope), and $T_1' = 20.30$ K/GPa and $T_2' = 25.93$ K/GPa. As the results of [40] show, by a pressure of 80–100 GPa, volume change at melt for fcc-Cu decreases by a factor of ~ 2. The literature data on δ at $P = 0$ span an interval of values from 0.045 [41] to 0.053 [42], or 0.049 ± 0.004. Hence, at the fcc-bcc-liquid triple point $\delta \sim 0.025$. For our estimate of T_{12}' we assume that at the triple point $0.02 \leq \delta \leq 0.03$. Then, Equation (6) gives $T_{12}' = 19.9 \pm 0.9$ K/GPa. Thus, the initial slope of the fcc-bcc solid–solid phase boundary is ~ 20 K/GPa. Now we can estimate the bcc-fcc entropy difference at the triple point: with the above values of V_1, V_2 and T'_{12}, it follows from (6) that $S_2 - S_1 = 0.127 \pm 0.006\ k_{\mathrm{B}}$ in good agreement with 0.12 k_{B} of Ref. [16].

4.2. Inverse-Z Simulations

Based on the above material of this work, we conclude that the the initial slope of the fcc-bcc solid–solid phase transition boundary is $\sim$20 K/GPa and that it is relatively flat. To further constrain the location of this boundary in the P-T plane, we carried out two sets of independent inverse Z runs (the inverse Z method is described in detail in [22]) to solidify liquid Cu and to check whether there is any solid–solid phase boundary so that liquid Cu solidifies into fcc on one side of this boundary and into bcc (or another solid structure) on the other side. We used a computational cell of 512 atoms prepared by melting a $8 \times 8 \times 8$ solid simple cubic (sc) supercell, which would eliminate any bias towards solidification into fcc or bcc, or any other solid structure. We used sc unit cells of 1.9446 and 1.8652 Å; the dimensions of bcc unit cells with the same volume as the sc ones are 2.45 and 2.35 Å, respectively, which corresponds to $\sim$200 and 360 GPa.

We carried out NVT simulations using the Nosé–Hoover thermostat with a timestep of 1 fs. Complete solidification typically required from 15 to 25 ps, or 15,000–25,000 timesteps. The inverse Z runs indicate that in each case, liquid Cu solidifies into fcc below $\sim$5000 K and into bcc above $\sim$5000 K, so that the fcc-bcc transition boundary at high-P is relatively flat at $\sim$5000 K. The final states of the solidification were identified as fcc and bcc from analyzing the corresponding radial distribution functions (RDFs). RDFs of the final solid states are noisy; upon fast quenching of the two structures to low T, where RDFs are more discriminating, we could compare them to the RDFs of fcc and bcc and properly identify.

The RDFs of the solidified states at $\sim$400 GPa below the transition boundary are shown in Figure 5, and of those solidified above the transition boundary in Figure 6. The 4500 K state lies very close to the transition boundary. We assign it to fcc, because its short-range order (smaller-R) peaks are definitely fcc-like, while long-range order (larger-R) peaks are smeared and may somewhat resemble those of bcc in Figure 6. Most likely, this 4500 K state is some mixture of bcc and fcc, so it must be very close to the transition boundary or even lie on the boundary itself.

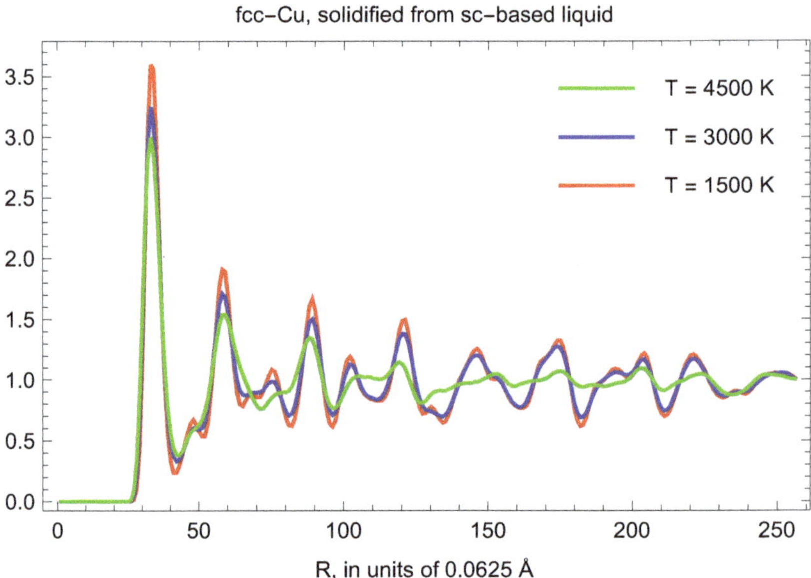

Figure 5. Radial distribution functions (RDFs) of the final states of the solidification of liquid Cu at $\sim$400 GPa at lower temperatures.

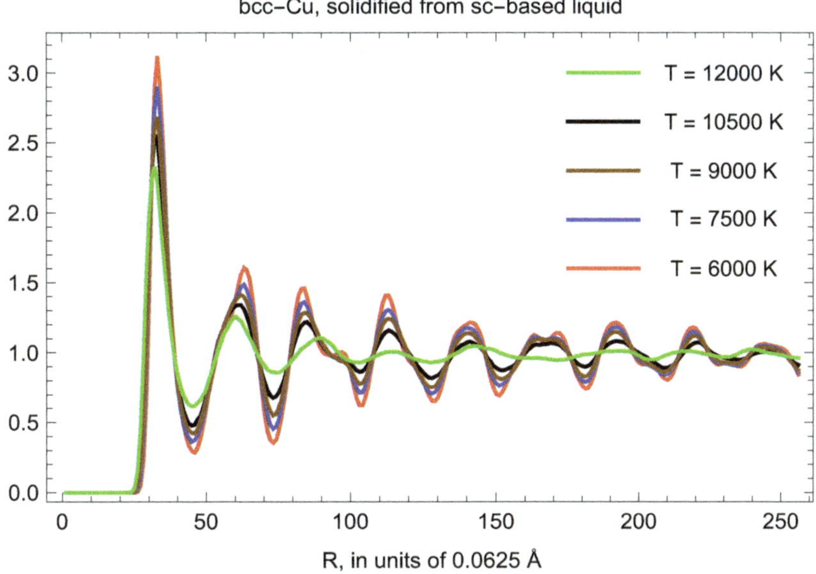

Figure 6. Radial distribution functions (RDFs) of the final states of the solidification of liquid Cu at ∼400 GPa at higher temperatures.

A few more comments are in order. The 12,000 K state at ∼500 GPa did not solidify, most likely for the reason of not being supercooled enough to initiate the solidification process [22]. Indeed, 12,000 K constitutes ∼0.92 of the corresponding T_m of ∼13,050 K (i.e., 8% of supercooling), while for the solidification process to occur, a supercooling of at least 15% is needed [22]. For the other set of points at ∼250 GPa, the highest solidification T of 7000 K constitutes ∼0.83 of the corresponding T_m of ∼8460 K (i.e., 17% of supercooling), which apparently allows for the solidification process to go through in this case.

5. Ab Initio Phase Diagram of Copper

Now we combine all the results of the previous sections of this work to construct the *ab initio* phase diagram of copper.

We take the P-T coordinates of the fcc-bcc-liquid triple point to be (79, 3460). A simple analytic form of the fcc-bcc solid–solid transition boundary which (i) crosses this triple point, (ii) has an initial slope of 19.9 K/GPa and (iii) takes into account the results of the inverse Z solidification simulations (i.e., to be between the 4500 K fcc and 6000 K bcc points at ∼260 GPa and to be close or even cross the 5000 K point at ∼430 GPa is

$$T(P) = 3460 + 19.9\,(P - 79) - 4.8\,(P - 79)^{1.2}. \tag{7}$$

This phase boundary and the two melting curves (3) and (4) define the topology of our *ab initio* phase diagram of Cu shown in Figure 7. This phase diagram is topologically similar to the *ab initio* phase diagram of Smirnov [18], except that their melting curve of bcc-Cu seems to be the continuation of the melting curve of fcc-Cu to higher P, that is, the melting curve of Cu does not change its slope at the fcc-bcc-liquid triple point. The results of the previous section demonstrate that at the triple point the slope of the Cu melting curve increases by ∼20%, from 20.3 K/GPa on the fcc side to 25.9 K/GPa on the bcc side. The experimental conditions of Ref. [43] correspond to melting from bcc phase, and indeed, the three data points of [43] appear to lie on our bcc-Cu melting curve, see Figure 7.

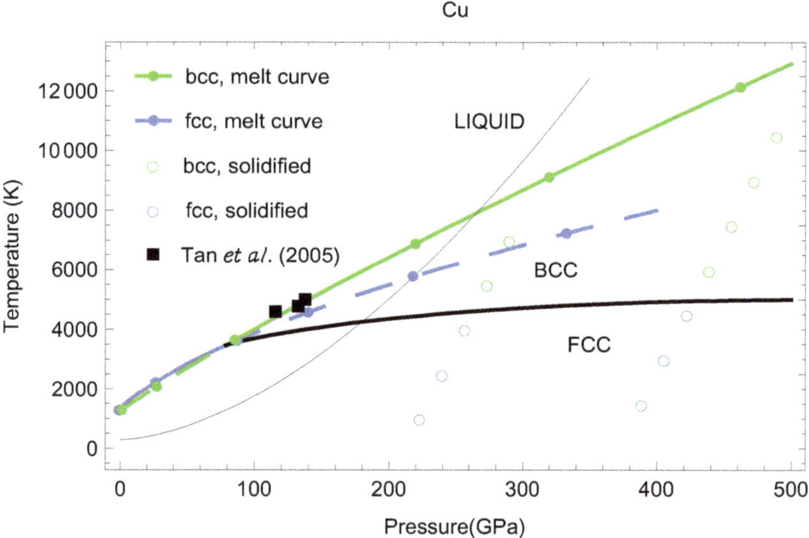

Figure 7. The *ab initio* phase diagram of Cu. Three experimental points (filled circles) are from ref. [43]. The fcc-bcc solid–solid phase transition boundary (black curve) is estimated based on the results of the inverse Z method simulations of the solidification of liquid Cu into fcc and bcc solid structures on other sides of the boundary. The principal Hugoniot of Cu is also shown as a thin black curve.

Shown also in Figure 7 is the Cu principal Hugoniot obtained from fitting the numerical data of Ref. [44] with a simple analytic form:

$$T_{\mathrm{H}}(P) = 293 + 0.617 \cdot P^{1.688}. \tag{8}$$

This principal Hugoniot crosses the fcc-bcc phase transition boundary at $(P, T) = (180.0, 4249.9)$, in agreement with Ref. [11], which claims the bcc-fcc transition on the Hugoniot at 180 GPa. Furthermore, in Ref. [5], a distinct kink was detected in the Poisson ratio as a function of P at $\sim$185 GPa such that the Poisson ratio data were fitted with two different straight segments below and above the kink. In view of our findings, this kink is naturally explained as that corresponding to the fcc-bcc transition in Cu on its principal Hugoniot.

The Hugoniot melting point corresponds to the intersection of the bcc-Cu melting curve (4) and the above Hugoniot: $(P, T) = (265.1, 7896.1)$, in agreement with Ref. [5], which claims the melting on the Hugoniot at 265 ± 6 GPa.

Figure 8 compares our *ab initio* melting curve of Cu to several melting curves among a few tens of those available in the literature, as it is not feasible to collect them all in one figure. Shown are the experimental melting curves of references [30,45] as well as the theoretical melting curves of refs. [38,40,46,47]. It is clearly seen that the best overall agreement of our melting curve of Cu, as a combination of both fcc and bcc segments, is with the theoretical melting curves of Belonoshko et al. [38] and Ghosh [39] which are virtually identical to each other. The reason for such a good agreement must be that in both [38,39] molecular dynamics simulations were done using embedded-atom model (EAM) for interatomic potentials, and the EAM parameters were obtained from fitting the model to *ab initio* data. It is interesting to note that in Ref. [43], their three experimental melting points are compared to the theoretical melting curve of [38], and excellent agreement is found. This fact provides mutual support to the validity of the experimental results of ref. [43] and to the computational methodologies of both Refs. [38,39] and the present study.

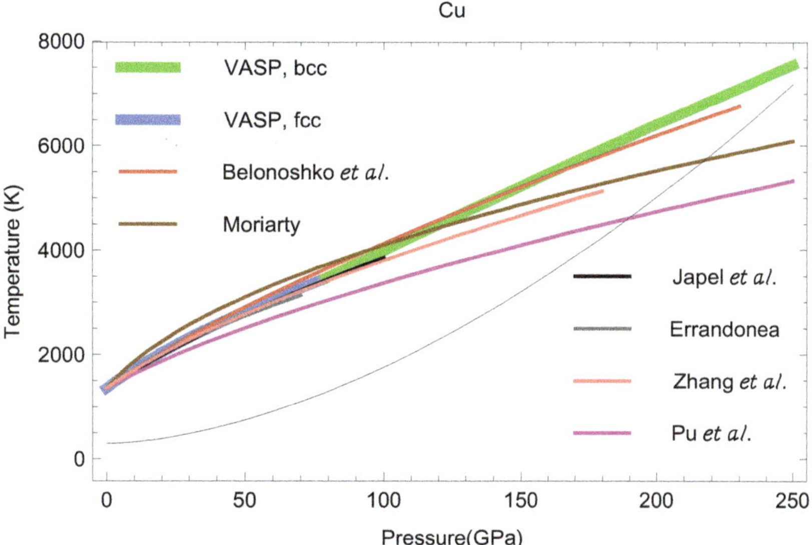

Figure 8. Comparison of the *ab initio* melting curve of Cu calculated in this work (which combines both lower-P fcc and higher-P bcc segments) to several melting curves of Cu available in the literature: Errandonea [30], Belonoshko [38], Pu et al. [40], Japel et al. [45], Moriarty [46], and Zhang et al. [47]. The Cu principal Hugoniot is also shown as a thin black line.

We note that the vast majority of the melting curves of fcc-Cu available in the literature converge to a unique analytic form similar to Equation (3). For example, in Figure 8, all the fcc-Cu melting curves except that of Ref. [40] can be effectively represented by that of ref. [47]. If only fcc-Cu were considered, the principal Hugoniot would have crossed it at $\sim$220 GPa, which would have been the P of the Hugoniot melting of Cu (P_m^{H}). In fact, this is in agreement with several papers which all predicted $P_m^{\mathrm{H}} \sim 220$ GPa: $P_m^{\mathrm{H}} = 220$–230 [17], $P_m^{\mathrm{H}} = 200$–220 [48], and $P_m^{\mathrm{H}} = 220$ [49], to name just a few. In the latter work, the Hugoniot T_m is 5560 K, in good agreement with 5843.0 from Equation (8).

6. Concluding Remarks

Let us now summarize the results of our theoretical study.

We have constructed the theoretical phase diagram of copper, using a suite of *ab initio* QMD simulations based on the Z methodology which combines both direct Z method for the simulation of melting curves and inverse Z method for the calculation of solid–solid phase boundaries. We determined that bcc-Cu becomes the thermodynamically stable solid structure of copper at high-PT and finds itself on the phase diagram of copper, along with fcc-Cu, its solid structure at ambient conditions. We have calculated the melting curves of both fcc-Cu and bcc-Cu, determined the location of the fcc-bcc-liquid triple point, and obtained an equation for the fcc-bcc solid–solid phase transition boundary. Last but not least, we have proposed thermal equations of state for both fcc-Cu and bcc-Cu, which appear to be in agreement with both experimental data and QMD simulations. Our theoretical phase diagram of copper represents the refinement of that of Smirnov for which the fcc-Cu and bcc-Cu melting curves as well as the fcc-bcc solid–solid phase boundary were estimated rather than being calculated using the DFT-based methodology as in the present work.

For a long time, copper has been considered both as a pressure and a shock-wave standard. It is worth dwelling on this point in light of our findings, which reaffirm the idea that copper is a multi-phase material that was put forward by Bolesta and Fomin [16,17] and confirmed in the subsequent experimental [11,12] and theoretical [18] studies.

As emphasized in [14], first-principles-based theoretical calculations do not find any other solid structure energetically competitive with fcc, at least to 10 TPa, and therefore Cu is not expected to undergo any P-induced structural phase transformations, which makes Cu an ideal choice as a pressure standard for all hydrostatic experiments expected in the near future, both isothermal and isentropic (ramp compression), in which T does not rise high enough to cross into the bcc-Cu phase stability region. In this respect, our thermal EOS of Cu, Equations (1) and (2), can be considered as an EOS standard.

The issue of Cu being a shock-wave standard deserves a somewhat more detailed discussion. Copper has been considered to be a reliable shock-wave standard because (i) it is a plastic material, so that the characteristics of its shock compression and static compression agree well with each other, (ii) when compressed, Cu has been supposed not to undergo any polymorphic transitions, and (iii) when melted, it undergoes a minor volume change, so that its shock compression characteristics change very little across the Hugoniot melting transition. Now, as it is firmly established, both experimentally and theoretically, that Cu is a polymorphic material, it is interesting to revisit the concept of Cu being a reliable shock-wave standard. First of all, the main shock-wave characteristics of a material depend on the values of a and b in the following (quasi)-linear relation between particle (U_p) and shock (U_s) velocities along the Hugoniot: $U_s = a + b\,U_p$. These values, in turn, depend on the parameters of the EOS [50]: $a = \sqrt{B_0/\rho_0}$, $b = (1 + B_0')/4$. As mentioned above, both fcc-Cu and bcc-Cu have very similar EOS parameters listed under Equation (1); in particular, the two values of a differ from each other by less than 0.5%. The two values of b are exactly the same. Hence, the shock-wave characteristics of both fcc-Cu and bcc-Cu are expected to be virtually identical to each other; in particular, T as a function of P along the Hugoniot should not change across the fcc-bcc transition, this is why Equation (8) was used as a common $T_{\mathrm{H}}(P)$ for both fcc and bcc. For this reason, experimental data on $U_s = U_s(U_p)$ for both fcc-Cu and bcc-Cu should be described by a common straight segment instead of two different ones, which is very clearly seen in, e.g., Figure 3 of Ref. [14]; a deviation of $U_s = U_s(U_p)$ from a single straight segment occurs at $U_p \sim 3.5$ km/s which corresponds to $P \sim 285$ GPa, above the Hugoniot melting point of 265 GPa; in other words, this deviation occurs in the P-T region of liquid Cu.

Thus, despite being a polymorphic material, both the thermal and shock-wave characteristics of two of its solid phases, fcc and bcc, are indeed virtually identical. This observation allow us to conclude that copper remains to be a reliable shock-wave standard, in addition to being reliable pressure calibrant and EOS standard.

Author Contributions: The authors contributed equally to this work. All authors have read and agreed to the published version of the manuscript.

Funding: This work was done under the auspices of the US DOE/NNSA. D.E. acknowledges financial support from Spanish Ministerio de Ciencia, Innovación y Universidades (Grants Nos. PID2019-106383GB-C41 and RED2018-102612-T), and Generalitat Valenciana (Prometeo/2018/123 EFIMAT).

Data Availability Statement: All relevant data that support the findings of this study are available from the corresponding author upon request.

Acknowledgments: The QMD simulations were performed on the LANL cluster Badger as part of the Institutional Computing project w20_phadiagurox.

Conflicts of Interest: The authors declare no conflict of interest.

References

1. Meyers, M.A.; Gregori, F.; Kad, B.K.; Schneider, M.S.; Kalantar, D.H.; Remington, B.A.; Ravichandran, G.; Boehly, T.; Wark, J.S. Laser-induced shock compression of monocrystalline copper: Characterization and analysis. *Acta Mater.* **2003**, *51*, 1211–1228. [CrossRef]
2. Murphy, W.J.; Higginbotham, A.; Kimminau, G.; Barbrel, B.; Bringa, E.M.; Hawreliak, J.; Kodama, R.; Koenig, M.; McBarron, W.; Meyers, M.A.; et al. The strength of single crystal copper under uniaxial shock compression at 100 GPa. *J. Phys. Cond. Mat.* **2010**, *22*, 065404. [CrossRef] [PubMed]

3. Kraus, R.G.; Davis, J.P.; Seagle, C.T.; Fratuono, D.E.; Swift, D.C.; Brown, J.L.; Eggert, J.H. Dynamic compression of copper to over 450 GPa: A high-pressure standard. *Phys. Rev. B* **2016**, *93*, 134105. [CrossRef]

4. Bell, P.M.; Xu, J.A.; Mao, H.K. Static compression of gold and copper and calibration of the ruby pressure scale to pressures to 1.8 megabars (static. RNO). In *Shock Waves in Condensed Matter*; Springer: Boston, MA, USA, 1986; pp. 125–130.

5. Hayes, D.; Hixson, R.S.; McQueen, R.G. High pressure elastic properties, solid–liquid phase boundary and liquid equation of state from release wave measurements in shock-loaded copper. *AIP Conf. Proc.* **2000**, *505*, 483.

6. Fratanduono, D.E.; Smith, R.F.; Ali, S.J.; Braun, D.G.; Fernandez-Pañella, A.; Zhang, S.; Kraus, R.G.; Coppari, F.; McNaney, J.M.; Marshall, M.C.; et al. Probing the solid phase of noble metal copper at terapascal conditions. *Phys. Rev. Lett.* **2020**, *124*, 015701. [CrossRef] [PubMed]

7. Wang, L.G.; Šob, M. Structural stability of higher-energy phases and its relation to the atomic configurations of extended defects: The example of Cu. *Phys. Rev. B* **1999**, *60*, 844. [CrossRef]

8. Friedel, J. On the stability of the body centred cubic phase in metals at high temperatures. *J. Physique Lett.* **1974**, *35*, 59–63. [CrossRef]

9. Anupam, N.; Mitra, N. A metastable phase of shocked bulk single crystal copper: An atomistic simulation study. *Sci. Rep.* **2017**, *7*, 7337.

10. Sims, M.; Briggs, R.; Coppari, F.; Coleman, A.L.; Gorman, M.G.; Panella-Fernandez, A.; Smith, R.F.; Eggert, J.H.; Wicks, J.K. Copper phase determination to 290 GPa using laser-shock compression. In Proceedings of the 2019 COMPRES Annual Meeting, Big Sky Resort, Big Sky, MT, USA, 2–5 August 2019.

11. Sharma, S.M.; Turneaure, S.J.; Winey, J.M.; Gupta, Y.M. Transformation of shock-compressed copper to the body-centered-cubic structure at 180 GPa. *Phys. Rev. B* **2020**, *102*, 020103(R). [CrossRef]

12. Mei, W.; Wen, Y.; Xing, H.; Ou, P.; Sun, J. Ab initio calculations of mechanical stability of bcc Cu under pressure. *Solid State Commun.* **2014**, *184*, 25. [CrossRef]

13. Xiong, Q.; Kitamura, T.; Li, Z. Transient phase transitions in single-crystal coppers under ultrafast lasers induced shock compression: A molecular dynamics study. *J. Appl. Phys.* **2019**, *125*, 194302. [CrossRef]

14. Greeff, C.W.; Boettger, J.C.; Graf, M.J.; Johnson, J.D. Theoretical investigation of the Cu EOS standard. *J. Phys. Chem. Solids* **2006**, *67*, 2033. [CrossRef]

15. Errandonea, D.; Burakovsky, L.; Preston, D.L.; MacLeod, S.G.; Santamaría-Perez, D.; Chen, S.; Cynn, H.; Simak, S.I.; McMahon, M.I.; Proctor, J.E.; et al. Experimental and theoretical confirmation of an orthorhombic phase transition in niobium at high pressure and temperature. *Commun. Mater.* **2020**, *1*, 60. [CrossRef]

16. Bolesta, A.V.; Fomin, V.M. Molecular dynamics simulation of shock-wave loading of copper and titanium. *AIP Conf. Proc.* **2017**, *1893*, 020008.

17. Bolesta, A.V.; Fomin, V.M. Molecular dynamics simulations of polycrystalline copper. *J. Appl. Mech. Tech. Phys.* **2014**, *55*, 800. [CrossRef]

18. Smirnov, N.A. Relative stability of Cu, Ag, and Pt at high pressures and temperatures from ab initio calculations. *Phys. Rev. B* **2021**, *103*, 064107. [CrossRef]

19. Wallace, D.C. *Statistical Physics of Crystals and Liquids*; World Scientific: Singapore, 2002; p. 191.

20. Dewaele, A.; Loubeyre, P.; Mezouar, M. Equations of state of six metals above 94 GPa. *Phys. Rev. B* **2004**, *70*, 094112. [CrossRef]

21. Zinov'ev, V.E. *Handbook of Thermophysical Properties of Metals at High Temperatures*; Nova Science Publishers: New York, NY, USA, 1996; p. 95.

22. Burakovsky, L.; Chen, S.P.; Preston, D.L.; Sheppard, D.G. Z methodology for phase diagram studies: Platinum and tantalum as examples. *J. Phys. Conf. Ser.* **2014**, *500*, 162001. [CrossRef]

23. Burakovsky, L.; Burakovsky, N.; Preston, D.L. Ab initio melting curve of osmium. *Phys. Rev. B* **2015**, *92*, 174105. [CrossRef]

24. Belonoshko, A.B.; Skorodumova, N.V.; Rosengren, A.; Johansson, B. Melting and critical superheating. *Phys. Rev. B* **2006**, *73*, 012201. [CrossRef]

25. Brand, H.; Dobson, D.P.; Vočadlo, L.; Wood, I.G. Melting curve of copper measured to 16 GPa using a multi-anvil press. *High Pres. Res.* **2006**, *26*, 185. [CrossRef]

26. Errandonea, D. The melting curve of ten metals up to 12 GPa and 1600 K. *J. Appl. Phys.* **2010**, *108*, 033517. [CrossRef]

27. Butuzov, V.P. The investigation of phase transformations as superhigh pressures. *Kristallografiya* **1957**, *2*, 536.

28. Mitra, N.R.; Decker, D.L.; Vanfleet, H.B. Melting Curves of Copper, Silver, Gold, and Platinum to 70 kbar. *Phys. Rev.* **1967**, *161*, 613. [CrossRef]

29. Mirwald, P.; Kennedy, G.C. The melting curve of gold, silver, and copper to 60-kbar pressure: A reinvestigation. *J. Geophys. Res.* **1979**, *84*, 6750. [CrossRef]

30. Errandonea, D. High-pressure melting curves of the transition metals Cu, Ni, Pd, and Pt. *Phys. Rev. B* **2013**, *87*, 054108. [CrossRef]

31. Cohen, L.M.; Klement, W.; Kennedy, G.C. Melting of Copper, Silver, and Gold at High Pressures. *Phys. Rev.* **1966**, *145*, 519. [CrossRef]

32. Akella, J.; Kennedy, G.C. Melting of Gold, Silver, and Copper - Proposal for a New High-Pressure Calibration Scale. *J. Geophys. Res.* **1971**, *76*, 4969. [CrossRef]

33. Górecki, T. Vacancies and a generalised melting curve of metals. *High Temp. High Press.* **1979**, *11*, 683.

34. Tam, P.D.; Tan, P.D.; Hoc, N.Q.; Phong, P.D. Melting of matals copper, silver and gold under pressure. *Proc. Nat. Conf. Theor. Phys.* **2010**, *35*, 148.
35. Tam, P.D.; Hoc, N.Q.; Tinh, B.D.; Tan, P.D. Melting curve of metals Cu, Ag and Au under pressure. *Mod. Phys. Lett. B* **2016**, *30*, 1550273. [CrossRef]
36. Vočadlo, L.; Alfè, D.; Price, G.D.; Gillan, M.J. Ab initio melting curve of copper by the phase coexistence approach. *J. Chem. Phys.* **2004**, *120*, 2872. [CrossRef] [PubMed]
37. Mazhukin, V.I.; Demin, M.M.; Aleksashkina, A.A. Atomistic modeling of thermophysical properties of copper in the region of the melting point. *Math. Montisnigri* **2018**, *41*, 99.
38. Belonoshko, A.B.; Ahuja, R.; Eriksson, O.; Johansson, B. Quasi ab initio molecular dynamic study of Cu melting. *Phys. Rev. B* **2000**, *61*, 3838. [CrossRef]
39. Ghosh, K. Melting curve of metals using classical molecular dynamics simulations. *J. Phys. Conf. Ser.* **2012**, *377*, 012085. [CrossRef]
40. Pu, C.; Yang, X.; Xiao, D.; Cheng, J. Molecular dynamics simulations of shock melting in single crystal Al and Cu along the principle Hugoniot. *Mater. Today Commun.* **2021**, *26*, 101990. [CrossRef]
41. Cahill, J.A.; Kirshenbaum, A.D. The density of liquid copper from its melting point (1356 °K) to 2500 °K. Furthermore, an estimate of its critical constants. *J. Phys. Chem.* **1962**, *66*, 1080. [CrossRef]
42. Blumm, J.; Henderson, J.B. Measurement of the volumetric expansion and bulk density of metals in the solid and molten regions. *High Temp. High Press.* **2000**, *32*, 109. [CrossRef]
43. Tan, H.; Dai, C.D.; Zhang, L.Y.; Xu, C.H. Method to determine the melting temperatures of metals under megabar shock pressures. *Appl. Phys. Lett.* **2005**, *87*, 221905. [CrossRef]
44. Kinslow, R. (Ed.) *High-Velocity Impact Phenomena*; Academic Press: New York, NY, USA, 1970; p. 532.
45. Japel, S.; Schwager, B.; Boehler, R.; Ross, M. Melting of copper and nickel at high rressure: The role of *d* electrons. *Phys. Rev. Lett.* **2005**, *95*, 167801. [CrossRef]
46. Moriarty, J.A. High-pressure ion-thermal properties of metals from ab initio interatomic potentials. In *Shock Waves in Condensed Matter*, Gupta, Y.M., Ed.; Plenum Press: New York, NY, USA, 1986.
47. Zhang, B.; Wang, B.; Liu, Q. Melting curves of Cu, Pt, Pd and Au under high pressures. *Int. J. Mod. Phys. B* **2016**, *30*, 1650013. [CrossRef]
48. Bringa, E.M.; Cazamias, J.U.; Erhart, P.; Stölken, J.; Tanushev, N.; Wirth, B.D.; Rudd, R.E.; Caturla, M.J. Atomistic shock Hugoniot simulation of single-crystal copper. *J. Appl. Phys.* **2004**, *96*, 3793. [CrossRef]
49. Gubin, S.A.; Maklashova, I.V.; Mel'nikov, I.N. The Hugoniot adiabat of crystalline copper based on molecular dynamics simulation and semiempirical equation of state. *J. Phys. Conf. Ser.* **2018**, *946*, 012098. [CrossRef]
50. Johnson, J.D. The features of the principal Hugoniot. *AIP Conf. Proc.* **1998**, *429*, 27.

MDPI

Article

P–V–T Equation of State of Iridium Up to 80 GPa and 3100 K

Simone Anzellini [1,*], Leonid Burakovsky [2], Robin Turnbull [3], Enrico Bandiello [3] and Daniel Errandonea [3]

[1] Diamond Light Source Ltd., Harwell Science & Innovation Campus, Diamond House, Didcot OX11 0DE, UK

[2] Los Alamos National Laboratory, Theoretical Divisions, Los Alamos, NM 87545, USA; burakov@lanl.gov

[3] Departamento de Física Aplicada-Instituto de Ciencia de Materiales, Matter at High Pressure (MALTA) Consolider Team, Universidad de Valencia, Edificio de Investigación, C/Dr. Moliner 50, Burjassot, 46100 Valencia, Spain; robin.turnbull@uv.es (R.T.); enrico.bandiello@uv.es (E.B.); daniel.errandonea@uv.es (D.E.)

* Correspondence: simone.anzellini@diamond.ac.uk

Abstract: In the present study, the high-pressure high-temperature equation of the state of iridium has been determined through a combination of in situ synchrotron X-ray diffraction experiments using laser-heating diamond-anvil cells (up to 48 GPa and 3100 K) and density-functional theory calculations (up to 80 GPa and 3000 K). The melting temperature of iridium at 40 GPa was also determined experimentally as being 4260 (200) K. The results obtained with the two different methods are fully consistent and agree with previous thermal expansion studies performed at ambient pressure. The resulting thermal equation of state can be described using a third-order Birch–Murnaghan formalism with a Berman thermal-expansion model. The present equation of the state of iridium can be used as a reliable primary pressure standard for static experiments up to 80 GPa and 3100 K. A comparison with gold, copper, platinum, niobium, rhenium, tantalum, and osmium is also presented. On top of that, the radial-distribution function of liquid iridium has been determined from experiments and calculations.

Keywords: iridium; equation of state; high pressure; X-ray diffraction; laser heating; density-functional theory; melting; radial-distribution function

Citation: Anzellini, S.; Burakovsky, L.; Turnbull, R.; Bandiello, E.; Errandonea, D. P-V-T Equation of State of Iridium Up to 80 GPa and 3100 K. *Crystals* **2021**, *11*, 452. https://doi.org/10.3390/cryst11040452

Academic Editor: Shujun Zhang

Received: 30 March 2021
Accepted: 15 April 2021
Published: 20 April 2021

Publisher's Note: MDPI stays neutral with regard to jurisdictional claims in published maps and institutional affiliations.

1. Introduction

Iridium (Ir) belongs to the family of the 5d transition metals. It exhibits a face-centered cubic (fcc) structure and its electronic structure is [Xe] $4f^{14}\,5d^7\,6s^2$. Similar to the other elements of the 5d family (such as Re, W, Pt, and Ta), Ir has always attracted considerable interest in the scientific community due to its outstanding mechanical and thermal properties. In particular, Ir is the most commonly used metal in high-temperature (HT) crucibles, thermocouples, and encapsulators of nuclear-powered electrical generators in space technology. Furthermore, thanks to its high shear modulus, chemical inertness, refractory nature, and phase stability, Ir is ideally suited as gasket material in static experiments in diamond-anvil cells (DAC) or as pressure standard for high-pressure (HP) experiments (in both static and dynamic experiments). As it exhibits excellent mechanical properties and high resistance to oxidation and corrosion at elevated temperature, Ir is also used in numerous applications as a static component at high T and/or in aggressive environments. In particular, Ir is notably inert in comparison to other transition metals [1], and it is largely immune to chemical reactions in comparison to refractory metals such as rhenium [2] or tungsten [3].

Despite the multiple technological applications of Ir, our current understanding of its mechanical properties is rather limited and knowledge of its HP–HT phase diagram is virtually non-existent. In particular, a solid–solid phase transition was observed in shock-wave experiments [4] to occur between 140 and 180 GPa. In a previous DAC experiment, performed under non-hydrostatic conditions using energy-dispersive X-ray diffraction

(XRD), a distortion of the fcc lattice was observed at room temperature (RT) and 59 GPa [5] and interpreted as a formation of a complex super-lattice. However, no phase transition was observed at RT in a recent DAC experiment performed under hydrostatic conditions (using helium as pressure medium) up to 140 GPa using angular-dispersive XRD [1]. A phase transition has been predicted to happen at 2000 K and 150 GPa from a fcc phase to a disordered hexagonal close-packed phase ($rhcp$) by a recent theoretical study [6] based on ab initio simulations.

An additional interest in iridium was generated by predicted structural anomalies attributed to an unusual electronic transition called core-level crossing (CLC). In fact, it is known that electronic transitions are expected to strongly affect material properties, such as elastic moduli, thermal expansion, resistivity and thermal conductivity. Such an electronic transition has been recently observed for the first time in a double-staged DAC study of osmium (another 5d metal), which compressed up to 770 GPa [7]. In particular, the CLC electronic transition was correlated to a discontinuity in the measured c/a axial ratio of the Os hcp structure at 440 GPa. Consequently, systematic theoretical studies have been conducted to predict possible CLC transitions for all of the 5d transition metals. In the case of Ir, such a transition was predicted to occur at 80 GPa [8]. However, a recent X-ray absorption near-edge structure (XANES) experiment that aimed to obtain information on the electronic structural evolution of Ir up to 90 GPa could not find any evidence of any CLC transition [1].

It is therefore evident that an in-depth characterization of the structural and physical properties of Ir at HP and HT conditions is still lacking. In the present work, the structural evolution of Ir has been characterized up to 80 GPa and 4260 K through a combination of in situ powder X-ray diffraction experiments and density-functional theory (DFT) calculations. In particular, the structural domain of solid Ir (always fcc) in the investigated P-T range, the melting point at 40.3 GPa, the radial-distribution function (RDF) of liquid Ir, and the corresponding thermal equation of state (EoS) have all been established.

2. Materials and Methods

2.1. Experimental Details

Two membrane DACs with diamonds of culet sizes of 400 µm and 300 µm were equipped with pre-indented and spark-erosion drilled Re gaskets. In order to prevent any oxidation, the sample preparation and loading was performed in a glovebox under an argon atmosphere. Samples were obtained from Ir powder sourced from Goodfellow (99.9% purity). Grains of the powder sample were compressed between two diamond anvils in order to create a foil. The obtained foil was then cut and loaded in the DAC high-pressure chambers (see Figure 1) between two magnesium oxide (MgO) disks (99.9% purity). The MgO, acting as insulating material (both thermally and chemically) as well as pressure gauge, was oven dried at 200 °C for two hours before being loaded in the DAC to remove any water or moisture.

Angular-dispersive powder XRD patterns were collected at the extreme conditions beamline, I15, of the Diamond Light Source [9]. The polychromatic beam of the I15's wiggler was tuned to 29.5 keV and focused down to $6 \times 9 \ \mu m^2$ (full-width at half maximum (FWHM)). A Pilatus 2M detector was used to ensure fast data collections with a good signal/noise ratio. The sample-to-detector distance was measured following the standard procedure from the diffraction rings of a CeO_2 sample.

The loaded DACs were mounted on the laser-heating (LH) system of I15 [9]. Each heating ramp was performed at a selected initial P. The sample inside the DAC was heated in double-sided mode [10]. Before each heating ramp, the two Nd:YAG fiber lasers were individually focused on the sample surface and their power was linearly increased until a clear hot spot was observed on the camera. During the laser alignment procedure, the exposure time of the cameras was maximized to be able to observe the hot spot at a relatively "low T" (around 1000 K). In this way, it was possible to avoid any unwanted damage of the sample. Furthermore, in order to prevent the X-ray beam from sampling a radial T gradient on the sample surface, the two lasers were intentionally slightly unfocused to maximize the size of the hot spot at uniform T. Finally, the lasers were coupled together and their relative positions were tuned to obtain a uniform hot spot of around 30 μm in diameter.

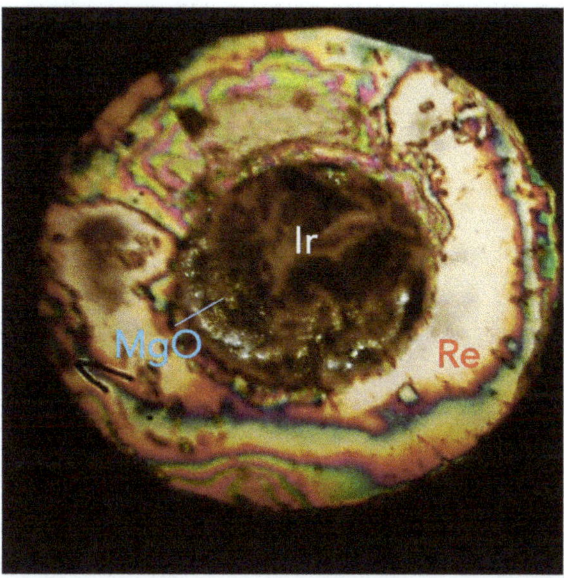

Figure 1. Photograph of the DAC high-pressure chamber after sample loading at ambient conditions. Despite the opacity of MgO, it is possible to observe the Ir embedded on it. The external ring is the Re gasket.

During the experiments, the T on both sides of the sample was measured by spectral radiometry following the procedure described in Benedetti and Loubeyre [11]. The error in the T measurements was calculated as the maximum values between the difference in the T measured on each side of the sample and the FWHM of the histogram of the two-colours pyrometry [10]. The P experienced by the sample was measured from the thermal EoS of MgO [12] (under the assumption of Ir and MgO being at the same T) using the measured volume of MgO and the average between the T measured from both sides of the sample. The errors in P measurement were estimated to be of the order of 5 GPa at 1500 K and of 7.5 GPa at 4260 K (the highest T reached in the present experiment). Such an estimation was obtained considering the adopted MgO thermal EoS [12] and the thermal gradient developed in the DAC high pressure chamber (therefore, inside the MgO), ranging from the measured T to 300 K at the interface with the diamonds. Before and after each heating ramp, the relative alignment of the X-rays with the lasers and the T reading were checked following the procedure described in Anzellini et al. [13]

In order to minimize the laser–sample interaction time (therefore, minimizing possible chemical reactions and sample damaging), the heating ramps were performed in a "trigger mode"; i.e., both lasers were set to a target power, after 0.3 s, a diffraction pattern and a temperature measurement were collected simultaneously. Then, 0.3 s after the XRD collection, both lasers were turned off synchronously. The XRD collection time was set to 1 s, whereas the acquisition time for the T measurements was adjusted according to the signal saturation.

During each run, the target power of the lasers was increased until a diffuse ring (characteristic of a liquid sample) was detected in the diffraction pattern or it was not possible to increase T any further. An accurate analysis of the diffraction patterns was performed in order to detect the appearance of the melting and to obtain structural and textural information about the sample and the insulating material. During the analysis procedure, masks were applied on a per-image basis using the DIOPTAS suite [14]. The images were azimuthally integrated and a LeBail analysis was performed using the TOPAS suite [15] software in order to identify crystal structures and determine unit–cell parameters at different P–T conditions. The EosFit suite [16] was used to determine the corresponding thermal EoS from the obtained data. The melting at HP was determined by the appearance of a diffuse scattering in the Ir diffraction signal.

2.2. Computational Methods

Calculations were performed using a quantum molecular dynamic (QMD) code and the Vienna Ab initio Simulation Package (VASP) based on DFT. In particular, for the present simulations of the equation of state of iridium, we used the local-density approximation (LDA). The success of LDA applied to materials as dense as Ir was demonstrated with the example of gold [17]. At low temperatures, the LDA error shifts the lattice constants towards smaller values, but this shift appears to be very negligible for high density materials. At these lattice constants the thermal pressure has a negligible volume dependence, and, therefore, the effect of the LDA error is small. This explains why LDA is very successful when applied to denser materials at low T. The absolute volume dependence of the thermal pressure is rather similar for both LDA and its alternative generalized-gradient approximation (GGA), and both show strongly increasing thermal pressure at larger volumes. However, as the volume decreases (with increasing pressure), the LDA increase of thermal pressure slows down, which causes LDA to deteriorate at smaller volumes and higher temperatures. This will be discussed when comparing DFT results on Ir with the experiments.

Ir was modeled by placing its nine outermost electrons ($5d^7\ 6s^2$) to the valence. These nine valence electrons were represented with a plane-wave basis set with cutoff energy of 300 eV, while the core electrons were represented by projector augmented-wave (PAW) pseudo potentials. A $2 \times 2 \times 2$ (32-atoms) super-cell was used in each case, with a very dense k-point mesh of $35 \times 35 \times 35$; this choice ensures full-energy convergence better than 0.1 meV/atom, which was checked for each run. For finite-T simulations, we used the algorithm of Nosé, which induces temperature fluctuations of approximately the same frequency as the typical phonon frequencies of the material at the simulated P–T conditions. The finite-T simulations typically require between 7500 and 10,000 time steps (of 1 fs) to achieve full energy convergence and to produce sufficiently long output for the extraction of reliable average for the value of pressure. An example of the time evolution of both temperature and pressure during a Nosé algorithm run is shown in Figure 2.

To calculate the theoretical RDF of liquid Ir, we first prepared a liquid by melting fcc-Ir. To this end, we used a $4 \times 4 \times 4$ (256 atoms) fcc-Ir super-cell with a lattice constant of 4.0 Å, which was subject to initial T of 20,000 K and NVE-run for 10 ps (NVE means we used a micro-canonical ensemble in which the total number of particles (N), the volume (V), and the total energy (E) are assumed to be constant in the ensemble). The emerging liquid that reached equilibrium at $T \approx 8000$ K was slowly cooled down to 4260 K. The size of the liquid super-cell was slightly decreased to correspond to $\approx$40 GPa. The adjusted liquid super-cell was then NVT-run for 5000 ps (N, V, and T constant in the ensemble) to extract its RDF, which was generated by VASP as part of the output of this run.

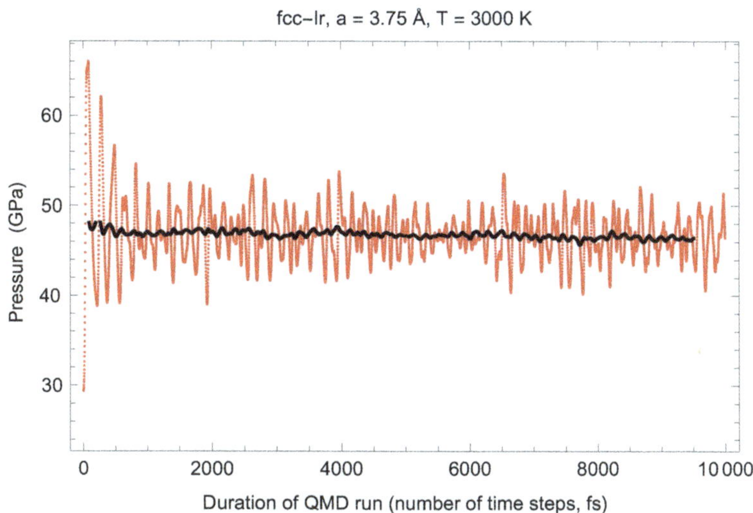

Figure 2. Time evolution of pressure during the 10,000 time step run of fcc-Ir with a lattice constant of 3.75 Å under a fixed temperature of 3000 K using the Nosé algorithm. The moving average is shown as a black line. In the vertical axis, pressure is in GPa.

3. Results

3.1. Experiments

During the experiment, several heating ramps were performed on Ir starting from two different initial pressures: 20 GPa and 30 GPa. The results obtained from the different ramps were completely comparable with each other and an example of the observed structural and textural evolution is reported in Figure 3 at selected T. In the investigated P–T range, only peaks belonging to fcc Ir and MgO were observed with no evidence of any solid–solid phase transitions or chemical reactions (such as IrO_2 or $MgIrO_3$). The texture of both MgO and Ir showed a similar temperature-induced evolution, starting with a high quality powder averaging at 300 K and showing increased recrystallization with the raising T (see Figure 3).

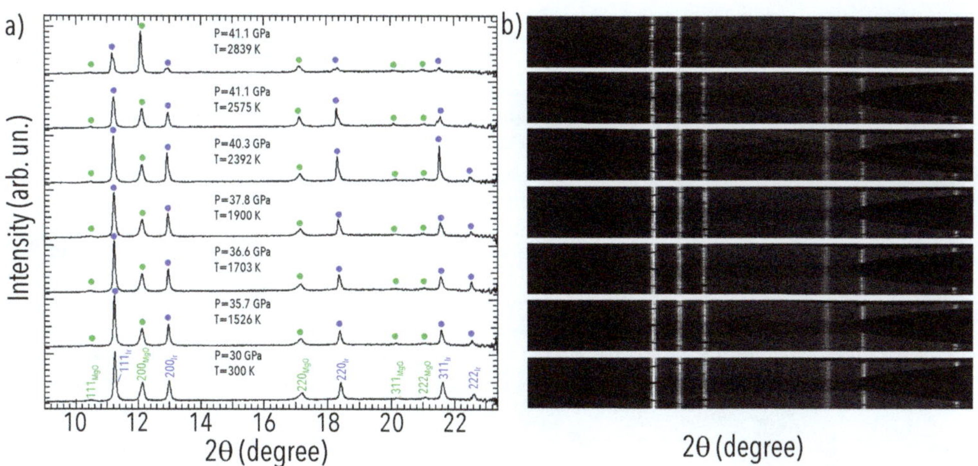

Figure 3. Selection of XRD patterns showing the (**a**) structural and the corresponding (**b**) textural evolution of an Ir sample embedded in MgO PTM between ambient T and 2900 K, in a pressure range between 30 GPa and 41 GPa.

During each heating ramp, a plateau in the temperature evolution was observed and, despite the continuous increase in the lasers power, it was impossible to reach higher *T*. During one particular heating ramp, together with the temperature plateau, it was possible to observe a massive diffuse XRD signal (see Figures 4 and 5), characteristic of the presence of molten material. An analysis of the textural evolution of this particular ramp (Figure 4) shows how the signal coming from the Ir sample evolves with the raising *T*: from a powder-like signal presenting several spots of higher intensities (at 1500 K) to a single-crystal-like signal caused by *T*-induced growth of the grains [18] (at 4260 K) and, finally, to a diffuse signal (at 4260 K) characteristic of the presence of liquid Ir. A better visualization of the observed behavior can be obtained from the integrated signals of the pattern reported in Figure 4. In fact, in Figure 5, it is possible to observe the amount of diffuse signal observed in the red pattern (at 2649 K) compared to the other two patterns obtained at lower *T*. Concerning the textural evolution of MgO, in Figure 4, it is possible to observe how it does not change much in the reported *T* range. This confirms that the measured liquid signal is actually from Ir and not from the insulating MgO material.

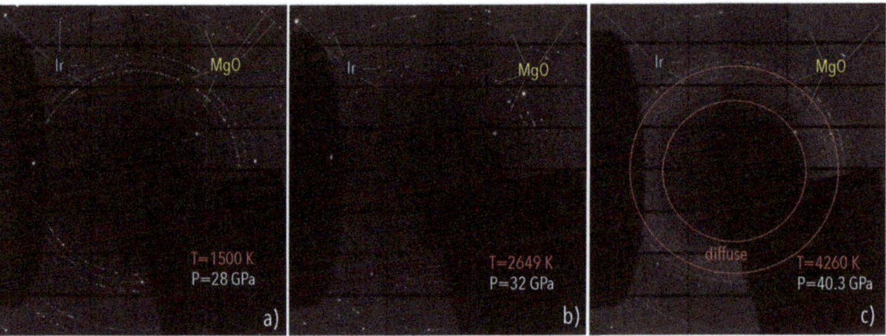

Figure 4. Textural evolution of Ir sample and MgO-insulating material during the heating ramp where melting was observed. (**a**) Both the sample and MgO are showing *T*-effects, characterized by the presence of smaller (more intense) spots withing the powder rings. (**b**) While the MgO texture remains quite similar to the previous case, Ir now exhibits the presence of single crystals caused by the growth of Ir grains [18]. (**c**) Ir crystal peaks have essentially disappeared and have now converted into an amorphous halo, whilst the XRD signal from the MgO is still similar to the previous case. In all the reported images, the peaks belonging to Ir and MgO are labelled and the corresponding *P–T* conditions indicated. In (**c**), the concentric red circles are used to assist the reader in identifying the presence of the diffuse halo. The shadows observed on each image are caused by the presence of the various optics of the LH system of I15 [9].

Due to the amount of liquid signal measured at 4260 K and 40.3 GPa, it was possible to extract the experimental RDF (reported as g(r) in Figure 6), obtaining confirmation of the melting of Ir and additional structural information. In particular, the two main oscillations in the RDF signal (r_1 and r_2) are found at 2.81 Å and 5.31 Å. Considering the volume expansion due to the solid/liquid transition, as expected in Ir from the slope of its Clausius–Clapeyron curve simulated by Burakovsky et al. [6,19], a value of 2.81 Å for the radius of the first coordination shell of liquid Ir is in good agreement with the first neighbour distance of 2.64 Å obtained from the last measured diffraction signal of solid Ir. Comparing the experimental and the theoretical RDF signals shown in Figure 6, it is possible to observe that both the r_1 and r_2 peaks obtained from the DFT simulation are slightly shifted towards lower r (by 0.01 nm) with respect to the experimental one. This is probably caused by the slight underestimation of volumes (therefore, of atomic distances) caused by the use of LDA (as mention above) compared to the experimental values. However, the experimental r_2/r_1 ratio of 1.87 is in good agreement with the theoretical value of 1.81 obtained from the DFT simulation. A similar value was also found for liquid Cu [20] (another *fcc* metal).

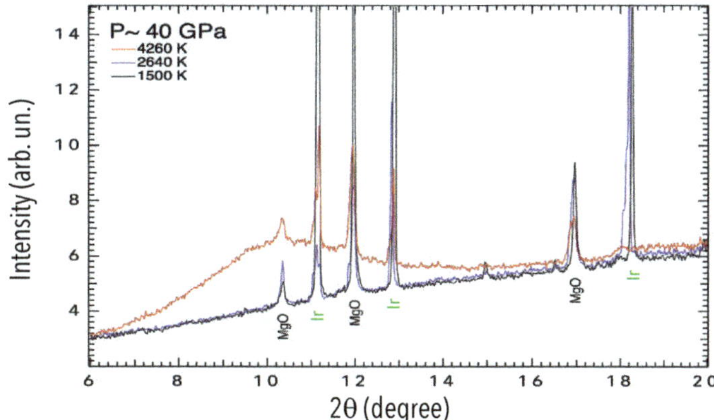

Figure 5. Integrated diffraction signal of Ir embedded in MgO at three different Ts (the same as those reported in Figure 4), before and after the melting of Ir. In particular, the appearance of the characteristic diffuse signal from a liquid is clearly visible in the red pattern. MgO and Ir peaks are labeled.

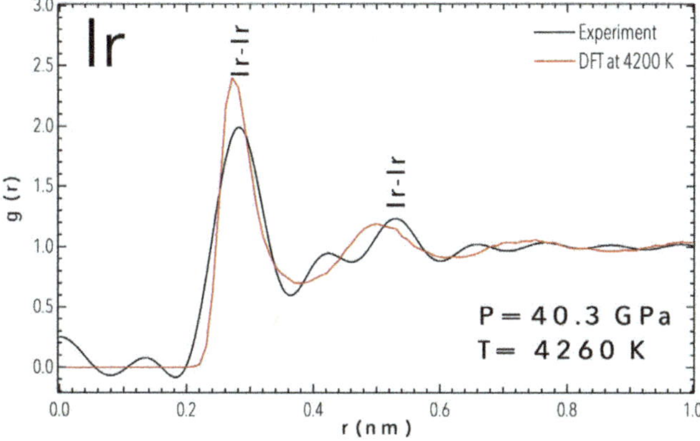

Figure 6. Comparison between the RDF signals of liquid Ir experimentally obtained at 40.3 GPa and 4260 K (black line) and the RDF signal obtained from DFT simulation at 4200 K and 40 GPa (red line).

In Figure 7, the experimental points obtained in the present study (solid and liquid) are compared to the phase boundaries calculated in Burakovsky et al. [6]. As it is possible to deduce from the figure, the present results (melting point included) are in good agreement with the previous calculation and far below the melting curve of MgO as reported in Kimura et al. [21] (confirming that the observed liquid signal is from Ir and not from the insulating material).

The unit–cell parameters of Ir at different P–T conditions were determined from a LeBail analysis of the measured XRD patterns. The present results corresponding to different isotherms are reported in Figure 8, together with the RT isotherms determined for experiments preformed using helium (He) [1], neon (Ne) [22], argon (Ar) [23] as pressure-transmitting media up to 140, 70, and 65 GPa, respectively. Results from HT experiments conducted at ambient pressure [24] are also reported in the figure for comparison.

The present RT results agree well with the experiments of Monteseguro et al. up to 10 GPa [1]. However, at higher pressures, they show a smaller compressibility, which becomes more similar to that determined from the experiments performed under Ar [23] and Ne [22].

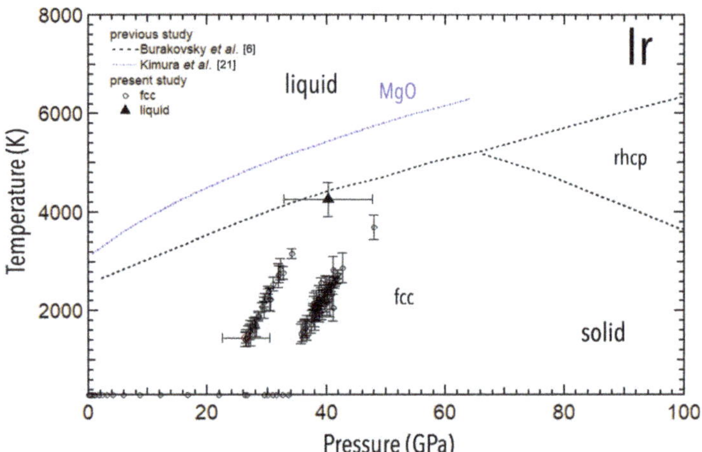

Figure 7. Ir phases as observed at the present experimental P–T conditions (solid symbols) compared to the DFT predicted phase diagram of Ir [6]. The melting line of MgO by Kimura et al. [21] is also reported for comparison. Errors in the T measurement are reported for all the experimental point obtained during laser heating. For simplicity, only the minimum and the maximum errors in P are reported for clarity reasons.

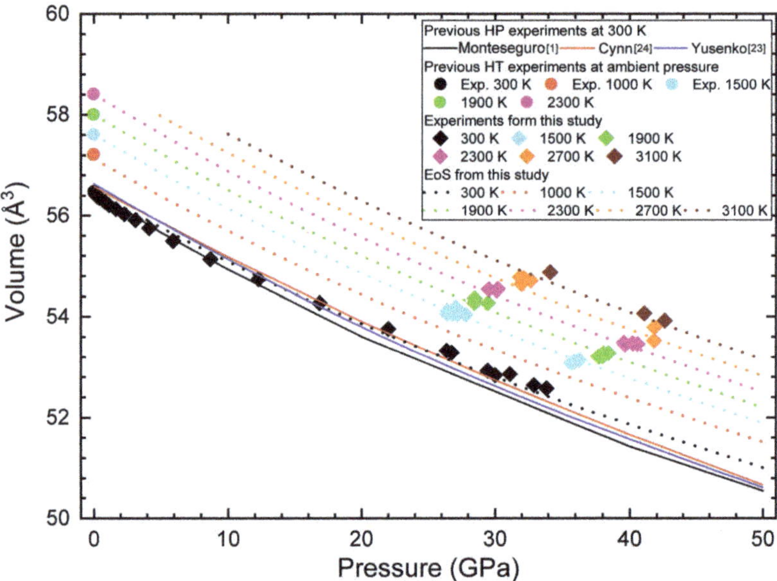

Figure 8. Unit–cell volume of Ir versus pressure for different temperatures. Diamonds correspond to the experiments of the present work. Circles are results from ambient-pressure experiments [24]. The black, blue, and red solid lines are the EoS determined from experiments carried out under He [1], Ne [22], and Ar [23]. The dashed lines are the isotherms obtained from the P–V–T EoS we determined.

A third-order Birch–Murnaghan (BM) equation of state [25] was fitted to the present data. The obtained volume at ambient pressure (V_0), bulk modulus (B_0), and its pressure derivative (B_0') are summarized in Table 1. They are compared with previous studies following the results from the present calculations.

Table 1. EoS parameters (at 300 K) determined from present and previous studies whose references are indicated in the center column. In the upper part, we show results from experiments indicating the pressure medium and the maximum P reached during the experiment. In the lower part, we show results from calculations.

V_0 [Å^3]	B_0 [GPa]	B_0'	Method	Pressure Medium	P Max
56.48	339 (3)	5.3 (1)	XRD [1]	Helium	140 GPa
56.58	383 (14)	3.1 (8)	XRD [23]	Argon	65 GPa
56.64 (24)	341(19)	4.7 (3)	XRD [22]	Neon	70 GPa
56.69	306 (23)	6.8 (1.5)	XRD [5]	MgO	65 GPa
56.48 (9)	360 (5)	6.0 (5)	XRD (this work)	MgO	35 GPa
56.58	366	5.0	DFT [6]		
56.17	377	5.3	DFT (this work)		

From the P–V–T data shown in Figure 8, it was possible to determine a thermal EoS taking advantage of the EosFit7 package [16]. For the analysis, we used diffraction data measured at 300 (2) K, 1500 (20) K, 1900 (20) K, 2300 (30) K, 2700 (50) K, and 3100 (50) K, as well as previous ambient-pressure results [24,26]. During the fitting procedure, the third-order BM EoS generated from the RT compression experiment was used as the isothermal part of the P–V–T EoS. In addition, a Berman equation was used as the thermal-expansion model [27], assuming a linear variation of B_0 with T. The pressure derivative of the bulk modulus was assumed to be P-independent. On the other hand, the thermal expansion was considered to be P-independent and to have a linear T-dependence. This simple model properly describes all the available experimental results, as can be seen in Figure 8. The obtained parameters are $dB_0/dT = -0.015(9)$ GPa/K, volumetric thermal expansion $\alpha = 1.6(2) \times 10^{-5}~K^{-1}$, and $d\alpha/dT = 8.0(7) \times 10^{-10}~K^{-2}$.

3.2. Computer Simulations

In Figure 9, the isotherms obtained from the present DFT calculations are presented and compared to the ones obtained from the experimental data (described in the previous section). According to the data reported in the figure, it is possible to observe how the present calculations underestimate the experimental volume at ambient P by less than 1%. Regarding the volumetric evolution with pressure, despite the underestimated value of V_0, at lower pressure the calculations give a slightly smaller compressibility than the experiments (larger bulk modulus). However, beyond 20 GPa and for temperatures up to 1000 K, calculations and experiment provide very similar volumetric compressions, with isotherms running nearly parallel to each other, as a consequence of the slightly smaller calculated B_0'. The RT EoS parameters obtained from the calculated isotherm at 300 K are summarized in Table 1. A detailed comparison with the experimental results is presented in the next section.

The calculated HT isotherms are well described using the same model employed to analyze the experiments. The parameters determined from the present computer simulations are $dB_0/dT = -0.03$ GPa/K, $\alpha = 1.6 \times 10^{-5}~K^{-1}$, and $d\alpha/dT = 8.0 \times 10^{-10}~K^{-2}$. The agreement with the experimental results is reasonably good. Only the effect of T on the bulk modulus is larger in the calculations than in the experiments, causing a stronger predicted compressibility for T higher than 1000 K. This is clearly visible in Figure 9, where the comparison of calculations with experiments shows that as both P and T go up, the theoretical p values start lagging behind their experimental counterparts; at the highest P and T of this work, the shift is as high as ≈20 GPa at 80 GPa and 3000 K.

Regarding the melting temperature, as previously discussed, the DFT calculations provide a melting curve in agreement (within the uncertainties) with the measured liquid Ir at 40 GPa. Furthermore, the calculated and the measured RDF are also in good agreement, with a difference in the position of the peaks of the first and second coordination spheres of ≈0.01 Å.

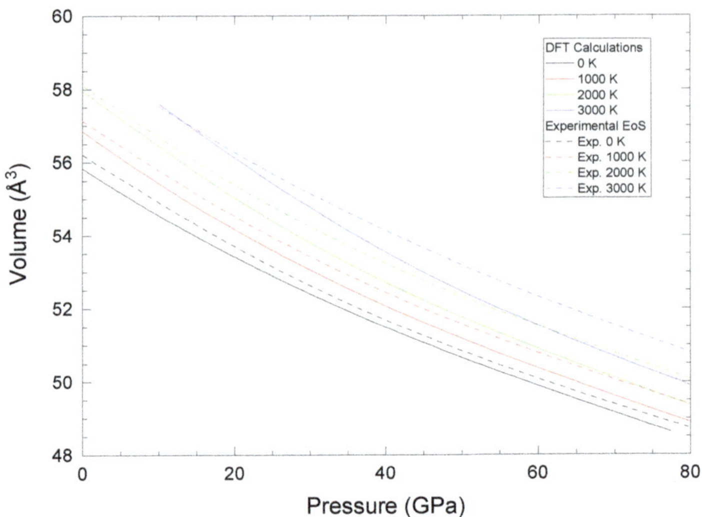

Figure 9. Calculated unit–cell volume of Ir versus P at different T (solid lines). The theoretical results are compared with the P–V–T EoS obtained from the experiments (dashed lines).

4. Discussion

Now we will compare the bulk modulus and its pressure derivative obtained from experiments and calculations in this work with those of the literature. Given the fact that B_0 and B_0' are correlated, a proper comparison can only be made by plotting B_0' versus B_0, as reported in Figure 10. Results from different works are included in the figure, including the error bars when available. Confidence ellipses are also represented in the figure to visualize the correlation between B_0 and B_0' of the present experiment.

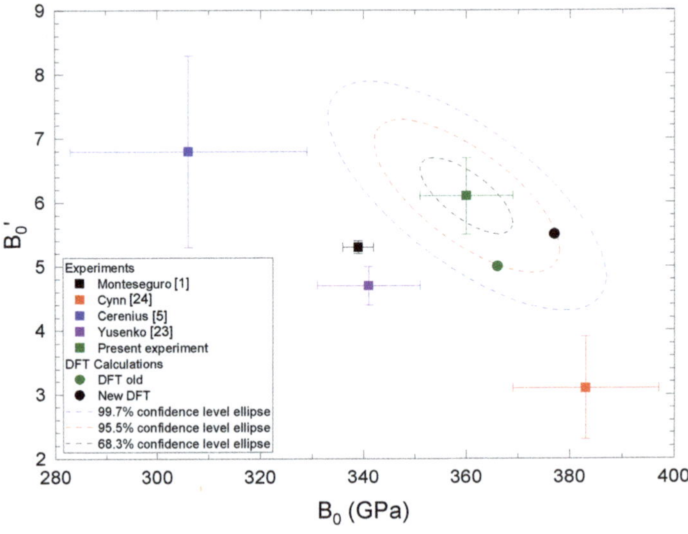

Figure 10. B_0' versus B_0. Values from different studies are included; see inset. Confidence ellipses from the present experiments are shown.

The theoretical values (from present and previous state-of-the-art calculations [6]) are within the 95.5% confidence level ellipse of the experimental ones. This means that they are only two sigmas away from the experimental results, which can be considered as a good match. The results from Monteseguro et al. [1] and Yusenko et al. [22] slightly underestimate the bulk modulus. Notice that the experiment of Monteseguro et al. [1] was done under hydrostatic conditions and in a larger pressure range. This can explain the small differences in the EoS parameters. Regarding other experiments, the results from Cynn et al. [23] overestimate B_0 and underestimate B_0'. On the other hand, the study by Cerenius et al. [5] reports an extremely low value for B_0 but a reasonable value for its pressure derivative. This could indicate a systematic error in the bulk modulus determination not affecting the pressure derivative, which could be related to the sample bridging between the two diamond anvils.

Regarding the calculations, in addition to the previous study of Burakovsky et al. [6], there were previous DFT and tight-binding calculations. Calculations using a Kleinman–Bylander separable non-local pseudopotential [28] give an overestimated bulk modulus ($B_0 = 385$ GPa) and the same happens for calculations using a Hartwigsen–Goedecker–Hutter semi-core potential ($B_0 = 399$ GPa) [29]. On the other hand, tight-binding calculations that underestimate the bulk modulus [30] give an underestimated bulk modulus ($B_0 = 341$ GPa).

In conclusion, Figure 10 and the comparison with previous studies show that the *RT* EoS reported here gives the most accurate description of the volumetric compression of Ir. This, and the agreement with previous studies and calculations of the thermal-expansion behavior of Ir at *HP*, reinforce confidence regarding the accurate determination of the present P–V–T EoS of Ir. In summary, the associated parameters are: $V_0 = 56.48$ (9) A^3, $B_0 = 360$ (5) GPa, $B_0' = 6.0$ (5), $dB_0/dT = -0.015(9)$ GPa/K, $\alpha = 1.6(2) \times 10^{-5}$ K^{-1}, and $d\alpha/dT = 8.0(7) \times 10^{-10}$ K^{-2}.

Compared to the other transition metals (such as Au, Cu, Pt, Nb, Re, Ta, etc.), which have been proposed as pressure standards in the literature [31], Ir results less compressible than most of them (with the exception of Os [32]). In fact, their bulk moduli are ranging from 136 GPa (Cu) to 320 GPa (Ru) [33–36]. However, Ir has several advantages compared to the other metals, for example, it has a much higher melting temperature than Au, Cu, and Pt [13,37,38], which makes it more suitable for *HP–HT* experiments. On the other hand, it is much more inert than Re, Ta, and W, making it a better candidate as standard for *HP–HT* experiments where chemical reactions are a sensitive issue [2,3]. In addition, the thermal expansion of Ir ($\alpha = 1.6(2) \times 10^{-5}$ K^{-1}) is one third of that of Au and Cu [39,40] and around 30% smaller than that of Pt and Ru [33,41]. This fact makes Ir a more interesting metal for pressure calibration at *HT*, since the smaller thermal expansion makes it more accurate, providing smaller thermal pressure (being it proportional to $\alpha \times B_0$ [10]). In the case of Ir $\alpha \times B_0 = 0.0058$ GPa K^{-1}, while in Au, Cu, Pt, and Ru, the same magnitude ranges from 0.0062 to 0.0074 GPa K^{-1}. In conclusion, Ir can be considered as an ideal standard used for pressure calibration in high P–T XRD experiments due to its simple structure, phase stability, strong signal, inert chemical properties, high melting temperature, and thermal expansion properties.

5. Conclusions

In the present study, a thermal pressure–volume equation of state for iridium valid up to 80 GPa and 3100 K has been obtained combining *HP–HT* laser-heating powder XRD experiments and molecular-dynamic DFT calculations. The stability of the *fcc* phase of iridium was also explored experimentally, confirming it as the only solid phase observed in the investigated P–T region (48 GPa and 3100 K). The melting temperature of Ir at 40 GPa was also determined to be 4260 (200) K. Results are compared with previous experimental and theoretical studies. The reliability of the present results is supported by the consistency between the values yielded for EoS parameters by the two methods. In particular, we found that the EoS parameters determined from the present experiments agree very well with the calculated ones (within 95.5% confidence level ellipse). The experimental parameters are $V_0 = 56.48\ (9)\ \text{Å}^3$, $B_0 = 360\ (5)\ \text{GPa}$, $B_0' = 6.0\ (5)$, $dB_0/dT = -0.015\ (9)\ \text{GPa/K}$, $\alpha = 1.6\ (2) \times 10^{-5}\ K^{-1}$, and $d\alpha/dT = 8.0\ (7) \times 10^{-10}\ K^{-2}$. The reported results will allow the use of Ir as calibration standard for high-pressure and high-temperature experiments. The comparison with other metals (Au, Cu, Pt, Nb, Re, Ta, and Os) shows that Ir is a reliable pressure standard for high-pressures and high-temperatures DAC experiments. Finally, the radial-distribution function of liquid iridium at HP and HT has been reported for the first time. Information on the first and second coordination shells of liquid iridium is reported.

Author Contributions: D.E. conceived the experiment. S.A., R.T. and E.B. conducted the experiment. S.A. and D.E. analyzed the results. L.B. performed the theoretical calculations. The manuscript is written through contributions of all authors. All authors have given approval to the final version of the manuscript.

Funding: This work was partially supported by the Spanish Ministry of Science, Innovation, and Universities under grants PID2019-106383GB-C41 and RED2018-102612-T (MALTA Consolider-Team Network) and by Generalitat Valenciana under grant Prometeo/2018/123 (EFIMAT). R.T. acknowledges funding from the Spanish MINECO via the Juan de la Cierva Formación program (FJC2018-036185-I).

Data Availability Statement: All relevant data that support the findings of this study are available from the corresponding authors upon request.

Acknowledgments: The authors acknowledge the DLS synchrotron facilities for provision of beamtime on the beamlines I15 (DLS Ref. CY21610).

Conflicts of Interest: The authors declare no conflict of interest.

Sample Availability: Samples were obtained from a commercial supplier of high-purity metals; Goodfellow.

References

1. Monteseguro, V.; Sans, J.A.; Cuartero, V.; Cova, F.; Abrikosov, I.A.; Olovsson, W.; Popescu, C.; Pascarelli, S.; Garbarino, G.; Jönsson, H.J.M.; et al. Phase stability and electronic structure of iridium metal at the megabar range. *Sci. Rep.* **2019**, *9*, 8940. [CrossRef]
2. Chulia-Jordan, R.; Santamaria-Perez, D.; Marqueno, T.; Ruiz-Fuertes, J.; Daisenberger, D. Oxidation of High Yield Strength Metals Tungsten and Rhenium in High-Pressure High-Temperature Experiments of Carbon Dioxide and Carbonates. *Crystals* **2019**, *9*, 676. [CrossRef]
3. Errandonea, D. Observation of chemical reactions between alkaline-earth oxides and tungsten at high pressure and high temperature. *J. Phys. Chem. Solids* **2009**, *70*, 1117–1120. [CrossRef]
4. Schock, R.; Johnson, K. Compression of iridium to 175 kbar. *Fiz. Metal. Met.* **1971**, *31*, 1100.
5. Cerenius, Y.; Dubrovinsky, L. Compressibility measurements on iridium. *J. Alloys Compd.* **2000**, *306*, 26. [CrossRef]
6. Burakovsky, L.; Burakovsky, N.; Cawkwell, M.; Preston, D.; Errandonea, D.; Simak, S. Ab initio phase diagram of iridium. *Phys. Rev. B* **2016**, *94*, 094112. [CrossRef]
7. Dubrovinsky, L.; Dubrovinskaia, N.; Bykova, E.; Bykov, M.; Prakapenka, V.; Prescher, C.; Glazyrin, K.; Liermann, H.P.; Hanfland, M.; Ekholm, M.; et al. The most incompressible metal osmium at static pressure above 750 gigapascal. *Nature* **2015**, *525*, 226. [CrossRef] [PubMed]
8. Tal, A.; Katsnelson, M.; Ekholm, M.; Jnsson, H.; Dubrovinsky, L.; Dubrovinskaia, N.; Abrikosov, I. Pressure-induced crossing of the core level in 5d metals. *Phys. Rev. B* **2016**, *93*, 205150. [CrossRef]
9. Anzellini, S.; Kleppe, A.; Daisenberger, D.; Wharmby, M.; Giampaoli, R.; Boccato, S.; Baron, M.; Miozzi, F.; Keeble, D.; Ross, A.; et al. Laser-heating system for high-pressure X-ray diffraction at the extreme condition beamline I15 at Diamond Light Source. *J. Synchrotron Radiat.* **2018**, *25*. [CrossRef]
10. Anzellini, S.; Boccato, S. A Practical Review of the Laser-Heated Diamond Anvil Cell for University Laboratories and Synchrotron Applications. *Crystals* **2020**, *10*, 459. [CrossRef]
11. Benedetti, L.; Loubeyre, P. Temperature gradients, wavelength-dependent emissivity, and accuracy of high and very-high temperatures measured in the laser-heated diamond cell. *High Press. Res.* **2004**, *24*, 423–445. [CrossRef]
12. Dorogokupets, P.I.; Oganov, A.R. Ruby, metals, and MgO as alternative pressure scales: A semiempirical description of shock-wave, ultrasonic, X-ray, and thermochemical data at high temperatures and pressures. *Phys. Rev. B* **2007**, *75*, 024115. [CrossRef]
13. Anzellini, S.; Monteseguro, V.; Bandiello, E.; Dewaele, A.; Burakovsky, L.; Errandonea, D. In situ characterization of the high pressure- high temperature meltig curve of platinum. *Sci. Rep.* **2019**, *9*, 13034. [CrossRef]

14. Prescher, C.; Prakapenka, V. DIOPTAS: A program for reduction of two-dimensional X-ray diffraction data and data exploration. *High Press. Res.* **2015**, *35*, 223. [CrossRef]

15. Coelho, A. TOPAS and TOPAS-Academic: An optimization program integrating computer algebra and crystallographic object written in C++. *J. Appl. Crystallogr.* **2018**, *51*, 210. [CrossRef]

16. Angel, R.J.; Gonzalez-platas, J.; Alvaro, M. EosFit7c and a Fortran module (library) for equation of state calculations. *Z. Kristallogr.* **2014**, *229*, 405–419. [CrossRef]

17. Grabowski, B.; Wippermann, S.; Glensk, A.; Hickel, T.; Neugebauer, J. Random phase approximation up to the melting point: Impact of anharmonicity and nonlocal many-body effects on the thermodynamics of Au. *Phys. Rev. B* **2015**, *91*, 201103. [CrossRef]

18. Hrubiak, R.; Meng, Y.; Shen, G. Microstructures define melting of molybdenum at high pressure. *Nat. Commun.* **2017**, *8*, 14562. [CrossRef] [PubMed]

19. Burakovsky, L.; Cawkwell, M.; Preston, D.; Errandonea, D.; Simal, S. Recent ab initio phase diagram studies: Iridium. *J. Phys. Conf. Ser.* **2017**, *950*, 042021. [CrossRef]

20. Sukhomlinov, S.; Müser, M. A mixed radial, angular, three-body distribution function as a tool for local structure characterization: Application to single-component structures. *J. Chem. Phys.* **2020**, *152*, 194502. [CrossRef]

21. Kimura, T.; Ohfuji, H.; Nishi, M.; Irifune, T. Melting temperatures of MgO under high pressure by micro-texture analysis. *Nat. Commun.* **2017**, *8*, 15735. [CrossRef]

22. Yusenko, K.; Khandarkhaeva, S.; Fedotenko, T.; Pakhomova, A.; Gromilov, S.; Dubrovinsky, L.; Dubrovinskaia, N. Equations of state of rhodium, iridium and their alloys up to 70 GPa. *J. Alloys Compd.* **2019**, *788*, 212. [CrossRef]

23. Cynn, H.; Klepeis, J.E.; Yoo, C.S.; Young, D.A. Osmium has the Lowest Experimentally Determined Compressibility. *Phys. Rev. Lett.* **2002**, *88*, 676. [CrossRef]

24. Halvorson, J.J.; Wimber, R.T. Thermal Expansion of Iridium at High Temperatures. *J. Appl. Phys.* **1972**, *43*, 2519–2522. [CrossRef]

25. Birch, F. Elasticity and constitution of the Earth's interior. *J. Geophys. Res.* **1952**, *57*, 227–286. [CrossRef]

26. Arblaster, J. Crystallographic properties of iridium. *Platin. Met. Rev.* **2010**, *54*, 93. [CrossRef]

27. Anzellini, S.; Errandonea, D.; MacLeod, S.; Botella, P.; Daisenberger, D.; DeAth, J.; Gonzalez-Platas, J.; Ibanez, J.; McMahon, M.; Munro, K.; et al. Phase diagram of calcium at high pressure and high temperature. *Phys. Rev. Mater.* **2018**, *2*, 083608. [CrossRef]

28. Grussendorff, S.; Chetty, N.; Dreysse, H. Theoretical studies of iridium under pressure. *J. Phys. Condens. Matter* **2003**, *15*, 4217. [CrossRef]

29. Fang, H.; Liu, B.; Gu, M.; Liu, X.; Huang, S.; Ni, C.; Wang, R. High-pressure lattice dynamic and thermodynamic properties of Ir by first-principles calculations. *Physica B* **2010**, *405*, 732. [CrossRef]

30. Gheribi, A.E.; Roussel, J.M.; Rogez, J. Phenomenological Hugoniot curves for transition metals up to 1 TPa. *J. Phys. Condens. Matter* **2010**, *19*, 476218. [CrossRef]

31. Sokolova, T.S.; Dorogokupets, P.I.; Dymshits, A.M.; Danilov, B.S.; Litasov, K.D. Microsoft excel spreadsheets for calculation of P–V–T relations and thermodynamic properties from equations of state of MgO, diamond and nine metals as pressure markers in high-pressure and high-temperature experiments. *Comput. Geosci.* **2016**, *94*, 162–169. [CrossRef]

32. Takemura, K. Bulk modulus of osmium: High-pressure powder X-ray diffraction experiments under quasihydrostatic conditions. *Phys. Rev. B* **2004**, *70*, 012101.

33. Anzellini, S.; Errandonea, D.; Cazorla, C.; MacLeod, S.; Monteseguro, V.; Boccato, S.; Bandiello, E.; Anichtchenko, D.D.; Popescu, C.; Beavers, C. Thermal equation of state of ruthenium characterized by resistively heated diamond anvil cell. *Sci. Rep.* **2020**, *10*, 7092. [CrossRef] [PubMed]

34. Dewaele, A.; Loubeyre, P.; Mezouar, M. Equations of state of six metals above 94 GPa. *Phys. Rev. B* **2004**, *094112*, 1–8.

35. Anzellini, S.; Dewaele, A.; Occelli, F.; Loubeyre, P.; Mezouar, M. Equation of state of rhenium and application for ultra high pressure calibration. *J. Appl. Phys.* **2014**, *115*, 043511, [CrossRef]

36. Errandonea, D.; Burakovsky, L.; Preston, D.E.A. Experimental and theoretical confirmation of an orthorhombic phase transition in niobium at high pressure and temperature. *Commun. Mater.* **2020**, *1*, 60, [CrossRef]

37. Weck, G.; Recoules, V.; Queyroux, J.; Datch, F.; Bouchet, J.; Ninet, S.; Garbarino, G.; Mezouar, M.; Loubeyre, P. Determination of the melting curve of gold up to 110 GPa. *Phys. Rev. B* **2020**, *101*, 014106. [CrossRef]

38. Dewaele, A.; Mezouar, M.; Guignot, N.; Loubeyre, P. Melting of lead under high pressure studied using second-scale time-resolved X-ray diffraction. *Phys. Rev. Condens. Matter Mater. Phys.* **2007**, *76*, 1–5. [CrossRef]

39. Dorogokupets, P.I.; Dewaele, A. Equations of state of MgO, Au, Pt, NaCl-B1, and NaCl-B2: Internally consistent high-temperature pressure scales. *High Press. Res.* **2007**, *27*, 431–446. [CrossRef]

40. Wang, Y.; Zhang, J.; Xu, H.; Lin, Z.; Daemen, L.L.; Zhao, Y.; Wang, L. Thermal equation of state of copper studied by high P-T synchrotron X-ray diffraction. *Appl. Phys. Lett.* **2009**, *94*, 071904. [CrossRef]

41. Zha, C.S.; Mibe, K.; Bassett, W.A.; Tschauner, O.; Mao, H.K.; Hemley, R.J. P-V-T equation of state of platinum to 80 GPa and 1900 K from internal resistive heating/X-ray diffraction measurements. *J. Appl. Phys.* **2008**, *103*, 054908. [CrossRef]

Article

Anomalous Behavior in the Atomic Structure of Nb₃Sn under High Pressure

Irene Schiesaro [1,*], Simone Anzellini [2], Rita Loria [1], Raffaella Torchio [3], Tiziana Spina [4], René Flükiger [5], Tetsuo Irifune [6], Enrico Silva [7] and Carlo Meneghini [1]

[1] Dipartimento di Scienze, Università Roma Tre, Via della Vasca Navale 79, 00146 Rome, Italy; ritaloria@gmail.com (R.L.); carlo.meneghini@uniroma3.it (C.M.)

[2] Diamond Light Source Ltd., Harwell Science & Innovation Campus, Diamond House, Didcot OX11 0DE, UK; simone.anzellini@diamond.ac.uk

[3] ESRF—European Synchrotron Radiation Facility, 38000 Grenoble, France; raffaella.torchio@esrf.fr

[4] Superconducting Radio Frequency (SRF) Materials and Research Department, Fermilab, Batavia, IL 60510, USA; tspina@fnal.gov

[5] Department of Quantum Matter Physics, University of Geneva, 1211 Geneva, Switzerland; rene.flukiger@unige.ch

[6] Geodynamics Research Center, 6 Ehime University, Matsuyama 790-8577, Japan; irifune@dpc.ehime-u.ac.jp

[7] Dipartimento di Ingegneria, Università Roma Tre, 00146 Roma, Italy; enrico.silva@uniroma3.it

* Correspondence: ire.schiesaro@gmail.com

Abstract: In the present study, the local atomic structure of a Nb₃Sn superconductor sample has been probed by X-ray absorption fine structure (XAFS) as a function of hydrostatic pressure (from ambient up to 26 GPa) using a diamond anvil cell set-up. The analysis of the Nb-K edge extended X-ray absorption fine structure (EXAFS) data was carried out combining standard multi shell structural refinement and reverse Monte Carlo method to provide detailed in situ characterization of the pressure-induced evolution of the Nb local structure in Nb₃Sn. The results highlight a complex evolution of Nb chains at the local atomic scale, with a peculiar correlated displacement of Nb–Nb and Nb–Nb–Nb configurations. Such a local effect appears related to anomalies evidenced by X-ray diffraction in other superconductors belonging to the same A15 crystallographic structure.

Keywords: Nb₃Sn; local atomic structure; high pressure; XAFS

Citation: Schiesaro, I.; Anzellini, S.; Loria, R.; Torchio, R,; Spina, T.; Flükiger, R.; Irifune, T.; Silva, E.; Meneghini, C. Anomalous Behavior in the Atomic Structure of Nb₃Sn under High Pressure. *Crystals* **2021**, *11*, 331. https://doi.org/10.3390/cryst11040331

Academic Editor: Sergio Brutti

Received: 18 February 2021
Accepted: 21 March 2021
Published: 25 March 2021

Publisher's Note: MDPI stays neutral with regard to jurisdictional claims in published maps and institutional affiliations.

1. Introduction

Nb₃Sn is a brittle intermetallic material belonging to the class of the A15 compounds (space group *Pm-3n*) that in 1954 reached a superconducting transition temperature of $T_c \approx 18$ K [1]. Even more important, is its capability to carry high current densities $J_c > 10^3$ A/mm² [2], allowing to make compact magnets reaching high critical fields B_{c2} up to 30 T [3,4], essential for high field superconductor applications. These properties contributed to make Nb₃Sn the most widely used high field superconductor in top science projects (CERN High luminosity LHC project [5], ITER [6] project) and industrial applications (NMR instruments, compact cyclotrons), with a record production in the period 2009–2014 of 150 Tons/Year for the ITER toroidal field magnets only [7]. The exceptional requirements of these magnet-based projects revamped the interest in this material and considerable efforts are now undertaken to further improve their critical performances and efficiencies during the applications [8–10]. In particular, the effects of strains (axial, transverse, hydrostatic) on J_c, T_c and the electrical resistivity, which may be caused by thermal contractions and strong Lorentz forces due to the high currents, were extensively explored [4,9,11–15]. However, less is known about the structural modifications induced by pressure, especially on the crystallographic and atomic scale [15–17]. These informations are crucial for achieving accurate model of the Nb₃Sn properties. As a matter of fact, density functional theory (DFT) calculations have shown that squeezing the structure actually affects the phonon spectra and electronic density of states $N(E_f)$ of Nb$_3$Sn [15]. However, accurate models require deep knowledge about the atomic structure at the crystallographic and atomic (local) scale.

Recently, structural changes have been highlighted in different A15 compounds as a function of pressure. High pressure X-ray diffraction (HP-XRD) experiments [15], carried out on technological Nb$_3$Sn wires, showed an anomaly in the unit cell volume compression (around 5 GPa) not associated with any structural phase transition. A similar effect, ascribed to isostructural transitions, was observed around 15 GPa for other A15 systems, e.g., Nb$_3$Ga [18] and Nb$_3$Al [19] . Whereas, low temperature (T = 10 K) HP-XRD measurements, performed on a slightly non stoichiometric Nb$_3$Sn$_{1-x}$ single crystal, have evidenced structural instabilities around 3 GPa, suggesting a dimerization of Nb–Nb chains, providing alternating shorter/longer Nb–Nb pairs [17]. Recent simulations on Nb$_3$Al suggested the *Pm-3n* phase to be energetically not favoured with respect to the C_2/c [20] phase, where the lower symmetry allows for distortions of the Nb chains.

The XRD technique provides accurate long-range order structural characterization, but misses details about the local atomic structure, i.e., the average relative atomic arrangement [21]. These details can be assessed exploiting the X-ray absorption fine structure (XAFS) spectroscopy, a chemically selective technique, sensitive to the local atomic structure. However, to the best of our knowledge, XAFS literature data on Nb$_3$Sn are quite rare, even at ambient pressure/temperature conditions [22,23].

In this work, Nb K-edge HP-XAFS was used to obtain further insight on the atomic structure of Nb$_3$Sn between ambient pressure (AP) and 26 GPa. XAFS data analysis was carried out combining well established multi-shell data refinement [24] and Reverse Monte Carlo (RMC) approach [25]. In this way, it was possible to build a 3D model of Nb$_3$Sn structural details around the Nb sites, and to extract further information on the pressure-dependence of the relative arrangements of Nb neighbors (many-body distribution functions) in the structure.

2. Materials and Methods

X-ray absorption spectroscopy (XAS) measurements were performed at the beamline BM23 [26] of the European Synchrotron Radiation Facility (ESRF) at the Nb K-edge (18.986 keV). The X-ray beam was focused down to 3×3 μm^2 using the micro-XAS facility the data of which were collected in transmission mode using two gas filled ionization chambers to measure both the incident (I_o) and transmitted (I_t) X-ray intensities. For sake of comparison, additional Nb K-edge XAS spectra were measured at the P65 beamline (Petra-III, DESY synchrotron radiation facility in Hamburg, Germany) [27] on powder of the same sample, in standard transmission geometry at ambient conditions.

A membrane diamond anvil cell (DAC) was equipped with nano-polycrystalline diamonds [28,29] with a culet size of 400 μm. The gasket was prepared from a pre-indented and laser drilled stainless steel foil. The Nb$_3$Sn samples were obtained from a polycrystalline bulk piece sintered by Hot Isostatic Pressure (HIP) technique (2 kbar Argon pressure at 1250 °C for 24 h) at the University of Geneva [30]. A grain size of about 20 μm and a composition very close to stoichiometry (24.8 at.% Sn) were determined from SEM/EDS analysis. A sharp superconducting transition was observed at 17.9 K by AC susceptibility. Finally, a Rietveld refinement yielded a lattice constant at ambient conditions of 5.291 Å and a Bragg–Williams [31] long-range order parameter S of 0.98. Further details on the synthesis procedure and preliminary characterization can be found in Ref. [32]. Samples from the same batch were characterized by means of microwave measurements [33], yielding a normal state resistivity $\rho_n \simeq 14.8$ μΩ · cm, a critical temperature T$_c$ = 17.8 K and an extrapolated upper critical field H_{c2}, giving $\mu H_{c2}(0)$ of $\simeq 27$ T [34], in full agreement with the accepted values [4] for pure Nb$_3$Sn. The sample for the XAS measurements was prepared by grinding a piece of Nb$_3$Sn bulk in an agate mortar to obtain a fine powder. The latter was then squeezed between two diamond anvils in order to obtain a thin and homogeneous pellet which was cut and loaded in the DAC's high pressure chamber. A ruby chip was placed a few μm away from the sample and used as pressure gauge. Once the good quality of the sample loading was checked by means of X-rays (providing an absorption jump of 0.3 at the Nb K-edge), the high pressure chamber of the DAC was filled with Ne gas to ensure good hydrostatic conditions for the investigated pressure range [35]. During the experiment, the pressure inside the DAC was measured, before and after each energy scan, using the ruby fluorescence method [36], following the calibration of Dorogokupets and Oganov [37]. High pressure measurements (HP) were carried out at room temperature at seven pressure values in the range between 0.3 GPa and 26 GPa. In order to get a good statistics, the data at each pressure point was obtained by collecting at least four energy spectra, which were checked for energy scale alignment and then averaged.

For each pressure point (AP and HP data), the experimental absorption signals $\alpha(E) = \log(I_o/I_t)$ were treated along the standard procedures for pre-edge background subtraction, edge jump normalization and extraction of the structural extended X-ray absorption fine structure (EXAFS) signal $\chi(k)$ using the ESTRA software suite [38]. The pre-edge background was modelled with a straight line calculated by fitting the data in the interval 18.788–18.908 keV before the edge. The obtained background was then subtracted from the total signal, obtaining $\alpha'(E)$. The edge energy E_0, defining the origin of the photo-electron wave-vector $k = \hbar^{-1}\sqrt{2m_e(E - E_0)}$ (m_e = electron mass), was selected at the first inflection point of the absorption edge (maximum of the first derivative) and refined during the fit. The post edge atomic absorption background (α_0) was calculated fitting a N-knots polynomial spline through the data. The structural EXAFS signal was finally calculated as: $\chi_{exp}(k) = (\alpha' - \alpha_0)/\alpha_0$. In order to avoid artefacts coming from the signal extraction, the same procedure and normalization parameters were used to extract the EXAFS signal from all the data at different pressures. The k^2-weighted experimental spectra at the measured pressures are presented in Figure 1 along with the moduli of their Fourier transforms (FT) and best fits (see below). The sample homogeneity is crucial for reliable XAFS analysis, the good agreement between HP and AP data underlining the good quality of the spectra Figure 1). Furthermore, the agreement between structural parameters obtained from the analysis of AP and 0.3 GPa data confirms the reliability and reproducibility of the analysis procedures.

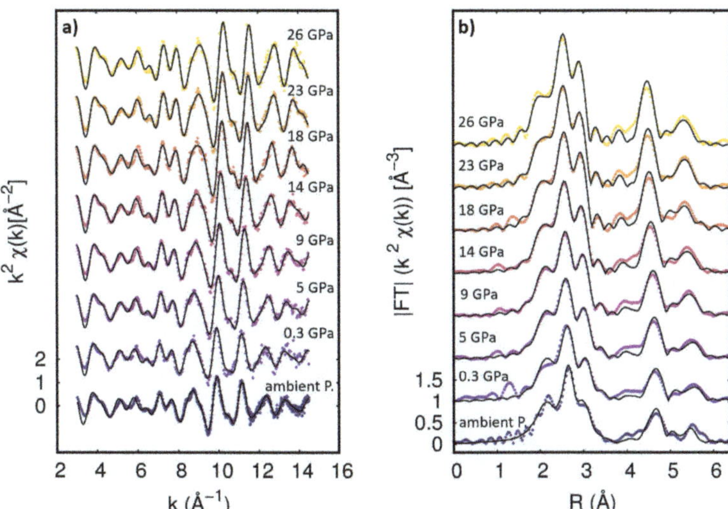

Figure 1. (a) Experimental (dots) and best fit (lines) Nb K-edge k^2-weighted extended X-ray absorption fine structure (EXAFS) spectra as a function of pressure. (b) Moduli of the FT of k^2-weighted EXAFS spectra (dots) and best fit curves (full lines). In each frame the corresponding pressures are marked and the curves are vertically shifted for sake of clarity.

2.1. Standard EXAFS Data Analysis

The quantitative EXAFS data analysis was firstly carried out following a standard multi-shell [24,39] data analysis procedure using the program FitEXA [38]. The k^2-weighted raw spectra were fitted in the 3–14.5 Å^{-1} k-range to the model function $k^2\chi_{th}$. The non-linear least-square data refinement and error analysis has been implemented using the MINUIT routine package [40]. The theoretical EXAFS function is given by a sum of partial contributions (coordination shells) $\chi_{th} = \sum_i \chi_i$, each χ_i being calculated from the standard EXAFS formula [38]:

$$\chi_i(k) = S_0^2 N_i \frac{A_i sin(2kR_i + \phi_i)}{kR_i^2} e^{-2k^2/\sigma_i^2} e^{-2R_i/\lambda_i}.\tag{1}$$

This model is valid in the case of small and harmonic (Gaussian) disorder [41] and assumes each contribution originating from a Gaussian shaped coordination shell defined by three structural parameters: the multiplicity number N_i, the average coordination distance R_i and the mean square relative displacement (MSRD) factor σ_i^2. The semi-empirical amplitude reduction factor S_0^2 takes into account for many body losses, after preliminary tests it was fixed to 0.9 for all the data analysis. The theoretical amplitude (A_i), phase (ϕ_i) and mean free path (λ_i) functions were calculated using the FEFF 8.2 program [42] using an atomic cluster based on the crystallographic Nb$_3$Sn structure within the *Pm-3n* space group [43]. The squeezing of the lattice parameters in the investigated pressure range [15] is less than 5% and it is expected to have a weak effect on the A_i, ϕ_i and λ_i; therefore they were not recalculated as a function of pressure.

In order to reduce the number of free parameters and the correlations among them (so improving the reliability of the obtained results) during the multi-shell EXAFS data fitting, constraints were imposed based on the crystallographic structure [39] of Nb$_3$Sn. The adopted unit cell schematized in Figure 2 was cubic (edge length a). The Sn atoms were located at the cube corners and center, forming a body-centred cubic (*bcc*) lattice, while the Nb atoms were arranged in pairs along chains parallel to the x, y, and z axis, at the center of each face of the cube.

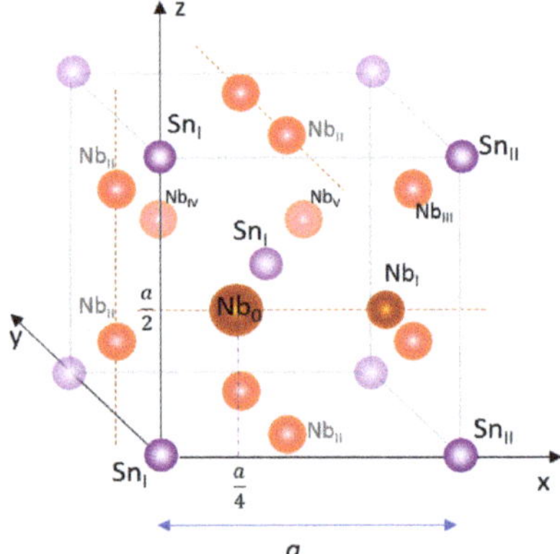

Figure 2. Representation of the Nb$_3$Sn structure, where Nb and Sn atoms are reported in light red and purple, respectively. For sake of clarity some of the atoms have been labelled following the nomenclature of Table 1, in order to identify the various paths used for the EXAFS analysis. The dimension of the Nb$_0$ atom has been enhanced to help the reader finding the absorbing atom.

The three main peaks observed between 1.8 Å and 3.2 Å in the FT (uncorrected for the phase shift) were partially overlapping (Figure 1b). They were assigned to the first three neighboring shells around a generic Nb$_0$ absorber in the A15 structure (Table 1), namely the two Nb_I along the Nb chain ($a/2$ away from Nb_0), the four Sn_I at the close cube corners and centers of adjacent cubes ($a\sqrt{5}/4$ away) and the eight Nb_{II} on the close faces ($a\sqrt{6}/4$ away). The features in the FT in the 3–6 Å region derived from single (SS) and multiple scattering (MS) contributions in the more distant region. For the analysis the SS and MS were selected on the basis of their amplitude and statistical significance in the fitting. In particular, the intense MS paths Nb_0-Nb_I-Nb_{IV} and Nb-Sn_I-Nb_V, being enhanced by collinear arrangements (forward and double forward scattering) were considered in the fitting model. For sake of clarity, the neighboring shells assignment used for the EXAFS data analysis is resumed in Table 1, giving the atomic position relative to the generic Nb_0 absorber at $(1/4, 0, 1/2)$ in the unit cell.

Table 1. Definitions of the seven neighboring shells used for the EXAFS data analysis calculated assuming the generic Nb_0 absorber located at $(\frac{1}{4}, 0, \frac{1}{2})$. The atom labels are those shown in Figure 2. For each shell, the labels of the neighboring atoms used in the text and the half path length R(a) as a function of the cube edge (lattice parameter) a are reported. For each atomic configuration contributing to the shell the multiplicity (N), the scattering model (single SS, or multiple MS, scattering) and the neighboring relative position respect to Nb_0 (in units of the lattice parameter a) are shown. The scattering paths are also reported for sake of completeness (path) along with the relative intensity respect to the Nb_I shell as given by FEFF, the MS terms include the three and four leg paths.

Shell	Atoms	R(a)	N	Scattering	Position	Path
I	Nb_I	$\frac{a\sqrt{4}}{4}$	2	SS	$(\pm\frac{1}{2},0,0)$	Nb_0-Nb_I (100%)
II	Sn_I	$\frac{a\sqrt{5}}{4}$	2+	SS	$(-\frac{1}{4},0,\pm\frac{1}{2})$	Nb_0-Sn_I (173%)
			2	SS	$(\frac{1}{4},\pm\frac{1}{2},0)$	
III	Nb_{II}	$\frac{a\sqrt{6}}{4}$	4+	SS	$(-\frac{1}{4},\pm\frac{1}{2},\pm\frac{1}{4})$	Nb_0-Nb_{II} (247%)
			4	SS	$(\frac{1}{4},\pm\frac{1}{4},\pm\frac{1}{2})$	
IV	Sn_{II}	$\frac{a\sqrt{13}}{4}$	2+	SS	$(\frac{3}{4},0,\pm\frac{1}{2})$	Nb_0-Sn_{II} (45%)
			2	SS	$(-\frac{3}{4},\pm\frac{1}{2},0)$	
V	Nb_{III}	$\frac{a\sqrt{14}}{4}$	4+	SS	$(\frac{3}{4},\pm\frac{1}{2},\pm\frac{1}{4})$	Nb_0-Nb_{III} (152%)
			4+	SS	$(\frac{1}{4},\pm\frac{3}{4},\pm\frac{1}{2})$	
			4+	SS	$(-\frac{3}{4},\pm\frac{1}{4},\pm\frac{1}{2})$	
			4	SS	$(-\frac{1}{4},\pm\frac{1}{2},\pm\frac{3}{4})$	
VI	Nb_{IV}	$\frac{a\sqrt{16}}{4}$	2+	MS	$(\pm1,0,0)$	Nb_0-Nb_1-Nb_{IV} (163%)
			2+	SS	$(0,\pm1,0)$	Nb_0-Nb_{IV} (46%)
			2	SS	$(0,0,\pm1)$	
VII	Nb_V	$\frac{a\sqrt{20}}{4}$	2+	MS	$(\frac{1}{2},\pm1,0)$	Nb_0-Sn_I-Nb_V (142%)
			2+	MS	$(-\frac{1}{2},0,\pm1)$	
			2	SS	$(-\frac{1}{2},\pm1,0)$	Nb_0-Nb_V (42%)

Based on the crystallographic structure, the multiplicity numbers (N_i) of the various contributions were kept fixed. All the shell distances R_i were constrained to the lattice parameter a (left free to vary) through the R(a) functions (Table 1), except the nearest neighbors R_{Nb_I} distance. Indeed, due to the correlated disorder [41,44], it may depart from the expected crystallographic length. Double and triple scattering contributions (MS) (as well as SS one), were considered for the VI and VII shells. The MSRD factors were refined independently for the various shells.

After some preliminary tests, it was found that seven shells were enough to fit the data reproducing the main structural features observed in the FT graph up to 6 Å. An example of the best fit obtained at 23 GPa is presented in Figure 3, together with the partial contributions of the seven shells used in the analysis. The edge energy shift E_o and the S_o^2 were kept fixed for all the spectra in order to improve the reliability of the observed trends on the structural parameters. During the EXAFS analysis, a total of 13 free parameters were used. The pressure-induced evolution of the Nb–Nb nearest neighbors (R_I) distance and the lattice parameter ($a/2$) are reported in Figure 4, together with the MSRD for the first (Nb$_I$) and the second (Sn$_I$) shells.

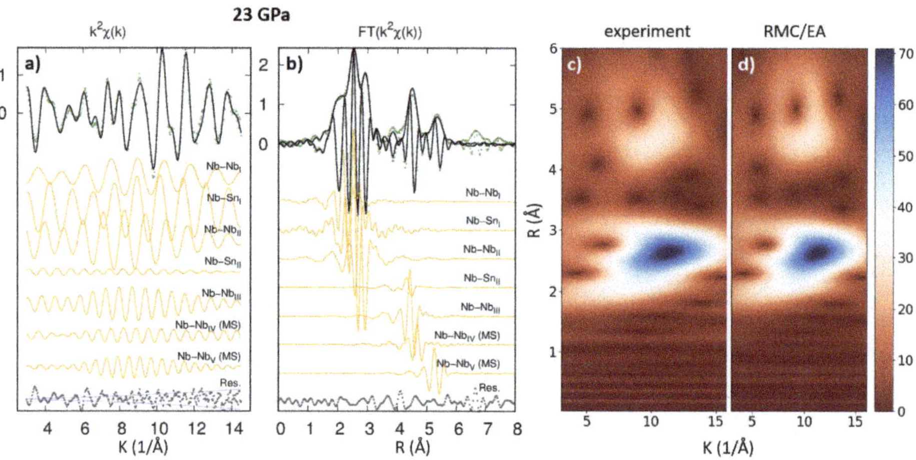

Figure 3. Example of the best fit obtained at 23 GPa for the EXAFS signal of Nb$_3$Sn at the Nb K-edge using the standard (left panels) and Reverse Monte Carlo (RMC) (right panels) analysis . In particular, (**a**,**b**) panels report the k^2-weighted EXAFS signal and the corresponding Fourier transform (FT) moduli |FT|, respectively. In both the panels, the experimental data are reported as green dots, whereas the best fit curves as black lines. The partial contribution used in the analysis are shown (yellow lines), vertically shifted for sake of clarity. Right panels (**c**,**d**) show the k^2-weighted Morlet wavelet transform for the experimental and CA-model from the RMC refinement. The Fourier transform (FT) moduli |FT| and the Back Fourier filtered (FF) curves obtained from the RMC refinement are shown in Figure S1 of the Supporting Information.

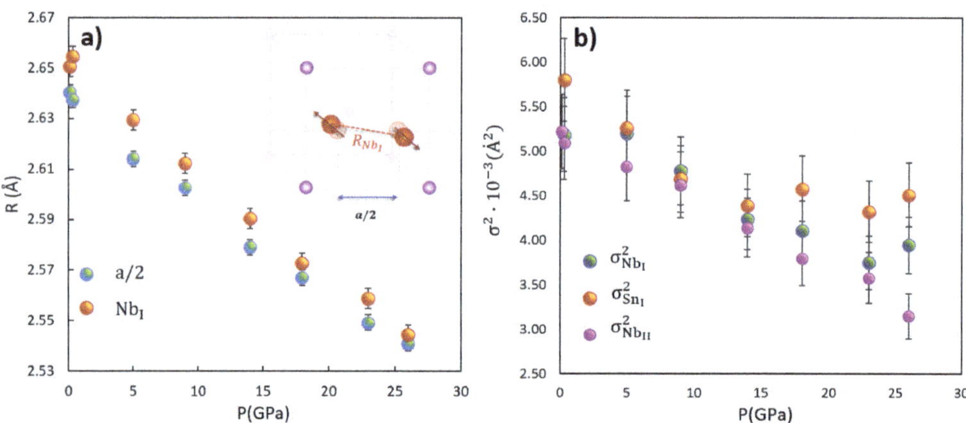

Figure 4. (**a**) R_{Nb_I} distance (orange symbols) and the $a/2$ parameters (light blue symbols) are shown. The R_{Nb_I} is systematically larger than $a/2$, signifying an anticorrelated Nb–Nb neighbor displacement perpendicularly to the Nb chain, as schematized in the inset. (**b**) The mean square relative displacement (MSRD) of the first ($\sigma^2_{Nb_I}$), second ($\sigma^2_{Sn_I}$) and third ($\sigma^2_{Nb_{II}}$) neighbors are shown as a function of pressure.

2.2. RMC Data Analysis

The RMC method is a simulation technique providing a 3D model of the atomic structure of a sample. This is obtained by minimizing the difference between the structure-related experimental signal and a configuration averaged (CA) theoretical curve calculated from the simulated atomic positions. The advantage of RMC with respect to other simulation techniques (such as standard Monte Carlo method or molecular dynamics), consists in the fact that it does not require any knowledge on the actual interatomic potentials of the material.

In the present study, the RMC analysis was performed using the EvAX suite [25]. This software combines an evolutionary algorithm (EA) with RMC, providing an optimized computational efficiency. EvAX was explicitly built for characterizing the local structural and thermal disorder in crystalline materials from the analysis of their experimental EXAFS signal within the multiple-scattering formalism [45]. It has already been successfully applied for the characterization of several systems [25,44,46].

In EvAX the configuration average EXAFS signal $\chi_{CA}(k)$ is obtained using the ab initio real space multiple scattering FEFF8.5L code [47], embedded in the EvAX distribution. The obtained signal is then compared with the experimental one ($\chi_{exp}(k)$). The initial atomic configuration is generally defined according to the space group and lattice parameter of the selected material (Pm-$3n$ in this case). It is then possible to choose the size of the structural unit (supercell), its shape and boundary conditions. During the RMC refinement procedure, the atomic positions are randomly displaced and a new $\chi'_{CA}(k)$ is calculated and compared to the $\chi_{exp}(k)$, the new atomic configuration are accepted or discarded following a modified METROPOLIS algorithm [45]. In the present study the comparison between $\chi_{CA}(k)$ and $\chi_{exp}(k)$ has been carried out looking at the weighted squared differences $\zeta_{k,R}$ between the configuration averaged theoretical and experimental spectra in both the k and R space, known as Morlet wavelet transform space (WT) [45]:

$$\zeta_{k,R} = \frac{||WT_{CA}(k,R) - WT_{exp}(k,R)||^2}{||WT_{exp}(k,R)||^2} \tag{2}$$

in this way, it was possible to improve the fitting constraints by taking into account the two-dimensional representation of the EXAFS signal with a simultaneous localization in the energy and frequency space domains. In order to implement the evolutionary algorithm, a population of several supercells is used whose structures are RMC refined and crossed at each generation using genetic rules so to massively improve the statistic of the simulation and makes the convergence of the procedure faster than standard RMC algorithms.

In the present study, EvAX simulations of the Nb K edge EXAFS data were performed using the Linux Cluster of the Department of Matematica e Fisica of the University Roma Tre. The minimization procedure has been performed in the k^2-weighted WT [48], considering k between 3–16 Å^{-1} and R between 0–6 Å. The Nb$_3$Sn crystallographic structure from Ref. [43] was used as the starting configuration for a population of 32 supercells, each one consisting of $3 \times 3 \times 3$ cells with periodic boundary conditions, containing a total of 216 atoms (162 Nb + 54 Sn) per supercell. The theoretical $\chi_{CA}(k)$ EXAFS spectra were calculated considering MS paths with up to 4 scattering legs with 6 Å maximum length. At each RMC iteration, new supercell configurations were generated by randomly displacing all the atoms of the simulation box by a maximum displacement of 0.2 Å. The simulated annealing approach was used to efficiently reach the global minimum. The acceptance ratio of the new atomic configuration was not fixed but it decreased slowly following a cooling scheme. The length of the cooling scheme, set by the number of iterations after which only the atomic displacements improving the agreement between experiment and theory were accepted, was set to 1500. Figure 3c,d show an example of the k^2-weight multi-shell data fit and RMC best fit in the WT space, obtained for the HP data at 23 GPa. The k^2-weighted configuration averaged curve $k^2\chi_{CA}(k)$ and its Fourier transform modulus, as obtained from the EvAX output for the HP data at at 23 GPa, are shown in the Supporting Material.

The n-body distribution functions ($g^{(n)}$) formalism [49] allows the description of the relative neighboring arrangements, providing a set of parameters containing their geometrical configurations. For example, the three-body configuration ($g^{(3)}$) is defined by two distances and the angle between them. The structural parameters defining the

$g^{(n)}$ (interatomic distances and bond angles) and the corresponding statistical parameters (mean, variance, and anharmonic terms), are calculated directly from the 3D atomic models using a specifically written Python code with the NumPy [50] library package. In order to evaluate the simulation's uncertainties, four independent RMC runs were performed using different seeds for the pseudo-random number generator [25]). The parameters and their uncertainties reported in Figure 5 correspond to the average values and standard deviations of those obtained from the different seeds.

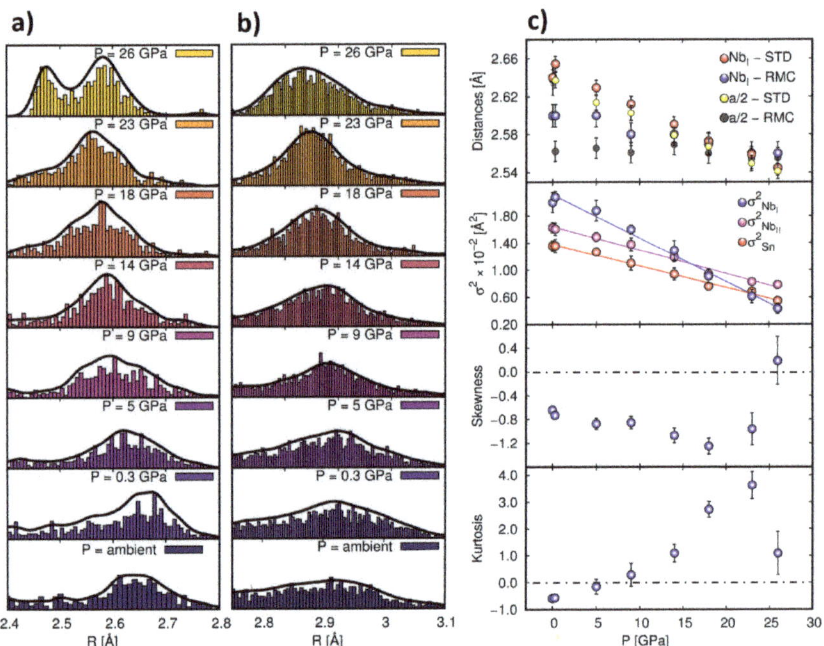

Figure 5. Synthesis of the main RMC results. (**a**) R_{Nb_I} and (**b**) R_{Sn_I} distributions histograms as a function of pressure. (**c**) Top panel (Distance): pressure dependence of R_{Nb_I} and $a/2$ compared to the one obtained from the standard analysis. (**c**) Second panel from the top (MSRD): pressure-induced evolution of the variance (σ^2) of the first three shells as obtained from the RMC simulations. (**c**) Third panel from the top (Skewness): pressure-induced evolution of the asymmetry of the distribution of the Nb–Nb nearest neighbor pair distribution function. (**c**) Bottom panel (Kurtosis): pressure-induced evolution of the Kurtosis of the distribution of the first shell

3. Results and Discussion

The obtained EXAFS signal (weighted by k^2) and the corresponding Fourier transforms (FT) are reported in Figure 1. The effect of pressure was evident, with the squeezing of the distances in the FT and the corresponding expansion of the EXAFS oscillations in the reciprocal space. The experimental EXAFS signals were analyzed by using both the multi-shell data analysis [24,39,51] and the RMC/EA methods [25], in order to obtain detailed local structural and topological information.

Due to its intrinsic nature, the XAFS signal probed the local atomic structure around the absorbing atoms, providing the relative neighboring arrangement and the corresponding disorder around the average absorber [52]. Furthermore, owing to the strong interaction between the photo-excited electrons and the potentials of the neighboring atoms, the XAFS signal was sensitive to multiple scattering processes [47]. Therefore, the MS analysis could provide details about the topology of the local atomic structure around the absorber (bond angles, shape and orientation of coordination polyhedra, etc.) through the many body distribution functions ($g^{(n)}$) [25,49]. Such information is hidden from other structural probes, such as neutron and X-ray diffraction techniques [53]. The MS analysis could be a difficult task using standard EXAFS methods based on the multi-shell data refinement due to Provost et al. [54]: i. increasing number of required parameters; ii. increasing correlations among the structural parameters (as more distant neighbors are taken into account); iii. the failure of the small harmonic disorder model [41] and the occurrence of correlated disorder in SS and MS shells [41,55]. Imposing appropriate constraints (as detailed above) is a valuable way to obtain highly reliable structural information [24] but topological details can be elusive, especially when dealing with ordered structures characterized by a large number of non negligible MS contributions. Therefore, the RMC-based atomistic simulations helped deepen the topological details through the direct access to the 3D structural models. Below, the results obtained combining multi-shell and RMC EXAFS data analysis are discussed.

The main structural parameters obtained from the EXAFS data analysis are shown in Figure 4. The average lattice parameter $a/2$ and R_{Nb_I} are shown for sake of comparison. The parameter a is the unique free parameter used to refine the further shells which are progressively less affected by the correlated disorder. It represents the analogous of the lattice parameter obtained by XRD and indeed, it matches well the a_{XRD} experimental behavior reported in the analysis of technological Nb_3Sn wires [15]. Noticeably, the Nb–Nb nearest neighbor distance R_{Nb_I} is significantly larger than $a/2$ at low pressures but converges to $a/2$ with increasing pressure. Interestingly, the values of $a/2$ and R_{Nb_I} found at 26 GPa coincide with the half of the lattice parameter obtained by HP-XRD at the same pressure, on Nb_3Sn powders from technological wire sample [15]. Such a behavior suggests a general anticorrelated displacement of the Nb–Nb neighbors perpendicularly to the average Nb chain directions at ambient conditions (as schematized in the inset of Figure 4a). The application of an external pressure acts against this anticorrelation, by increasing the alignment of the Nb–Nb bonds with the cell axis. The average tilting angle of the Nb–Nb bond with respect to the Nb chain, calculated assuming a perpendicular anticorrelated displacement of the Nb–Nb pairs, decreases from 6.5(5)° at 0.3 GPa to 3(1)° at 26 GPa.

In order to take into account possible anharmonic effects, the additional cumulant expansion of the EXAFS formula was attempted for the data analysis of the Nb_I shell [41]. However, we found the correlations between the first cumulant C_1 (distance) and the third one C_3 (skewness), taking into account the asymmetry of the distribution, to be higher than 90% . The same is true for the correlations between C_2, being the MSRD, and C_4 (Kurtosis) describing the tailedness of the distribution. This gave large uncertainties on the parameters making the results less reliable. Therefore, we did not use the cumulant expansion in the standard analysis but exploited the RMC analysis to obtain deeper structural details.

The plots in Figure 5 show the Nb_I (panel a) and Sn_I (panel b) pair distribution functions as obtained from the RMC models. The Nb_0 and Sn_I distributions are broad and asymmetric at low pressures. Raising the pressure makes the distributions narrower and shortens the interatomic distances, visually showing the "squeezing" and the overall ordering of the structure. To obtain quantitative information about the local atomic structure around Nb atoms, we calculated the parameters characterizing the neighboring distributions (mean values, variances, higher moments of the distributions) as a function of pressure directly from the RMC structural models (see Figure 5c). The results obtained by RMC refinement of the EXAFS data measured at 0.3 GPa in DAC were fully consistent with those independently measured at AP on a standard set-up. This reinforces the confidence on the reproducibility of the data and the reliability of RMC analysis on independent data sets. However, from this data it is evident that at low pressures, the Nb_I distances, $a/2$ and the MSRD parameters calculated from the RMC atomic models differed from those obtained using the standard EXAFS analysis (Figure 5c, Distances and σ^2).

To explain this discrepancy we must point out that, as discussed in the EvAX manual (http://www.dragon.lv/evax/), the structural parameters characterizing the atomic distribution (average interatomic distances, MSRD, and so on) estimated from the atomic configurations, cannot directly be compared with the results of standard EXAFS if the atomic distributions are asymmetric. This is due to the fact that Equation (1) is blind to non-Gaussian contributions [41], while the RMC model converges to the most disordered solution consistent with the experimental data [56]. In our case, absolute differences are expected between the two methods, especially for the low pressure data, where the atomic distributions appear broader and asymmetric. As the pressure squeezes the atomic distributions, the results of RMC and standard EXAFS analysis converge to similar values.

In order to prove the consistency between the present RMC and standard analysis, we considered the $\chi_{CA}^{(AP)}$ obtained from the RMC structure at AP and the corresponding $\chi_{exp}^{(AP)}$ experimentally obtained and we fitted both signals following the standard EXAFS formula. The obtained results are reported in Figure 6 and Table 2. The good agreement between the structural parameters obtained fitting the experimental and the RMC signal confirmed how the above mentioned discrepancy was liked due to the different sensibility of the two methods. In fact, while the absolute values obtained from RMC and the standard EXAFS analysis differed, the overall structural behavior was confirmed. In particular, the R_{Nb_I} distances (Figure 5c—Distance panel) at low pressures result was larger than the $a/2$ calculated from the average next neighbor Nb distances. However, such a difference decreases with the increasing pressure and, similarly to the standard analysis case, $R_{Nb_I} \simeq a/2$ at high pressures. This confirms the model of anticorrelated Nb–Nb displacement along directions perpendicular to the Nb chains.

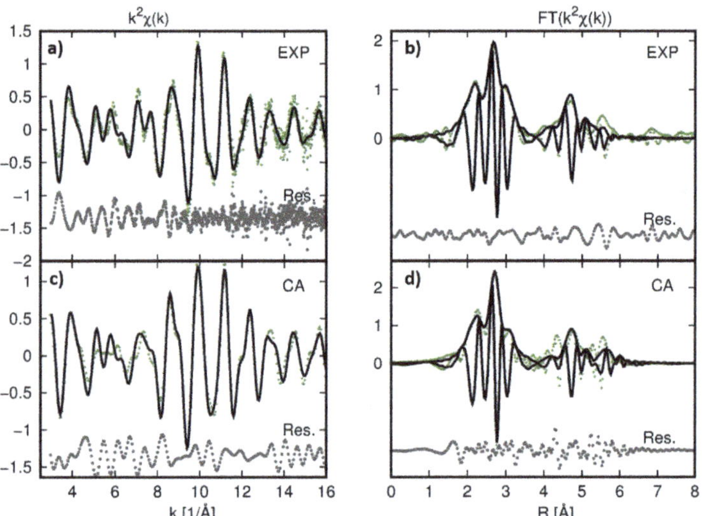

Figure 6. Best fit of the AP data comparing the $k^2\chi_{exp}^{(AP)}$ (top figures: (**a**,**b**)) and $k^2\chi_{CA}^{(AP)}$ (bottom figures: (**c**,**d**)). In particular, (**c**,**d**) panels report the k^2-weighted EXAFS signal and (**b**,**d**) panels their corresponding Fourier Trasform (FT) moduli |FT|. In both the panels, the experimental data are reported as green dots, the best fit curves as black lines and the residuals are reported as grey dots at the bottom of the figures.

Table 2. Comparison between the structural parameters obtained from the EXAFS and RMC fit of the experimental AP data $k^2\chi_{exp}^{(AP)}$ and from the EXAFS fit of the $\chi_{CA}^{(AP)}$ curve.

AP Data	R_{Nb_I} [Å]	$a/2$ [Å]	$\sigma_{Nb_I}^2$
EXAFS fit of χ_{exp}	2.644(4)	2.638(3)	0.0064(4)
RMC fit of χ_{exp}	2.60(1)	2.56(1)	0.020(2)
EXAFS fit of χ_{CA}	2.649(4)	2.644(4)	0.0064(4)

Considering the pair distribution functions obtained from the RMC model in Figure 5, it was possible to observe a bimodal distribution appearing at 26 GPa, with two sharp peaks with $\Delta R/R \sim 3.8\%$. We therefore tried to improve the standard EXAFS analysis of the data at 26 GPa considering a bimodal distribution for the Nb_I shell. During the fit, two $R_{Nb_I}(a,b)$ distances were refined with the same multiplicity numbers ($N(a,b) = 1$) and MSRD ($\sigma_{Nb_I}^2(a) = \sigma_{Nb_I}^2(b)$). The obtained fit is shown in Figure 7 and the corresponding results are reported in Table 3 along with the square residual function R_w^2, a statistical indicator representing the best fit quality [38]. The $R_{Nb_I}(a,b)$ distances (Table 3) matched well those in Figure 5a) and $\sigma_{Nb_I}^2(a,b)$ were half of the MSRD of the single shell model. This demonstrated the consistency between RMC and the standard EXAFS analysis and reinforces the reliability of the analysis procedures. It is important to notice that, as the double shells model slightly improved the fitting by about 5%, the number of free parameters increased by one. To evaluate the statistical significance of the best fit improvement we evaluated the associated Fisher F function [57]:

$$F = \frac{R_{w_1}^2 - R_{w_2}^2}{R_{w_2}^2} \frac{N_i - n_2}{n_2 - n_1} \simeq 1.7 \tag{3}$$

where n_1 and n_2 are the number of free parameters in the fit with single or double Nb_I shell, $R_{w_1}^2$ and $R_{w_2}^2$ the corresponding square residual functions and $N_i \simeq 51$ is the number of independent experimental points in the fit [57]. The experimental $F \simeq 1.7$ corresponded to a p-value $\simeq 0.2$, which established that the double shell model was not statistically justified in the standard analysis despite the improvement of the residual function.

Table 3. Quantitative parameters obtained from the EXAFS fit of the data at 26 GPa. The results obtained with a single shell model and a bimodal distribution model are compared. Parameters indicated with * were constrained during the analysis. The multiplicity number N was fixed to 2 for the single shell model while for the bimodal distribution model it was constrained to 1 for each shell. In the bimodal distribution model the R_{Nb_I} parameters was left free to vary for each shelll and the $\sigma_{Nb_I}^2$ of the two contributions was constrained to be the same. In the last column of the table the R_W^2 parameters indicates the best fit quality.

26 GPa	N	R_{Nb_I} [Å]	$\sigma_{Nb_I}^2$	R_W^2
single shell model	2 *	2.545(4)	0.0044(4)	0.0615
bimodal distribution model	1 *	2.498(5) 2.607(4)	0.0022(2) *	0.0587

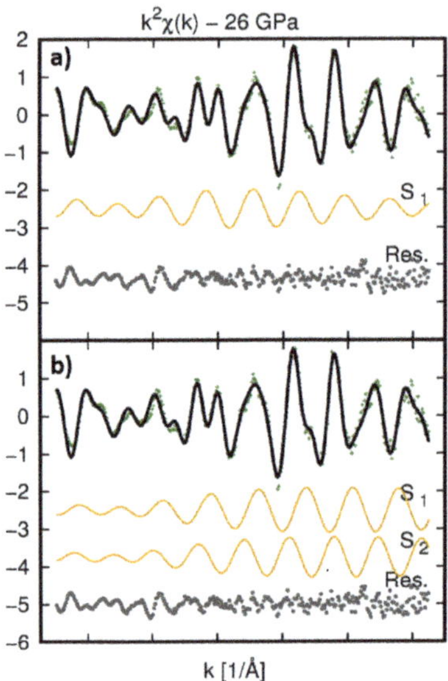

Figure 7. Best fit of the $k^2\chi(k)$ data at 26 GPa. In panel (**a**) the best fit obtained with a single shell model is reported while panel (**b**) shows the fit obtained with a bimodal distribution model of the Nb_I shell. In both the panels, the experimental data are reported as green dots and the best fit curves as a black line. The contributions used to fit the Nb_I shell (S_1 for the single shell model, S_1 and S_2 for the bimodal distribution model) are reported by orange lines vertically shifted for clarity. The contributions from the other coordination shells are not shown for sake of clarity. The best fit residuals ($k^2(\chi_{exp} - \chi_{fit})$) are reported as grey dots at the bottom of the panels.

The MSRD calculated for the first three shells ($\sigma^2_{Nb_I}$, $\sigma^2_{Sn_I}$, $\sigma^2_{Nb_{II}}$) are shown in Figure 5c—MSRD. They were all larger than those found in Figure 4. As discussed above such a discrepancy must be attributed to the different sensitivity of the standard EXAFS formula and RMC refinement to the atomic distribution functions. However, the three MSRD data decreased under raising pressure, pointing out to an overall ordering of the structure. At lower pressure, the $\sigma^2_{Nb_I}$ was the largest of the three but it decreased upon pressure increase as $\frac{\delta\sigma^2}{\delta P} \simeq -7.0(2) \times 10^{-4}$ Å^2/GPa. Whereas, the $\sigma^2_{Sn_I}$ and $\sigma^2_{Nb_{II}}$ were both decreasing with the same (slower) rate of $\frac{\delta\sigma^2}{\delta P} \simeq -3.5(2) \times 10^{-4}$ Å^2/GPa. To understand this effect we note that the MSRD measured by EXAFS for an hypothetical pair of atoms $A - B$ was [41,44]:

$$\sigma^2_{AB} = \sigma^2_A + \sigma^2_B - 2\gamma_{AB}\sigma_A\sigma_B$$

where σ^2_i is the average atomic displacement of the i-th atom around its equilibrium position in the crystallographic structure, γ_{AB} is the atomic displacement correlation function ($-1 \leq \gamma_{AB} \leq 1$). In this formula $\gamma_{AB} = 0$ means the A and B atomic displacements were uncorrelated, whereas $\gamma_{AB} > 0$ ($\gamma_{AB} < 0$) means that A and B atoms were displaced in the same (opposite) direction(s). The atomic displacement of distant neighbor shells was likely uncorrelated ($\gamma_{AB} \sim 0$) [52], so that the more rapid decrease of $\sigma^2_{Nb_I}$ was in agreement with the reduction of anti-correlated displacement of Nb atoms perpendicular to the Nb chain discussed above, and even suggested a positive correlation when raising the pressure above 23 GPa.

The higher moments of the Nb_I distribution were calculated directly from the 3D atomic models (Figure 5c—Skewness and Kurtosis). At ambient pressure, the Nb_I pair distribution is strongly asymmetric (Skewness ≈ -1), with a broad tail at low R. Under applied pressures (up to 20 GPa), the asymmetry slightly increased (Skewness absolute value), then it suddenly decreased and vanished at 26 GPa. The pressure-induced evolution of the Kurtosis showed an initial negative value at AP and 0.3 GPa. When raising the pressure, its value increased systematically up to 23 GPa then suddenly dropped down once 26 GPa are reached. These findings established a complex pressure-induced evolution of the Nb–Nb pair distribution functions and pointed out an anomaly above 23 GPa. Noticeably, no structural changes were found by XRD in the same pressure range, underlining the local nature of these changes.

Further details about the local atomic arrangement around the Nb atoms are obtained by looking at the three body distribution functions $g^{(3)}$, in particular by looking at the order along the Nb chains. The sensitivity of EXAFS to the $g^{(3)}$ of the Nb–Nb–Nb triangles was high because the MS signal was enhanced by the focusing effect on the central atom.

The RMC atomic model allowed the geometrical parameters defining the Nb–Nb–Nb arrangement to be directly calculated, in particular, the two Nb–Nb shorter bonds R_1, R_2, and the angle between them θ_{Nb}. In Figure 8 (panels b) the θ_{Nb} distribution is presented. It is possible to observe how its average value increased from 174.9° in the low pressure RMC models (AP and 0.3 GPa) up to 176.4° for the highest pressure model (26 GPa). This trend was in agreement with the standard EXAFS analysis and confirmed that a higher pressure not only squeezed the structure (interatomic distances) but also reduced the anticorrelated displacement of the Nb neighbors perpendicular to the Nb chains. Additional qualitative information can be derived by looking at the R_2 vs. R_1 plots (Figure 8a: panels) providing a view of the Nb–Nb–Nb displacement correlation along the chains. At low pressures (AP and 0.3 GPa) the R_2 vs. R_1 points were quite randomly scattered. Raising the pressure to 5 GPa established a peculiar trend in the R_2 vs. R_1 distribution with two different atomic displacement models: *i*. A fraction of (R_1, R_2) configurations were arranged close to the center of the panels ($R_2 \simeq R_1$) and were weakly correlated; *ii*. The remaining fraction of (R_1, R_2) point roughly aligned along the plot's diagonal, suggesting a configuration where R_1 and R_2 varied in an anti-correlated way: larger R_1 correspond to shorter R_2 and *vice versa*, with the relative difference between R_1 and R_2 being $\Delta R/R \sim 9\%$. We emphasize how this finding established an evolution of the Nb–Nb–Nb $g^{(3)}$ up to 5 GPa, in the same pressure region where anomalies in the compressibility of the Nb_3Sn (technological wires) have been reported [15].

In order to represent the Nb–Nb–Nb $g^{(3)}$ behavior above 5 GPa, we built a qualitative model as schematized in the right panels of Figure 8c: i. a fraction of Nb–Nb–Nb configuration (highlighted in red) corresponds to Nb-Nb_c-Nb isosceles triangles ($R_1 \approx R_2$) with Nb_c randomly displaced, preferentially in the direction perpendicular to the Nb chains (Figure 8c, top scheme). ii. The other fraction (highlighted in green), was associated to the asymmetric arrangement of Nb_c (with $R_1 < R_2$ or $R_2 < R_1$) where the Nb_c was randomly displaced, preferentially in the direction parallel to the Nb chains, thus providing a local dimerization of the Nb–Nb bonds (Figure 8c, bottom scheme). The results of such a simulation are shown for the 14 GPa analysis in Figure 8c (center panel) for sake of qualitative comparison. Under compression, the fraction of anticorrelated configurations decreased also reducing $\Delta R = |R_2 - R_1|$. The fraction of dimerized configurations disappeared at 23 GPa. Upon further compression (up to 26 GPa) the R_2 vs R_1 distribution results were more squeezed. In particular, the value of the average distribution of R_{Nb_1} became very close to $a/2$ (Figures 4 and 5). Interestingly, while a fraction of the configurations was localized at the points $R_1 \sim R_2$, at such a pressure a novel R_2 vs R_1 correlation mode was established, where shorter (around 2.5 Å) and sharper $R_1(R_2)$ distances were associated to longer (around 2.6 Å) and wider $R_2 (R_1)$ distances, corresponding to the bimodal R_{Nb_1} distribution highlighted in Figure 5.

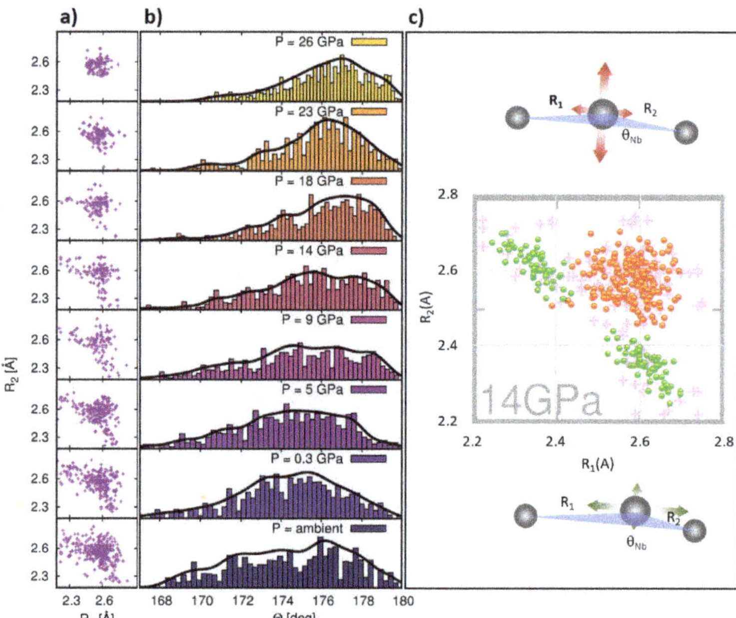

Figure 8. Pressure induced evolution of the $R_1 - R_2$ distances (**a**) and corresponding angles (**b**). Explanatory model (**c**) of the correlated disorder for $R_1 - R_2$ distribution compared with the experimental data obtained from RMC model at 14 GPa (background). The points are calculated by randomly displacing Nb$_c$. For the red points the average distances are the same $\bar{R}_1 = \bar{R}_2$ and the random displacement is preferentially perpendicular to the Nb chain (as schematized in the top scheme). For the green points the average distances are different $\bar{R}_1 > (<)\bar{R}_2$ and the random displacement is preferentially parallel to the Nb chain (as schematized in the bottom scheme).

Briefly the analysis of the Nb K-edge EXAFS data carried out combining traditional and RMC approaches provided a deep insight on the evolution of the local atomic structure around the Nb atoms in Nb$_3$Sn as a function of pressure. In particular, the EXAFS data analysis showed an anticorrelated displacement of Nb atoms perpendicularly to the Nb chains which decreased when raising the applied pressure. Our results also suggested the loss of the cubic (Pm-$3n$) symmetry at the local scale. Noticeably, recent ab initio simulations [58] showed that lower symmetry structures were energetically favoured with respect to the Pm-$3n$ in other A15 systems such as Nb$_3$Al. In particular, the C_2/c symmetry allowing anticorrelated Nb atom displacements perpendicularly to the Nb chains, was energetically favored with respect to the Pm-$3n$ one at ambient conditions and in a wide range of pressures. This work suggested that a similar behavior is observed in Nb$_3$Sn at the local scale.

The RMC analysis provided a further insight into the local Nb$_3$Sn structure thanks to the direct access to the 3D atomistic models. The analysis of the Nb$_I$ nearest neighbors distributions pointed out non-Gaussian contributions (Figure 5), in particular, the negative Skewness (~ -1) being associated to a fractions of closer Nb–Nb pairs. The Kurtosis parameter increased with raising pressure. Raising the pressure above 23 GPa sudden reduced the non-Gaussian contributions. This behavior is better understood looking at the parameters defining the three body distribution function $g^{(3)}$ of the the Nb–Nb–Nb chains (Figure 8). The RMC atomistic model suggested a bimodal behavior for the Nb–Nb–Nb configurations along the Nb chains: some of them depicted the symmetric displacement of the central Nb perpendicularly to the chain (isosceles triangles) coherently with the anticorrelated Nb–Nb displacement discussed above. Nevertheless a fraction of the Nb–

Nb–Nb triangles had $R_1 \neq R_2$, with $\Delta R/R \sim 9\%$. These scalene triangles produced the partial dimerization of the Nb chains. It is noticeable that recent single crystal HP-XRD measurements performed at low temperatures (10 K) [17] on a slightly non-stoichiometric Nb_3Sn_{1-x} demonstrated a pressure induced symmetry lowering to the $P4_2/mmc$ space group at around 3 GPa where some of the Nb–Nb chains were dimerized in agreement with the present RMC EXAFS analysis. Our results coherently recognized that the anticorrelated displacement along the Nb–Nb–Nb chains was locally stabilized in that low pressure room temperature region . Finally, our EXAFS analysis showed that raising the pressure up to 26 GPa suppressed most of the local disorder.

4. Conclusions

In this study, the local atomic structure around Nb in Nb_3Sn has been characterized in situ between ambient pressure and 26 GPa by XAS. The Nb K-edge EXAFS data were analysed using a standard multi-shell method, based on small Gaussian disorder model and a RMC method, providing a 3D atomic structural model. The results obtained from analysis carried out independently on the same sample, using different experimental set-up (AP and 0.3 GPa) are completely consistent. This demonstrates the reproducibility of the adopted procedures. The results of AP multishell EXAFS data analysis are in good agreement with previous literature [23]. The results obtained with RMC and the standard EXAFS analysis differ at low pressure, likely due to the different sensitivity of the two data analysis techniques to the structural disorder, and the presence of alarge non-Gaussian contributions. In fact, they become fully consistent at high pressures, where the sharper radial distributions are better described by the small Gaussian disorder model at the basis of the standard EXAFS formula. This strengthens the consistency between standard and RMC EXAFS analysis.

The results reported here provide a deep insight on the Nb local atomic structure, highlighting a complex behavior of Nb arrangement at the atomic scale which is not detectable by long range order probes such as X-ray diffraction. Noticeably, our EXAFS findings are coherent with novel details about the atomic structure of A15 systems which recently came to light [15,17–19].

It is worthwhile to note that the details of the phononic spectra and the electron density of states (DOS), both intimately related to the details of the atomic structure, are dominant in determining the stress-induced critical performance degradation of Nb_3Sn [59–61]. The present results represent a new knowledge, relevant for providing accurate physical models in view of a better understanding of the superconductive properties, thus helping to individuate a route towards improving their critical properties.

Supplementary Materials: The following are available online at https://www.mdpi.com/2073-4352/11/4/331/s1.

Author Contributions: C.M. and S.A. conceived the experiment. S.A., R.L., R.T. and C.M. conducted the experiment. I.S. and C.M. analyzed the results. T.S. and R.F. prepared the sample. T.I. provided the nano-polycrystalline diamonds. E.S. collaborated in understanding the results. The manuscript is written through contributions of all authors. All authors have given approval to the final version of the manuscript.

Funding: This work has been carried out within the framework of the EUROfusion Consortium and has received funding from the Euratom Research and Training Programme 2014–2018 and 2019–2020 under grant agreement No 633053.The views and opinions expressed herein do not necessarily reflect those of the European Commission. S.A. acknowledges the support from the Natural Environment Research Council of Great Britain and Northern Ireland via grants NE/M000117/1 and NE/M00046X/1.

Institutional Review Board Statement: Not applicable.

Informed Consent Statement: Not applicable.

Data Availability Statement: Data are provided in the figures of the article.

Acknowledgments: The authors acknowledge the ESRF for provision of beamtime on the beamlines BM23 (proposal number: MA-3442) and J. Jacobs for DAC preparation.

Conflicts of Interest: The authors declare no conflict of interest.

Abbreviations

The following abbreviations are used in this manuscript:

AP	Ambient Pressure
DAC	Diamond Anvil Cell
DFT	Density Functional Theory
DOS	Density of State
EA	Evolutionary Algorithm
EXAFS	Extended X-ray Absorption Fine Structures
FT	Fourier Transform
HP	High Pressure
MS	Multiple Scattering
MSRD	Mean square relative displacement
RMC	Reverse Monte Carlo
SS	Single Scattering
WT	Morlet wavelet transform space
XAFS	X-ray Absorption Fine Structures
XAS	X-ray Absorption Spectroscopy
XRD	X-ray Diffraction

References

1. Matthias, B.T.; Geballe, T.H.; Geller, S.; Corenzwit, E. Superconductivity of Nb_3Sn. *Phys. Rev.* **1954**, *95*, 1435. [CrossRef]
2. Kunzler, J.E.; Buehler, E.; Hsu, F.S.L.; Wernick, J.H. Superconductivity in Nb_3Sn at High Current Density in a Magnetic Field of 88 kgauss. *Phys. Rev. Lett.* **1961**, *6*, 89–91. [CrossRef]
3. Parrell, J.A.; Field, M.B.; Zhang, Y.; Hong, S. Advances in Nb_3Sn strand for fusion and particle accelerator applications. *IEEE Trans. Appl. Supercond.* **2005**, *15*, 1200–1204. [CrossRef]
4. Godeke, A. A review of the properties of Nb_3Sn and their variation with A15 composition, morphology and strain rate. *Supercond. Sci. Technol.* **2006**, *19*, R68. [CrossRef]
5. Bottura, L.; Rijk, G.; Rossi, L.; Todesco, E. Advanced accelerator magnets for upgrading the LHC. *IEEE Trans. Appl. Supercond.* **2012**, *22*, 4002008. [CrossRef]
6. Vostner, A.; Salpietro, E. Enhanced critical current densities in Nb_3Sn superconductors for large magnets. *Supercond. Sci. Technol.* **2006**, *19*, S90. [CrossRef]
7. ITER Home Page. Available online: http://https://www.iter.org/factsfigures (accessed on 1 February 2021).
8. Zhang, R.; Gao, P.; Wang, X.; Zhou, Y. First-principle study on elastic and superconducting properties of Nb_3Sn and Nb_3Al under hydrostatic pressure. *AIP Adv.* **2015**, *5*, 107233. [CrossRef]
9. Ren, Z.; Gamperle, L.; Fete, A.; Senatore, C.; Jaccard, D. Evolution of T^2 resistivity and superconductivity in Nb_3Sn under pressure. *Phys. Rev. B* **2017**, *95*, 184503. [CrossRef]
10. Quiao, L.; He, Y.; Wang, H.; Shi, Z.; Li, Z.; Xiao, G.; Yang, L. Effect on grain boundary deformation on the critical temperature degradation of superconducting Nb_3Sn under hydrostatic pressure. *J. Alloys Compd.* **2021**, *864*, 158116. [CrossRef]
11. Nishijima, G.; Watanabe, K.; Araya, T.; Katagiri, K.; Kasaba, K.; Miyoshi, K. Effect of transverse compressive stress on internal reinforced Nb_3Sn superconducting wires and coils. *Cryogenics* **2005**, *45*, 653–658. [CrossRef]
12. Lu, J.; Han, K.; Walsh, R.P.; Miller, J.R. I_C Axial Strain Dependence of High Current Density Nb_3Sn Conductors. *IEEE Trans. Appl. Supercond.* **2007**, *17*, 2639–2642. [CrossRef]
13. Nijhuis, A.; van Meerdervoort, R.P.P.; Krooshoop, H.J.G.; Wessel, W.A.J.; Zhou, C.; Rolando, G.; Sanabria, C.; Lee, P.J.; Larbalestier, D.C.; Devred, A.; et al. The effect of axial and transverse loading on the transport properties of ITER Nb_3Sn strands. *Supercond. Sci. Technol.* **2013**, *26*, 084004. [CrossRef]
14. Zhang, W.J.; Liu, Z.Y.; Liu, Z.L.; Cai, L.C. Melting curves and entropy of melting of iron under Earth's core conditions. *Phys. Earth Planet. Inter.* **2015**, *244*, 69–77. [CrossRef]
15. Loria, R.; Marzi, G.D.; Anzellini, S.; Muzzi, L.; Pompeo, N.; Gala, F.; Silva, E. The Effect of hydrostatic pressure on the superconducting and structural properties of Nb_3Sn: Ab-initio modeling ans SR-XRD investigation. *IEEE Trans. Appl. Supercond.* **2017**, *27*, 8400305. [CrossRef]
16. Chu, C.W. Pressure-Enhanced Lattice Transformation in Nb_3Sn Single Crystal. *Phys. Rev. Lett.* **1974**, *33*, 1283–1286. [CrossRef]
17. Svitlyk, V.; Mezouar, M. Pressure-Induced Symmetry Lowering in Nb_3Sn_{1-x} Superconductor. 2020. Available online: http://xxx.lanl.gov/abs/2011.14982 (accessed on 23 March 2021).

18. Mkrtcheyan, V.; Kumar, R.; Baker, J.; Connolly, A.; Antonio, D.; Cornelius, A.; Zhao, Y. High pressure transport and structural studies on Nb_3Ga superconductor. *Physica B* **2015**, *459*, 21–23. [CrossRef]

19. Yu, Z.; Li, C.; Liu, H. Compressibility anomaly in the superconducting material Nb_3Al under high pressure. *Physica B* **2012**, *407*, 3635–3638. [CrossRef]

20. Mao, J.; Chen, Y. Ground-state crystal structures of superconducting Nb_3Al and the phase transformation under high pressures. *J. Appl. Phys.* **2018**, *124*, 173902. [CrossRef]

21. Bunker, G. *Introduction to XAFS. A Practical Guide to X-ray Absorption Fine Structure Spectroscopy*; Cambridge University Press: Cambridge, UK, 2010.

22. Sakashita, H.; Kamon, K.; Terauchi, H.; Kamijo, N.; Maeda, H.; Toyota, N.; Fukase, T. EXAFS Study on Premartensitic Phase in Nb_3Sn. *J. Phys. Soc. Jpn.* **1987**, *56*, 4183–4187. [CrossRef]

23. Heald, S.; Tarantini, C.; Lee, P.; Brown, M.; Sung, Z.; Ghosh, A.; Larbalestier, D. Evidence from EXAFS for different Ta/Ti site occupancy in high critical current density Nb_3Sn superconductor wires. *Sci. Rep.* **2018**, *8*, 4798. [CrossRef]

24. Battocchio, C.; Meneghini, C.; Fratoddi, I.; Venditti, I.; Russo, M.V.; Aquilanti, G.; Maurizio, C.; Bondino, F.; Matassa, R.; Rossi, M.; et al. Silver Nanoparticles Stabilized with Thiols: A Close Look at the Local Chemistry and Chemical Structure. *J. Phys. Chem. C* **2012**, *116*, 19571–19578. [CrossRef]

25. Timoshenko, J.; Kuzmin, A.; Purans, J. EXAFS study of hydrogen intercalation into ReO_3 using the evolutionary algorithm. *J. Phys. Condens. Matter* **2014**, *26*, 055401. [CrossRef] [PubMed]

26. Mathon, O.; Beteva, A.; Borrel, J.; Bugnazet, D.; Gatla, A.; Hino, R.; Kantor, I.; Mairs, T.; Munoz, M.; Pasternak, S.; et al. The Time-resolved and Extreme-conditions XAS (TEXAS) facility at the European Synchrotron Radiation Facility: The energy-dispersive X-ray absorption spectroscopy beamline ID24. *J. Synchrotron Radiat.* **2015**, *22*, 1548–1554. [CrossRef] [PubMed]

27. Welter, E.; Chernikov, R.; Herrmann, M.; Nemausat, R. A beamline for bulk sample x-ray absorption spectroscopy at the high brilliance storage ring PETRA III. *AIP Conf. Proc.* **2019**, *2054*, 040002. [CrossRef]

28. Ishimatsu, N.; Matsumoto, K.; Maruyama, H.; Kawamura, N.; Mizumaki, M.; Sumiya, H.; Irifune, T. Glitch-free X-ray absorption spectrum under high pressure obtained using nano-polycrystalline diamond anvils. *J. Synchrotron Radiat.* **2012**, *19*, 768–772. [CrossRef]

29. Irifune, T.; Kurio, A.; Sakamoto, S.; Inoue, T.; Sumiya, H. Ultrahard polycrystalline diamond from graphite. *Nature* **2003**, *421*, 599–600. [CrossRef]

30. Spina, T. Proton Irradiation Effects on Nb_3 Sn Wires and Thin Platelets in View of High Luminosity LHC Upgrade. Ph.D. Thesis, Deparement de Physique de la Matiere Quantique (DQMP), Universite de Geneve, Geneva, Switzerland , 2015.

31. Bragg, W.L.; Williams, E. J. The effect of thermal agitation on atomic arrangement in alloys. *Proc. R. Soc. A* **1934**, *145*, 699–730.

32. Flükiger, R.; Spina, T.; Cerutti, F.; Ballarino, A.; Scheuerlein, C.; Bottura, L.; Zubavichus, Y.; Ryazanov, A.; Svetogovov, R.D.; Shavkin, S.; et al. Variation ofTc, lattice parameter and atomic ordering in Nb_3Sn platelets irradiated with 12 MeV protons: Correlation with the number of induced Frenkel defects. *Supercond. Sci. Technol.* **2017**, *30*, 054003. [CrossRef]

33. Alimenti, A.; Pompeo, N.; Torokhtii, K.; Spina, T.; Flükiger, R.; Muzzi, L.; Silva, E. Surface Impedance Measurements on Nb_3Sn in High Magnetic Fields. *IEEE Trans. Appl. Supercond.* **2019**, *29*, 3500104. [CrossRef]

34. Alimenti, A.; Pompeo, N.; Torokhtii, K.; Spina, T.; Flükiger, R.; Muzzi, L.; Silva, E. Microwave measurements of the high magnetic field vortex motion pinning parameters in Nb_3Sn. *Supercond. Sci. Technol.* **2021**, *34*, 014003. [CrossRef]

35. Klotz, S.; Chervin, J.C.; Munsch, P.; Le Marchand, G. Hydrostatic limits of 11 pressure. *J. Phys. D Appl. Phys.* **2009**, *42*, 075413. [CrossRef]

36. Syassen, K. Ruby under pressure. *High Press. Res.* **2008**, *28*, 75–126. [CrossRef]

37. Dorogokupets, P.I.; Oganov, A.R. Ruby, metals, and MgO as alternative pressure scales: A semiempirical description of shock-wave, ultrasonic, x-ray, and thermochemical data at high temperatures and pressures. *Phys. Rev. B* **2007**, *75*, 024115. [CrossRef]

38. Meneghini, C.; Bardelli, F.; Mobilio, S. Estra-Fitexa: A Software Package for Exafs Data Analysis. *Nucl Instrum. Methods Phys. Res. Sect. B Beam Interact. Mater. Atoms* **2012**, *285*, 153–157. [CrossRef]

39. Meneghini, C.; Matteo, S.D.; Monesi, C.; Neisius, T.; Paolasini, L.; Mobilio, S.; Natoli, C.R.; Metcalf, P.A.; Honig, J.M. Antiferromagnetic–paramagnetic insulating transition in Cr-doped V_2O_3 investigated by EXAFS analysis. *J. Phys. Condens. Mat.* **2009**, *21*, 355401. [CrossRef] [PubMed]

40. James, F. MINUIT: Function Minimization and Error Analysis Reference Manual Version 94.1. CERN Program Library D506:1994. Available online: http://cdssls.cern.ch/record/2296388/files/minuit.pdf (accessed on 23 March 2021).

41. Fornasini, P.; Monti, F.; Sanson, A. On the cumulant analysis of EXAFS in crystalline solids research papers. *J. Synchrotron Radiat.* **2001**, *8*, 1214–1220. [CrossRef] [PubMed]

42. Rehr, J.; Kas, J.; Vila, F.; Prange, M.; Jorissen, K. Parameter-free calculations of X-ray spectra with FEFF9. *Phys. Chem. Chem. Phys.* **2010**, *12*, 5503–5513. [CrossRef]

43. Shirane, G.; Axe, J.D. Neutron Scattering Study of the Lattice-Dynamical Phase Transition in Nb_3Sn. *Phys. Rev. B* **1971**, *4*, 2957–2963. [CrossRef]

44. Kuzmin, A.; Timoshenko, J.; Kalinko, A.; Jonane, I.; Anspoks, A. Treatment of disorder effects in X-ray absorption spectra beyond the conventional approach. *Radiat. Phys. Chem.* **2020**, *175*. [CrossRef]

45. Timoshenko, J.; Kuzmin, A.; Purans, J. Reverse monte carlo modeling of thermal disorder in crystalline materials from EXAFS spectra. *Comput. Phys. Commun.* **2012**, *183*, 1237–1245. [CrossRef]

46. Jonane, I.; Cintis, A.; Kalinko, A.; Chernikov, R.; Kuzmin, A. Low temperature X-ray absorption spectroscopy study of $CuMoO_4$ and $CuMo_{0}.90W_{0}.10O_4$ using reverse monte-carlo method. *Radiat. Phys. Chem.* **2020**, *175*, 108411. [CrossRef]

47. Rehr, J.; Albers, R.C.Theoretical approaches to x-ray absorption fine structure. *Rev. Mod. Phys.* **2000**, *72*, 621. [CrossRef]

48. Timoshenko, J.; Kuzmin, A. Wavelet data analysis of EXAFS spectra. *Comput. Phys. Commun.* **2009**, *180*, 920–925. [CrossRef]

49. Filipponi, A.; Di Cicco, A.; Natoli, C.R. X-ray-absorption spectroscopy and n-body distribution functions in condensed matter. I. Theory. *Phys. Rev. B* **1995**, *52*, 15122–15134. [CrossRef]

50. Harris, C.R.; Millman, K.J.; van der Walt, S.J.; Gommers, R.; Virtanen, P.; Cournapeau, D.; Wieser, E.; Taylor, J.; Berg, S.; Smith, N.J.; et al. Array programming with NumPy. *Nature* **2020**, *585*, 357–362. [CrossRef]

51. Meneghini, C.; Ray, S.; Liscio, F.; Bardelli, F.; Mobilio, S.; Sarma, D.D. Nature of "Disorder" in the Ordered Double Perovskite Sr_2FeMoO_6. *Phys. Rev. Lett.* **2009**, *103*, 046403. [CrossRef]

52. Timoshenko, J.; Anspoks, A.; Cintins, A.; Kuzmin, A.; Purans, J.; Frenkel, A.I. Neural Network Approach for Characterizing Structural Transformations by X-Ray Absorption Fine Structure Spectroscopy. *Phys. Rev. Lett.* **2018**, *120*, 225502. [CrossRef]

53. Egami, T.; Billinge, S. (Eds.) *Underneath the Bragg Peaks*, 2nd ed.; Pergamon Materials Series; Elsevier: Pergamon, Turkey, 2012; Volume 16.

54. Provost, K.; Beret, E.; Muller, D.; Marcos, E.S.; Michalowicz, A. Impact of the number of fitted Debye-Waller factors on EXAFS fitting. *J. Phys. Conf. Ser.* **2013**, *430*, 012015. [CrossRef]

55. Filipponi, A.; Di Cicco, A. X-ray-absorption spectroscopy and n -body distribution functions in condensed matter. II. Data analysis and applications. *Phys. Rev. B* **1995**, *52*, 15135–15149. [CrossRef] [PubMed]

56. Tucker, M.G.; Keen, D.A.; Dove, M.T.; Goodwin, A.L.; Hui, Q. RMCProfile: Reverse Monte Carlo for polycrystalline materials. *J. Phys. Condens. Matter* **2007**, *19*, 335218. [CrossRef]

57. Michalowicz, A.; Provost, K.; Laruelle, S.; Mimouni, A.; Vlaic, G. F-test in EXAFS fitting of structural models. *J. Synchr. Radiat.* **1999**, *6*, 233–235. [CrossRef]

58. Mao, H.K.; Chen, X.J.; Ding, Y.; Li, B.; Wang, L. Solids, liquids, and gases under high pressure. *Rev. Mod. Phys.* **2018**, *90*, 015007. [CrossRef]

59. Markiewicz, W. Elastic stiffness model for the critical temperature Tc of Nb_3Sn including strain dependence. *Cryogenics* **2004**, *44*, 767–782. [CrossRef]

60. Valentinis, D.F.; Berthod, C.; Bordini, B.; Rossi, L. A theory of the strain-dependent critical field in Nb_3Sn, based on anharmonic phonon generation. *Supercond. Sci. Technol.* **2013**, *27*, 025008. [CrossRef]

61. Godeke, A.; Hellman, F.; ten Kate, H.H.J.; Mentink, M.G.T. Fundamental origin of the large impact of strain on superconducting Nb_3Sn. *Supercond. Sci. Technol.* **2018**, *31*, 105011. [CrossRef]

Article

Mechanical Properties of Commercial Purity Aluminum Modified by Zirconium Micro-Additives

Ahmad Mostafa [1,*], Wail Adaileh [1], Alaa Awad [2] and Adnan Kilani [2]

[1] Department of Mechanical Engineering, Tafila Technical University, Tafila 66110, Jordan; wadaileh@ttu.edu.jo
[2] Department of Industrial Engineering, University of Jordan, Amman 11942, Jordan; aawad12@yahoo.com (A.A.); adnan_kilani@yahoo.com (A.K.)
* Correspondence: a.omar@ttu.edu.jo; Tel.: +962-32-250-326; Fax: +962-32-250-002

Abstract: The mechanical properties and the fractured surfaces of commercial purity aluminum modified by zirconium micro-additives were investigated by means of experimental examination. A commercial purity Al specimen was used as a reference material and seven Al-Zr alloys in the 0.02–0.14 wt.% Zr composition range (with 0.02 wt.% Zr step) were prepared by microalloying methods. Optical microscopy was used to examine the microstructures and to calculate the grain sizes of the prepared specimens. The phase assemblage diagrams were plotted and the relative amounts of solid phases were calculated at room temperature using FactSage thermochemical software and databases. Proof stress, strength coefficient and strain hardening exponent were measured from the stress-strain curves obtained from tensile experiments and Charpy impact energy was calculated for all specimens. The experiments showed that the grain size of commercial purity Al was reduced by adding any Zr concentration in the investigated composition range, which could be due to the nucleation of new grains at Al_3Zr particle sites. Accordingly, the microhardness number, tensile properties and Charpy impact energy were improved, owing to the large grain-boundary areas resulted from the refining effect of Zr, which can limit the movement of dislocations in the refined samples. The basic fracture mode in all specimens was ductile, because Al has an FCC structure and remains ductile even at low temperatures. The ductile fractures took place in a transgranular manner as could be concluded from the fractured surface features, which include voids, ridges and cavitation.

Keywords: grain refinement; mechanical properties; commercial purity aluminum; zirconium

check for updates

Citation: Mostafa, A.; Adaileh, W.; Awad, A.; Kilani, A. Mechanical Properties of Commercial Purity Aluminum Modified by Zirconium Micro-Additives. *Crystals* **2021**, *11*, 270. https://doi.org/10.3390/cryst11030270

Academic Editor: Evgeniy N. Mokhov

Received: 1 March 2021
Accepted: 8 March 2021
Published: 9 March 2021

Publisher's Note: MDPI stays neutral with regard to jurisdictional claims in published maps and institutional affiliations.

1. Introduction

Characterizing the mechanical properties of metals and alloys is fundamental for multiple technological applications. It is essential to depict the responses of structural components to external mechanical loadings [1,2]. Of particular interest, the investigation of titanium [3] and aluminum [4] alloys' behaviors is essential, because they are broadly used by the aerospace industry. Aluminum and its alloys are among the most in demand engineering materials for structural applications in many industries, because of their various positive attributes, such as high strength-to-weight ratio, good corrosion resistance and excellent thermal and electrical conductivities [5]. Aluminum properties are governed by the grain size, which is one of the important microstructural features as described by the Hall–Petch [6,7] relationship. Fine-grained structures usually show high yield strength, high toughness, good formability, improved machinability and uniform distribution of the secondary phases [8,9]. One way to modify the Al grain structure is by introducing small amounts of alloying elements, the so-called grain refiners, such as rare earth elements, transition metals and binary alloys (Ti+B) [5,10]. The main role of the grain refiners is to develop fine equiaxed Al solid solution (FCC_Al) grains in the cast structure either by increasing the number of nucleation sites or by grain multiplications [11]. The addition of transition metals, such as Zr [12], can modify the cast Al structure effectively by forming the Al_3Zr primary phase particles [13,14]. It has been reported that the grain refinement

127

effect on the parent metal is often associated with peritectic systems [14,15]. According to Murray et al. [16] and Wang et al. [17], FCC_Al solid solution solidifies from the liquid by L + Al_3Zr ↔ FCC_Al peritectic reaction at 660.8 °C and 0.28 wt.% (0.083 at.%) Zr. The maximum liquid solubility of Zr in Al was reported as 0.11 wt.% (0.030 at.%) Zr at 661 °C [18]. The refinement occurs during the cooling of the primary phase crystals, which react peritectically with the liquid. Upon further cooling, the peritectic reaction progresses and transforms the primary crystals into secondary phase crystals, which then act as nuclei for solidification of the remaining melt [15]. The role of the new nucleation sites is to block the grain boundary or sub-boundary migration, which refine the cast grain structure [19,20].

The effect of Zr on the mechanical properties of commercial purity Al [12] and other Al alloys [13,20–22] has been investigated. Gao et al. [12] found that addition of Zr with 0.2 wt.% to commercial purity Al refined the cast structure significantly by precipitating $L1_2$ structured Al_3Zr primary phase. Furthermore, the hardness and peak stress of Al-0.2Zr alloy were increased by 20% and 15.6%, respectively, as compared to that of pure Al. By adding ~0.12 wt.% Zr, Toschi et al. [13] noticed an increase in the tensile strength (*TS*) and yield strength (*YS*) of age hardenable aluminum alloy containing magnesium and copper (AA2618) at room temperature (by 7% and 22%, respectively) and after aging at 250 °C (by 10% and 12%, respectively). The improved tensile properties were due to the presence Al_3Zr primary phase particles. In an attempt to stabilize the fine-grained structure and improve the superplasticity behavior, Duan et al. [20] added 0.1 wt.% Zr to Al-Zn-Mg alloys. Seyed Ebrahimi et al. [21] found that addition of 0.3 wt.% Zr to the new super high strength Al–12.24Zn–3.25Mg–2.46Cu alloy decreased the average grain size by 20% and enhanced the tensile strength, yield strength and elongation values by 34%, 25% and 1850%, respectively.

The abovementioned literature data showed that Zr has significantly improved the microstructural features and mechanical behaviors of commercial purity Al and its alloys due to the formation of Al_3Zr primary phase. However, the wt.% Zr was varying from one study to another. This work aims at investigating the effects of minor Zr additives in the 0.02–0.14 wt.% Zr range (with 0.02 wt.% Zr step) on the microstructure, microhardness, tensile properties, and Charpy impact energy of commercial purity aluminum. Microalloying technique was used to control the chemical composition of different Al-Zr alloys.

2. Materials and Methods

Commercial purity aluminum of about 99.8 wt.% was used as a reference material and as a major component in the master alloys and microalloys. Seven microalloys with different Zr concentrations (0.02, 0.04, 0.06, 0.08, 0.10, 0.12 and 0.14 wt.%) were prepared by diluting the binary Al-5.73Zr wt.% master alloy in a commercial purity aluminum. The chemical compositions of the commercial purity Al is listed in Table 1.

Table 1. Chemical compositions of commercial purity Al (wt.%).

	Fe	Si	Mg	Ti	B	V	Zn	Others [1]	Al
Commercial purity Al	0.11	0.05	0.004	0.004	0.0005	0.008	0.005	0.015	Bal.

[1] Cu, Na and Mn.

The Al-5.73Zr wt.% master alloy was prepared by melting the predetermined quantity of Al in a graphite crucible at 850 °C for 15 min with a cryolite flux on the melt top to prevent oxidation. The furnace temperature was raised up to 1000 °C and a capsule of a pure zirconium powder wrapped in an Al foil was added to the melt. The melt was stirred for 30 s using a graphite rod and kept inside the furnace for another 5 min. The melt was then poured in a brass die with a cylindrical cavity.

The microalloys were prepared similarly by adding the predetermined quantities of Al and master alloy in a graphite crucible and melted at 800 °C for 15 min with the cryolite flux. The melt was stirred for 30 s to enhance mixing the master alloy with the parent metal and solidified inside brass dies of two forms. The first form was a 10 mm diameter and 100 mm long cylinder for making the tensile test specimens according to ASTM 557-15 [23] standard shown in Figure 1a. The other form was to prepare the 10 mm × 10 mm × 100 mm impact test specimens according to ASTM E23-18 [24] standard shown in Figure 1b.

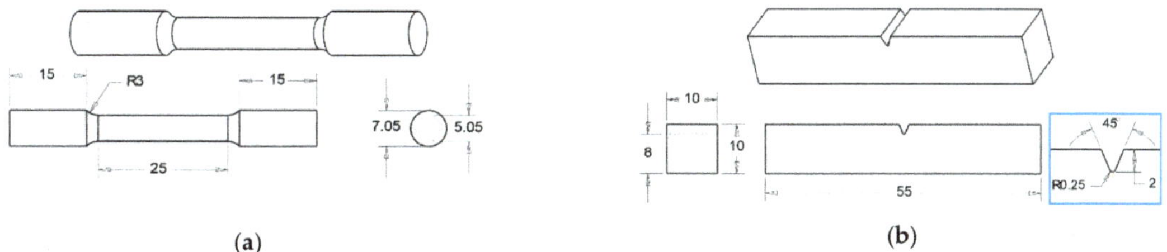

(a)   **(b)**

Figure 1. Standard test specimens (in mm) for (**a**) tensile and (**b**) impact experiments.

Slices from the prepared microalloys were taken using a slow cutting machine in order to minimize heat generated during cutting process and to prevent any microstructural changes. The slices were mounted in cold epoxy blocks, ground gradually from 240 up to 1200 grit using SiC sand papers and polished using 1 μm diamond paste. The polished specimens were chemically etched using a solution of 1.5% HF + 2% HNO_3 + 1.5% HCl + 95% distilled water for 90 s and cleaned under running water to perform metallurgical examination. A Quanta 200 scanning electron microscope (SEM, FEI, Hillsboro, OR, USA) equipped with an energy dispersive X-ray spectrometer (EDS) was used to determine the chemical composition of the master alloy and to examine the features of the fractured surfaces.

Microhardness tests were carried out using a HWDM-3 Vickers hardness tester (TTS Unlimited Inc., Kita-Ku, Osaka, Japan) equipped with a pyramid head indenter of 136° angle between each face and 100 g force. Seven different values were taken at different locations on each specimen, from which the average HV number for each alloy was determined. The grain sizes were measured according to the standard test methods for determining average grain size using the intercept method (ASTM E112-13) [25]. Five lines at different directions were used and the average grain size was determined for each specimen.

Tensile properties were acquired using a 250 KN capacity universal testing machine (Shimadzu, Kyoto, Japan) at a cross head speed of 10 mm/min. The tensile test was performed on circular samples, shown in Figure 1a, with a gradual axial loading until failure. The load-extension data were recorded by a data acquisition system in order to evaluate true stress-strain data. The impact toughness was determined with a Charpy impact test, also known as the CVN test [26]. The Charpy test allows an estimation of the total impact work needed for crack initiation and the work necessary for crack propagation from a V-notched specimen shown in Figure 1b. The energy absorbed versus Zr concentration relationship was plotted.

3. Results and Discussion

3.1. Microstructure and Grain Size Analyses

The optical micrographs of the commercial purity Al and modified alloys with different Zr concentrations are shown in Figure 2. It can be seen that the microstructure of reference material was dramatically changes by adding different Zr wt.% concentrations. In Figure 2a, commercial purity Al shows a mixture of grain structures dominated by large columnar grains. The columnar structure in Al was altered after adding Zr element and other grain and sub-grain structures were formed. Figure 2b,c illustrate the microstructure of Al-0.02Zr and Al-0.04Zr microalloys, respectively, which contain grains with determined boundaries and internal sub-grains colonized by the main grain boundaries. The grain boundaries in Figure 2b are well defined and the sub-grain boundaries are small. On the other hand, the main grain boundaries are less defined in Figure 2c due to the growth of sub-grains. In case of Al-0.06Zr microalloy, it seems that the sub-grains grew more than that in Al-0.04Zr microalloy and thus the Al columnar microstructure turned into large equiaxed grains, shown in Figure 2d, due to the merge grains along boundaries. Figure 2e shows a unique microstructure, differs than others, where the sub grains clustered within the equiaxed primary grain structure after adding 0.08 wt.% Zr. The increased Zr concentration leads to precipitate more of Al_3Zr second phase upon cooling, which breaks the coarse structure, seen in Al-0.06Zr microalloy, into smaller grains by nucleating several sub-grains. Evidence of the refining effect of high Zr concentrations (up to 0.14 wt.%) could be seen in Figure 2f–h. The grains are becoming more resistant to coarsening if the diffusivity and solubility of the added metal are small [27]. Among the transition metals, Zr has the smallest diffusion flux [28] and low solubility in Al.

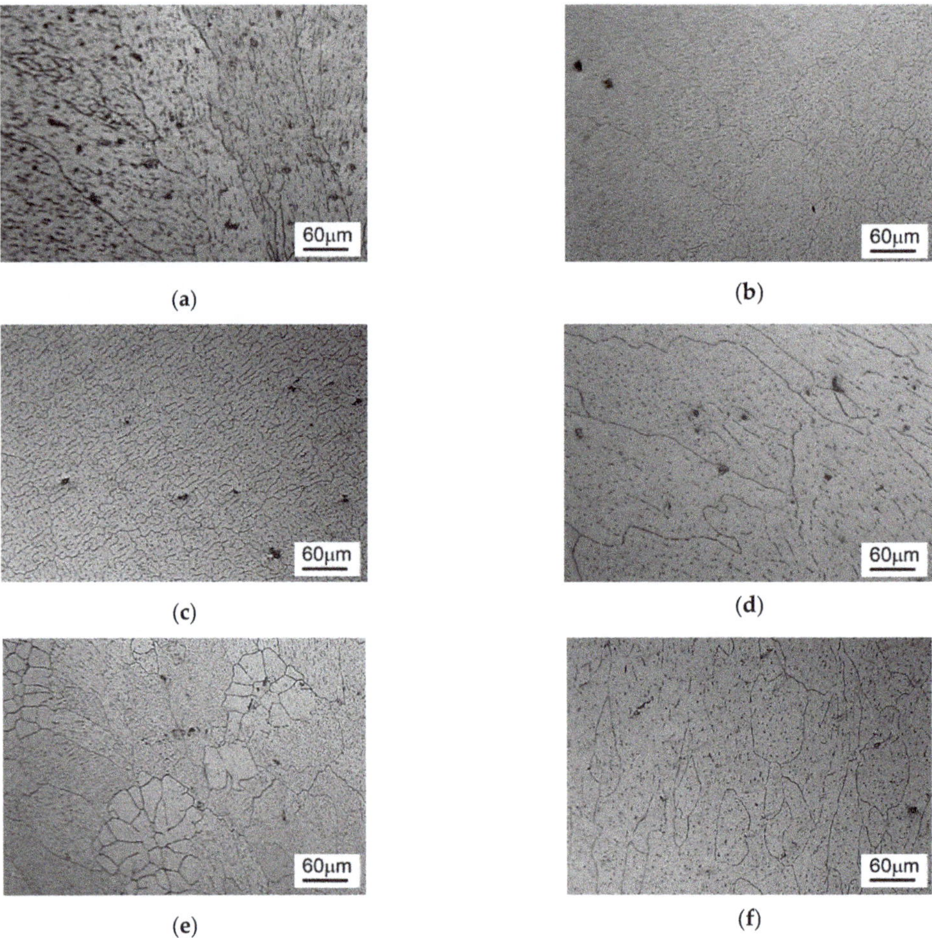

Figure 2. *Cont.*

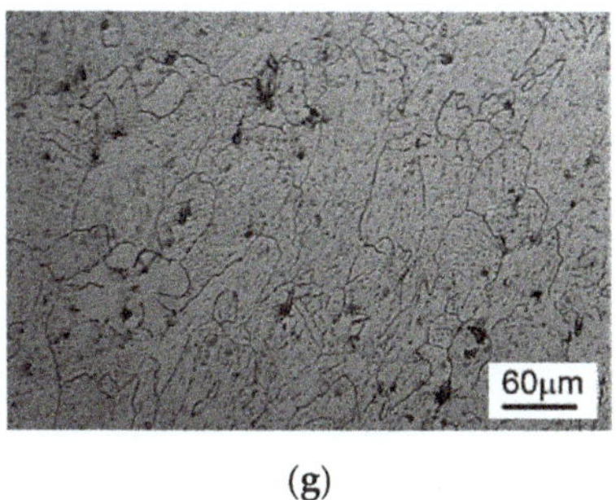
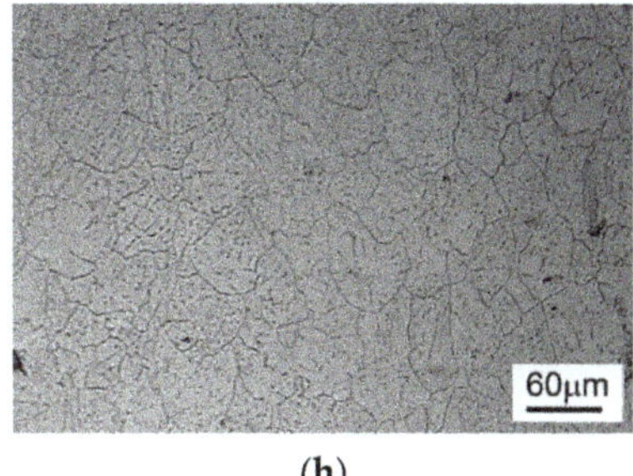

(**g**)   (**h**)

Figure 2. Micrographs of experimental aluminum alloys: (**a**) commercial purity Al, (**b**) Al-0.02Zr, (**c**) Al-0.04Zr, (**d**) Al-0.06Zr, (**e**) Al-0.08Zr, (**f**) Al-0.10Zr, (**g**) Al-0.12Zr and (**h**) Al-0.14Zr.

The Al-rich side of the Al-Zr binary phase diagram in Figure 3 was calculated using the FactSage thermochemical software and database [25] to better understand the relationship between the alloy composition and the microstructure. The dashed lines represent the composition of Al-Zr microalloys prepared in this study. It can be noted that Al_3Zr forms in 0.10, 0.12 and 0.14 wt.% Zr microalloys during cooling when the solidification path crosses the L + Al_3Zr two-phase field. The amount of Al_3Zr precipitates increase by increasing the Zr concentration and thus more nucleation sites in the melt are expected. The remaining liquid solidifies upon cooling, in the 550–661 °C temperature range, to form FCC_Al solid solution according to L + Al_3Zr ↔ FCC_Al peritectic reaction. At room temperature, Al_3Zr solute particles precipitate from the super saturated FCC_Al solid solution and disperse in FCC_Al matrix. This explains the formation of a uniform fine structures in 0.10–0.14 wt.% Zr microalloys.

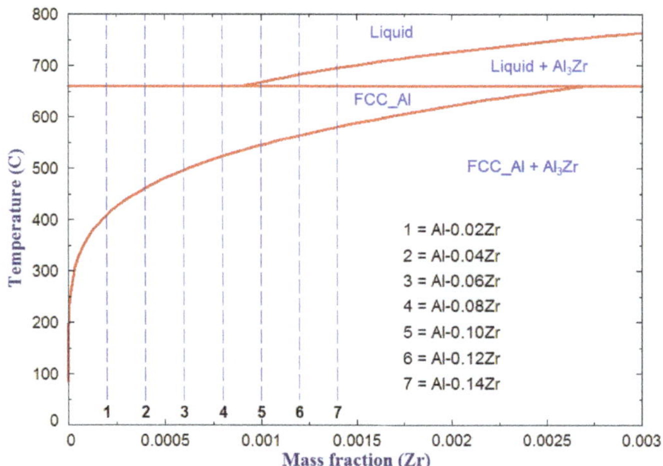

Figure 3. The calculated Al-rich side of the Al-Zr binary phase diagram.

The corresponding average grain size versus Zr wt.% concentration in commercial purity Al relationship is plotted in Figure 4. The grain size of pure aluminum specimen was measured by taking the equivalent diameter of equiaxed grains to be 160.3μm. It is noted that the average grain size of commercial purity aluminum has been reduced after 0.02–0.14 wt.% Zr addition. The average grain size has dropped from 160.3 μm to 88.8 μm after 0.02 wt.% Zr addition and then gradually increased up to 148.4 μm when 0.06 wt.%

Zr was added. Another drop in the average grain size, down to 66.1 μm, was observed after adding 0.08 wt.% Zr, which later gradually reduced down to 31.4 μm when 0.14 wt.% Zr was added. Gao et al. [12] observed the grain refinement effect of Zr on Al when the large Al dendrites were altered to fine equiaxed grains after adding 0.2 wt.% Zr.

Mahmoud et al. [29] added 0.1 to 0.3 wt.% Zr (with 0.05 wt.% Zr step) to pure Al and found that the grain structure has been refined in the experimental range. However, further Zr addition (>0.3 wt.%) slightly increased the average grain size, whereas excess Zr addition did not show any considerable effect on the Al structure [29].

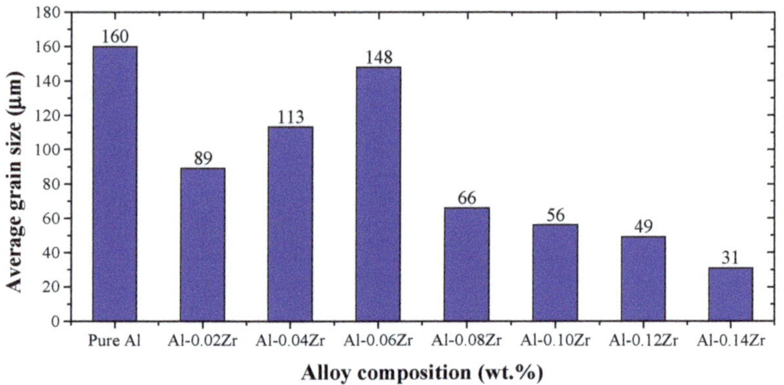

Figure 4. Average grain size (μm) vs. Zr concentration plot.

It was reported by Rohrer [30] that precipitation of a second phase is one mechanism that can occur to partition excess solute in the microstructure. The more solute on the grain boundary increases the grain boundaries and thus finer grain structures may result. It is mentioned earlier in this work that the maximum solid solubility of Zr in Al at 660.8 °C was reported as 0.28 wt.%. However, the amount of Zr solute in Al reduces when the alloy cools down to room temperature. Under such condition, the alloy becomes supersaturated with Zr solute and Al_3Zr primary phase precipitates peritectically [27]. The higher the Zr concentration in the alloy contains the more Al_3Zr phase particle precipitates.

Table 2 summarizes the relative amounts of solid phases (in gram) for each microalloy composition at room temperature. The primary Al_3Zr phase serves as an effective heterogeneous nuclei for Al and refine the grains, because of the lattice parameter mismatch between $L1_2$ structured Al_3Zr and FCC_Al matrix [31].

Table 2. Relative amounts of the solid phases in the studied alloys at room temperature (wt.%).

Microalloy	FCC_Al	Al_3Zr
Al-0.02Zr	99.962	0.037 (7)
Al-0.04Zr	99.925	0.075 (4)
Al-0.06Zr	99.887	0.113 (2)
Al-0.08Zr	99.849	0.150 (9)
Al-0.10Zr	99.811	0.188 (7)
Al-0.12Zr	99.774	0.226 (4)
Al-0.14Zr	99.736	0.264 (2)

The FactSage thermochemical software and database [32] was also used to calculate the relative amounts of solid phases (in gram) through plotting the phase assemblage diagram for each microalloy composition as illustrated in Figure 5.

It is important to mention that the phase assemblage diagrams are also used to determine the phase formation and/or decomposition temperatures. For instance, FCC_Al decomposes at 410, 470, 500, 530, 550, 570 and 580 °C to form Al₃Zr solid phase in Al-0.02Zr, Al-0.04Zr, Al-0.06Zr, Al-0.08Zr, Al-0.10Zr, Al-0.12Zr and Al-0.14Zr microalloys, respectively. The jumps in Al₃Zr amounts appeared in Figure 5a, at 650–700 °C temperature range, correspond to Al-0.12Zr and Al-0.14Zr microalloys and represent the precipitation of Al₃Zr phase from liquid in the Liquid + Al₃Zr two-phase field. The amount dropped down to zero again, because both Al₃Zr and the remaining liquid were consumed to form FCC_Al solid solution by the Liquid+Al₃Zr ↔ FCC_Al peritectic reaction. The uniform dispersion of Al₃Zr solid particles in the liquid explains the occurrence of fine equiaxed grains shown in Figure 2g,h.

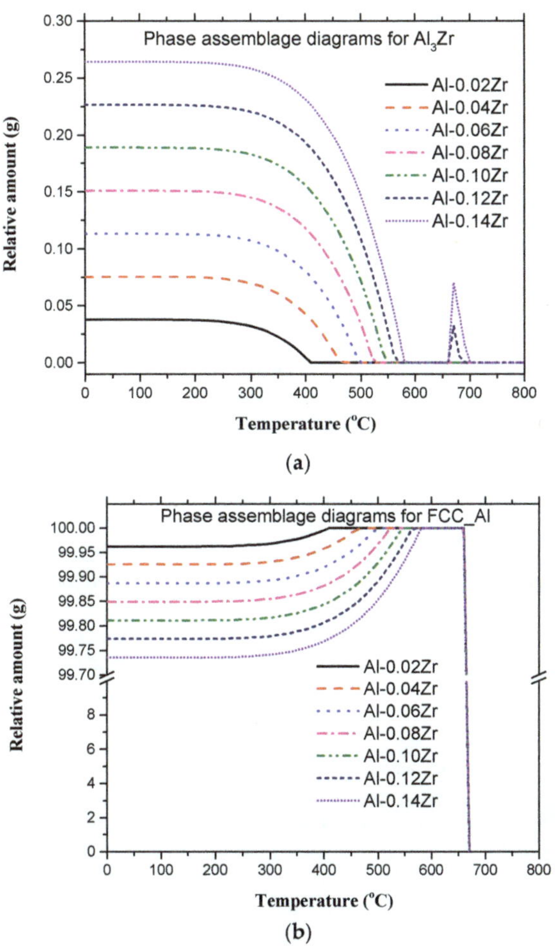

Figure 5. Phase assemblage diagrams for (**a**) Al₃Zr and (**b**) FCC_Al.

3.2. Microhardness Analysis

Figure 6 shows the average microhardness number for commercial purity Al and Al-Zr microalloys. The values varied from one composition to another, which clearly show an increasing trend with increased Zr concentration. It is remarkable that the microhardness numbers are inversely proportional to the grain size results, illustrated in Figure 4, where the hardness increases with decreasing the grain size. The average microhardness number obtained for pure Al was about 26.1 HV and increased up to 30.7 HV for Al-0.14Zr microalloy. In the work of Souza et al. [33], the microhardness number of both cast 0.22 and 0.32 wt.% Zr alloys was about 25.5 HV and increased up to 30 and 40 HV, respectively, after aging for 10 h. The alloys in the current work were slowly cooled in a brass die and thus aging was highly probable to occur.

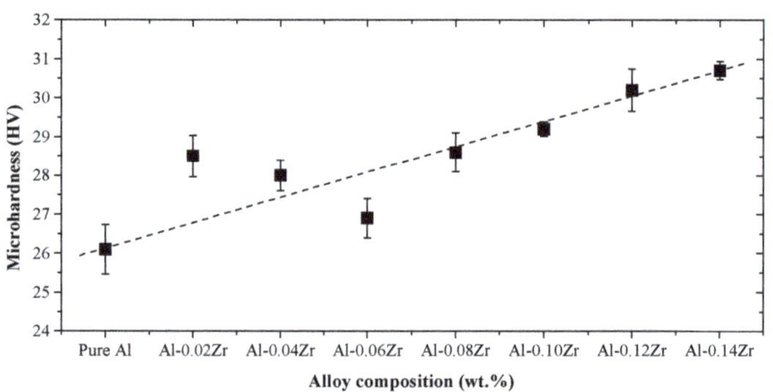

Figure 6. Microhardness number vs. Zr concentration plot.

The improved microhardness of Al-Zr microalloys stems from the formation of thermally stable second phase on the grain boundaries such as Al_3Zr [34]. The chemical driving force for Al_3Zr precipitation becomes higher for alloys with high Zr concentrations, because they have high solute contents as discussed in Section 3.1. The microhardness number may also increase due to the increased grain boundary areas in Al-Zr microalloys, which can limit the movement of dislocations in the refined samples [11,21].

3.3. Tensile Characteristics

The mechanical behaviors of commercial purity aluminum and Al-Zr microalloys are presented by the calculated true stress-true strain curves in Figure 7. Accordingly, the proof stress at 0.2% strain and mechanical behavior parameters, i.e., strength coefficient (k) and strain hardening exponent (n), were derived from the tensile experiments for all examined specimens and summarized in Table 3. The error percentages in the tensile properties could be due to the structural defects in commercial purity Al or due to the inhomogeneous distribution of the fine particles in Al-Zr microalloys.

The proof stress of all Al-Zr microalloys increased from 19.67 to 48.18 MPa as compared to 16.87 MPa for commercial purity aluminum. The increased proof stress value is attributed to the presence of Zr in all microalloys. With reference to the phase assemblage diagrams (Figure 5), the amount of Al_3Zr phase increases directly with the increase of Zr concentration in the microalloys, which has a notable effect on the mechanical properties. The formation of nanosized Al_3Zr particles within the FCC_Al matrix plays a significant role in blocking the dislocations' movement under mechanical loading [12]. Hence, the strength of Al-Zr microalloys increases with increased Zr concentration. Although Zr-0.06Zr microalloy showed higher proof stress value of 19.67 MPa as compared to 16.87 MPa for commercial purity aluminum, its value still lower than that of Al-0.02Zr and Al-0.04Zr microalloys, which is about 29.32 and 28.27 MPa, respectively. The improved mechanical performance of Al-0.02Zr and Al-0.04Zr microalloys could be attributed to the smaller average grain sizes of about 88.8 and 112.6 μm, respectively, as compared to 148.4 μm for Al-0.06Zr microalloy.

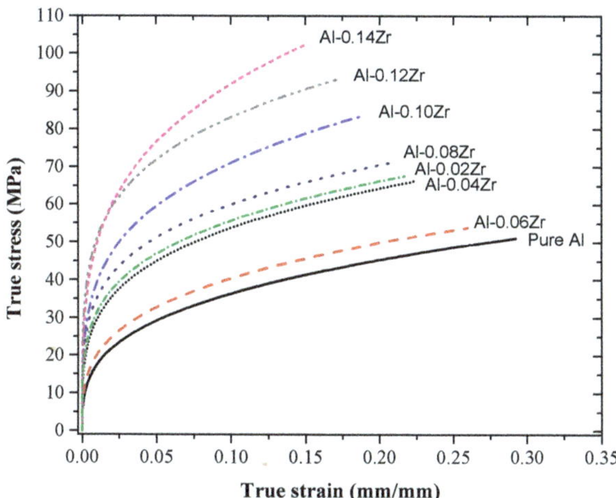

Figure 7. True stress-true strain curves for pure Al and Al-Zr microalloys.

The contribution of grain size to the improved mechanical properties can be expressed by Hall-Petch relationship [6,7]:

$$\sigma_y = \sigma_o + \left(k_y \times d^{-0.5}\right),\qquad(1)$$

where, σ_y is the yield stress, σ_o is the material constant for the dislocation movement at the starting stress, k_y is the material strengthening coefficient and d is the average grain diameter. In accordance to Hall-Petch relationship, the addition of Zr improved the mechanical properties of commercial purity Al as shown in Figure 7.

Table 3. Mechanical behavior parameters as extracted from the tensile experiments for all examined specimens.

Material	Proof Stress ($\sigma_{0.2}$) MPa	Strain Hardening Exponent (n)	Strength Coefficient (k) MPa
Pure Al	$16.87 \pm 9.6\%$	0.318	75.60
Al-0.02Zr	$29.32 \pm 5.1\%$	0.251	99.04
Al-0.04Zr	$28.27 \pm 7.0\%$	0.260	97.79
Al-0.06Zr	$19.67 \pm 6.1\%$	0.304	81.26
Al-0.08Zr	$32.63 \pm 8.4\%$	0.232	102.42
Al-0.10Zr	$36.14 \pm 7.3\%$	0.256	128.08
Al-0.12Zr	$48.18 \pm 4.6\%$	0.211	135.08
Al-0.14Zr	$47.74 \pm 3.3\%$	0.267	169.94

The strain hardening exponent (n) and strength coefficient (k) were calculated from ln σ vs. ln ε curves for all examined specimens. It is important to mention that ln σ vs. ln ε curves for all specimens were plotted in the plastic deformation region to fulfill Hollomon's equation, which is a power law relating the true strain to the true stress [11] as follows:

$$\sigma_T = k\varepsilon_T^n,\qquad(2)$$

where, σ_T is the true stress, k is the strength coefficient, ε_T is the true strain and n is the strain hardening exponent. It can be noticed from Table 3 that both n and k numbers are in compatible trends to the yield strength values. The increased strain hardening exponent values for all microalloys could be attributed to the decreased average grain size, which dramatically influenced the mechanical behavior of the Al-Zr microalloys. The value of n

indicates the capability to uniformly distribute the deformation. In other words, *n* evaluates the strain- hardening capability of the material [35].

3.4. Fractography Analysis of Tensile Specimens

The fractured surfaces of tensile specimens and their features were observed using SEM as shown in Figure 8. The basic fracture mode in all specimens is ductile, because Al has an FCC structure and remains ductile even at low temperatures. The ductile fracture took place in transgranular manner, which can be recognized by the flat surfaces of the specimen halves after fracture. This type of fracture occurs by void coalescence of a polycrystalline material during tension experiment [36]. The presence of ridges (stepped structure) are evident of crack propagation under progressive tensile loading and indicates the material resistance to failure upon loading. The hollow points (voids) are usually form at the locations of the second phase particles, i.e., Al_3Zr, in Al-Zr microalloys specifically. The crack-like voids are formed by the grain-boundary cavitation, by which a full grain is detached from the observed surface along its boundaries during the tensile test.

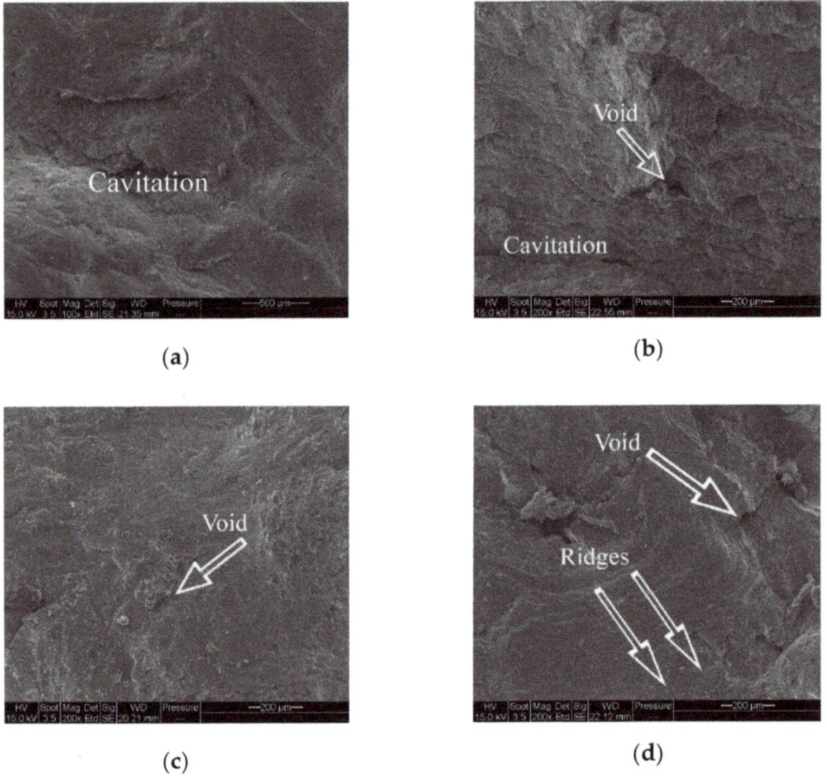

(a)

(b)

(c)

(d)

Figure 8. *Cont.*

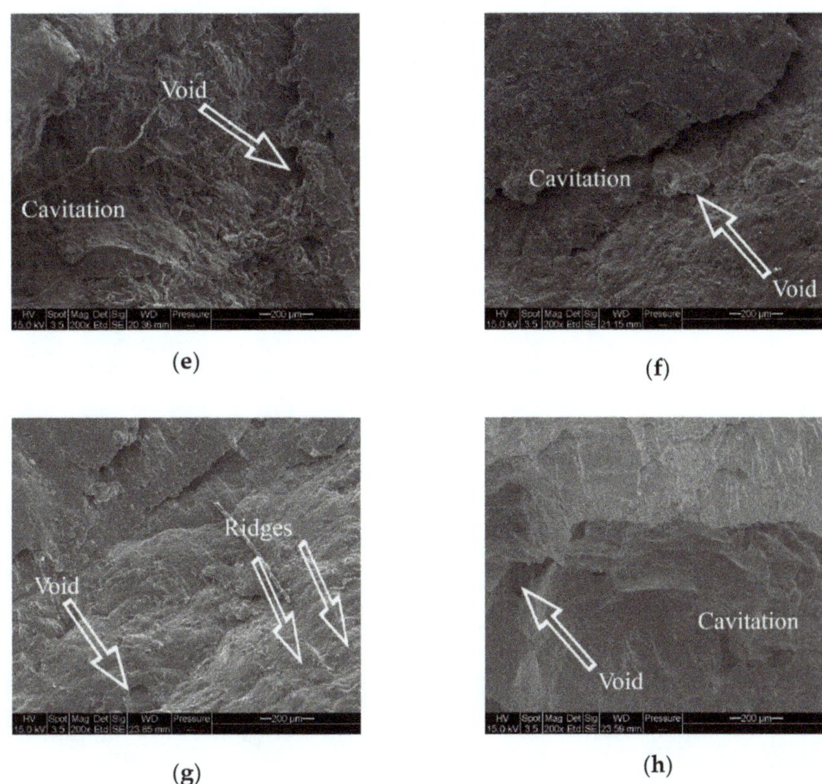

(e)

(f)

(g)

(h)

Figure 8. SEM images for the fractured tensile specimens: (**a**) commercial purity Al, (**b**) Al-0.02Zr, (**c**) Al-0.04Zr, (**d**) Al-0.06Zr, (**e**) Al-0.08Zr, (**f**) Al-0.10Zr, (**g**) Al-0.12Zr and (**h**) Al-0.14Zr.

3.5. Impact Fracture Analysis

The fracture toughness of a material is generally represented by its Charpy impact energy, which is a high strain-rate test (1×10^{-3}—1×10^{-4} s^{-1}) that determines the amount of energy absorbed by a material during fracture [26]. The bar chart in Figure 9 is the Charpy impact energy plotted for commercial purity Al and Al-Zr microalloys. The measured impact energy increased steadily from 36.6 J for commercial purity Al to 41.2, 47.09, 43.19, 46.11, 49.05, 51.99 and 47.07 J for Al-Zr microalloys containing Zr concentrations from 0.02 wt.% to 0.14 wt.% (with 0.02 wt.% Zr step), respectively. The increase in the impact energy was not significant and this could be due to alloying of commercial purity Al with minute Zr additions. It is highly recommended that the improvement in the fracture energy was due to the formation of Al$_3$Zr particles in Al-Zr microalloys, which also played a significant role in initiating new nucleation sites to refine the coarse Al grains.

Miyahara et al. [37] reported that the validated impact energy was independent of the grain size in alumina, whereas, Tarpani and Spinelli [38] plotted a direct relationship between Charpy impact energy and equivalent grain size, by which the fracture energy increases with increasing grain size. The relationship [38] was extracted indirectly from the effect of annealing temperature on the equivalent grain size, which in turn improved the fracture toughness of the material. In the current work, the purpose of grain refinement of commercial purity Al was to improve its mechanical properties and thus the specimens were not heat treated. The grains of commercial purity Al were refined by adding Zr inclusions, as discussed earlier in this work. Figure 10 shows an inverse relationship between the Charpy impact energy and average grain size for the examined specimens. In this case, the impact energy is inversely proportional to the grain boundary area that provide prominent resistance to impact fracture due to the increased number of dislocations.

In fact, the fracture energy is lowered if yield strength heightens for the same material [39]. However, in the current experiments, the chemical composition of the reference material is different than that of the microalloys. Thus, the Al-Zr microalloys lead to both high yield strength and high fracture energy as could be concluded from Figure 10.

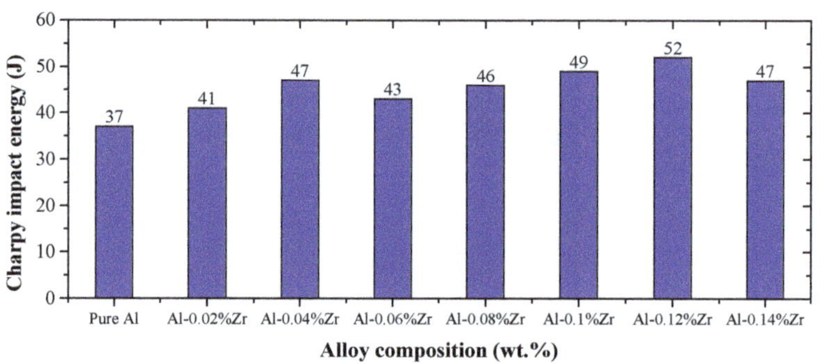

Figure 9. Charpy impact energy for commercial purity Al and Al-Zr microalloys.

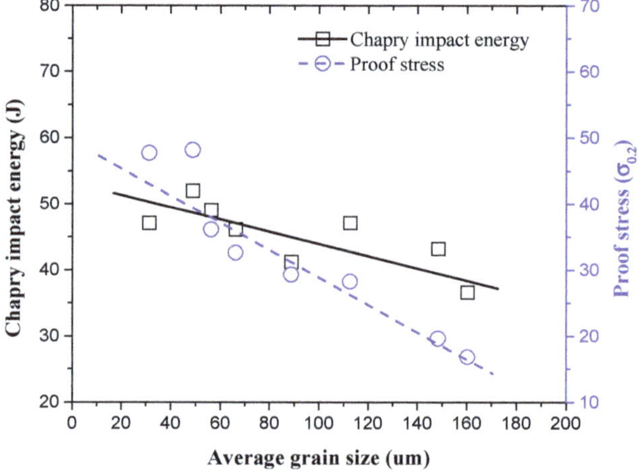

Figure 10. Charpy impact energy and proof stress for commercial purity Al and Al-Zr microalloys as functions of grain size.

4. Conclusions

In this work, the mechanical properties of commercial purity aluminum modified by Zr micro-additives have been investigated. The experiments showed that addition of Zr in the 0.02–0.14 wt.% composition range has reduced the grain size of commercial purity aluminum from 160.3 μm down to 31.4 μm for Al-0.14Zr microalloy, due to the formation of Al_3Zr particles which served as nucleation sites for the growth of new grains. The relative amount of Al_3Zr was calculated at room temperature as 0.0377–0.2647 wt.% for Al-Zr microalloys using FactSage thermochemical software and databases. As a result of this reduction, the microhardness number was improved due to the improved solute content with increased concentrations of Zr. The inverse relationship between the average grain size and Charpy impact energy indicated that the resistance to fracture was due to the increased grain boundary areas in the microalloys. Both Charpy impact energy and proof stress were improved as a result of increased number of dislocations in the fine-structured specimens. The presence of voids, fracture ridges and cavitation on the fractured surfaces were evidences that failure mode was ductile fracture in all examined specimens. Furthermore, the ductile fracture took place in a transgranular manner as could be noticed from the flat nature of the fractured surfaces.

Author Contributions: Conceptualization, A.K.; Formal analysis, A.M., A.A. and A.K.; Investigation, A.M. and A.A.; Methodology, A.M. and A.A.; Resources, A.M. and A.K.; Software, A.M.; Supervision, A.K.; Validation, A.M. and A.K.; Visualization, A.M.; Writing–original draft, A.M. and W.A.; Writing–review & editing, A.M. All authors have read and agreed to the published version of the manuscript.

Funding: This research received no external funding.

Data Availability Statement: Data is contained within the article.

Acknowledgments: The authors thankfully acknowledge the financial support provided by the University of Jordan and thanks are due to Tafila Technical University for granting necessary permissions to experimental facilities.

Conflicts of Interest: The authors declare no conflict of interest.

References

1. Errandonea, D.; Burakovsky, L.; Preston, D.L.; MacLeod, S.G.; Santamaría-Perez, D.; Chen, S.; Cynn, H.; Simak, S.I.; McMahon, M.I.; Proctor, J.E.; et al. Experimental and theoretical confirmation of an orthorhombic phase transition in niobium at high pressure and temperature. *Commun. Mater.* **2020**, *1*, 60. [CrossRef]
2. Straumal, B.; Korneva, A.; Kilmametov, A.; Lityńska-Dobrzyńska, L.; Gornakova, A.; Chulist, R.; Karpov, M.; Zięba, P. Structural and Mechanical Properties of Ti–Co Alloys Treated by High Pressure Torsion. *Materials* **2019**, *12*, 426. [CrossRef] [PubMed]
3. Smith, D.; Joris, O.P.J.; Sankaran, A.; Weekes, H.E.; Bull, D.J.; Prior, T.J.; Dye, D.; Errandonea, D.; Proctor, J.E. On the high-pressure phase stability and elastic properties of β -titanium alloys. *J. Phys. Condens. Matter* **2017**, *29*, 155401. [CrossRef] [PubMed]
4. Dorward, R.C.; Pritchett, T.R. Advanced aluminium alloys for aircraft and aerospace applications. *Mater. Des.* **1988**, *9*, 63–69. [CrossRef]
5. Li, H.; Li, D.; Zhu, Z.; Chen, B.; Chen, X.; Yang, C.; Zhang, H.; Kang, W. Grain refinement mechanism of as-cast aluminum by hafnium. *Trans. Nonferrous Met. Soc. China* **2016**, *26*, 3059–3069. [CrossRef]
6. Hall, E.O. The Deformation and Ageing of Mild Steel: III Discussion of Results. *Proc. Phys. Soc. Sect. B* **1951**, *64*, 747–753. [CrossRef]
7. Petch, N.J. The cleavage strength of polycrystal. *J. Iron Steel Inst.* **1953**, *174*, 25–28.
8. Zhang, Y.; Ma, N.; Yi, H.; Li, S.; Wang, H. Effect of Fe on grain refinement of commercial purity aluminum. *Mater. Des.* **2006**, *27*, 794–798. [CrossRef]
9. Mohd Syukry, Z.; Ahmad Badri, I. Effect of Aging Time to the Commercial Aluminum Alloy Modified with Zirconium Addition. *Mater. Sci. Forum* **2015**, *819*, 50–56. [CrossRef]
10. Zaid, A.I.O.; Mostafa, A.O. Effect of hafnium addition on wear resistance of zinc-aluminum 5 alloy: A three-dimensional presentation. *Adv. Mater. Lett.* **2017**, *8*, 910–915. [CrossRef]
11. Mostafa, A.O. Mechanical Properties and Wear Behavior of Aluminum Grain Refined by Ti and Ti+B. *Int. J. Surf. Eng. Interdiscip. Mater. Sci.* **2019**, *7*, 1–19. [CrossRef]
12. Gao, Z.; Li, H.; Lai, Y.; Ou, Y.; Li, D. Effects of minor Zr and Er on microstructure and mechanical properties of pure aluminum. *Mater. Sci. Eng. A* **2013**, *580*, 92–98. [CrossRef]
13. Toschi, S.; Balducci, E.; Ceschini, L.; Mørtsell, E.; Morri, A.; Di Sabatino, M. Effect of Zr Addition on Overaging and Tensile Behavior of 2618 Aluminum Alloy. *Metals* **2019**, *9*, 130. [CrossRef]
14. Wang, F.; Qiu, D.; Liu, Z.-L.; Taylor, J.A.; Easton, M.A.; Zhang, M.-X. The grain refinement mechanism of cast aluminium by zirconium. *Acta Mater.* **2013**, *61*, 5636–5645. [CrossRef]
15. Crossley, F.A.; Mondolfo, L.F. Mechanism of Grain Refinement in Aluminum Alloys. *JOM* **1951**, *3*, 1143–1148. [CrossRef]
16. Murray, J.; Peruzzi, A.; Abriata, J.P. The Al-Zr (aluminum-zirconium) system. *J. Phase Equilib.* **1992**, *13*, 277–291. [CrossRef]
17. Wang, T.; Jin, Z.; Zhao, J.-C. Thermodynamic Assessment of the Al-Zr Binary System. *J. Phase Equilib.* **2001**, *22*, 544–551. [CrossRef]
18. Wang, F.; Eskin, D.G.; Khvan, A.V.; Starodub, K.F.; Lim, J.J.H.; Burke, M.G.; Connolley, T.; Mi, J. On the occurrence of a eutectic-type structure in solidification of Al-Zr alloys. *Scr. Mater.* **2017**, *133*, 75–78. [CrossRef]
19. Lü, X.Y.; Guo, E.J.; Rometsch, P.; Wang, L.J. Effect of one-step and two-step homogenization treatments on distribution of Al3Zr dispersoids in commercial AA7150 aluminium alloy. *Trans. Nonferrous Met. Soc. China* **2012**, *22*, 2645–2651. [CrossRef]
20. Duan, Y.L.; Xu, G.F.; Peng, X.Y.; Deng, Y.; Li, Z.; Yin, Z.M. Effect of Sc and Zr additions on grain stability and superplasticity of the simple thermal-mechanical processed Al-Zn-Mg alloy sheet. *Mater. Sci. Eng. A* **2015**, *648*, 80–91. [CrossRef]

21. Seyed Ebrahimi, S.H.; Emamy, M.; Pourkia, N.; Lashgari, H.R. The microstructure, hardness and tensile properties of a new super high strength aluminum alloy with Zr addition. *Mater. Des.* **2010**, *31*, 4450–4456. [CrossRef]
22. Fang, H.C.; Chen, K.H.; Chen, X.; Huang, L.P.; Peng, G.S.; Huang, B.Y. Effect of Zr, Cr and Pr additions on microstructures and properties of ultra-high strength Al–Zn–Mg–Cu alloys. *Mater. Sci. Eng. A* **2011**, *528*, 7606–7615. [CrossRef]
23. ASTM International. B557-15. Standard Test Methods for Tension Testing Wrought and Cast Aluminum- and Magnesium-Alloy Products. In *Annual Book of ASTM Standards*; ASTM International: West Conshohocken, PA, USA, 2016.
24. ASTM International. E23-18. Standard Test Methods for Notched Bar Impact Testing of Metallic Materials. In *ASTM Book of Standards*; ASTM International: West Conshohocken, PA, USA, 2018; ISBN 435493654.
25. ASTM International ASTM E112-13. Standard Test Methods for Determining Average Grain Size. In *ASTM Book of Standards*; ASTM International: West Conshohocken, PA, USA, 2013.
26. Senčič, B.; Šolić, S.; Leskovšek, V. Fracture toughness–Charpy impact test–Rockwell hardness regression based model for 51CrV4 spring steel. *Mater. Sci. Technol.* **2014**, *30*, 1500–1505. [CrossRef]
27. Mahmudi, R.; Sepehrband, P.; Ghasemi, H.M. Improved properties of A319 aluminum casting alloy modified with Zr. *Mater. Lett.* **2006**, *60*, 2606–2610. [CrossRef]
28. Ryum, N. Precipitation and recrystallization in an A1-0.5 wt.% Zr-alloy. *Acta Metall.* **1969**, *17*, 269–278. [CrossRef]
29. Mahmoud, A.E.; Mahfouz, M.G.; Elrab, H.G.G.-; Doheim, M.A. Grain refinement of commercial pure aluminium by zirconium. *J. Eng. Sci.* **2014**, *42*, 1232–1241.
30. Rohrer, G.S. The role of grain boundary energy in grain boundary complexion transitions. *Curr. Opin. Solid State Mater. Sci.* **2016**, *20*, 231–239. [CrossRef]
31. Kaibyshev, O.A.; Faizova, S.N.; Hairullina, A.F. Diffusional mass transfer and superplastic deformation. *Acta Mater.* **2000**, *48*, 2093–2100. [CrossRef]
32. Bale, C.W.; Bélisle, E.; Chartrand, P.; Decterov, S.A.; Eriksson, G.; Gheribi, A.E.; Hack, K.; Jung, I.-H.; Kang, Y.-B.; Melançon, J.; et al. FactSage thermochemical software and databases, 2010–2016. *Calphad* **2016**, *54*, 35–53. [CrossRef]
33. Souza, P.H.L.; de Oliveira, C.A.S.; do Vale Quaresma, J.M. Precipitation hardening in dilute Al–Zr alloys. *J. Mater. Res. Technol.* **2018**, *7*, 66–72. [CrossRef]
34. Knipling, K.E.; Dunand, D.C.; Seidman, D.N. Nucleation and Precipitation Strengthening in Dilute Al-Ti and Al-Zr Alloys. *Metall. Mater. Trans. A* **2007**, *38*, 2552–2563. [CrossRef]
35. Silva, R.; Pinto, A.; Kuznetsov, A.; Bott, I. Precipitation and Grain Size Effects on the Tensile Strain-Hardening Exponents of an API X80 Steel Pipe after High-Frequency Hot-Induction Bending. *Metals* **2018**, *8*, 168. [CrossRef]
36. Pineau, A.; Benzerga, A.A.; Pardoen, T. Failure of metals I: Brittle and ductile fracture. *Acta Mater.* **2016**, *107*, 424–483. [CrossRef]
37. Miyahara, N.; Mutoh, Y.; Yamaishi, K.; Uematsu, K.; Inoue, M. The Effects of Grain Size on Strength, Fracture Toughness, and Static Fatigue Crack Growth in Alumina. In *Grain Boundary Controlled Properties of Fine Ceramics*; Springer: Dordrecht, The Netherlands, 1992; pp. 125–136.
38. Tarpani, J.R.; Spinelli, D. Grain size effects in the charpy impact energy of a thermally embrittled RPV steel. *J. Mater. Sci.* **2003**, *38*, 1493–1498. [CrossRef]
39. Nakai, M.; Itoh, G. The Effect of Microstructure on Mechanical Properties of Forged 6061 Aluminum Alloy. *Mater. Trans.* **2014**, *55*, 114–119. [CrossRef]

Article

High-Pressure Spectroscopy Study of $Zn(IO_3)_2$ Using Far-Infrared Synchrotron Radiation

Akun Liang [1], Robin Turnbull [1], Enrico Bandiello [1], Ibraheem Yousef [2], Catalin Popescu [2], Zoulikha Hebboul [3] and Daniel Errandonea [1,*]

[1] Departamento de Física Aplicada-ICMUV, Universitat de València, Dr. Moliner 50, 46100 Burjassot, Spain; akun2.liang@uv.es (A.L.); robin.turnbull@uv.es (R.T.); enrico.bandiello@uv.es (E.B.)
[2] CELLS-ALBA Synchrotron Light Facility, 08290 Barcelona, Spain; iyousef@cells.es (I.Y.); cpopescu@cells.es (C.P.)
[3] Laboratoire Physico-Chimie des Matériaux (LPCM), Université Amar Telidji de Laghouat, BP 37G, Laghouat 03000, Algeria; z.hebboul@lagh-univ.dz
* Correspondence: daniel.errandonea@uv.es

Abstract: We report the first high-pressure spectroscopy study on $Zn(IO_3)_2$ using synchrotron far-infrared radiation. Spectroscopy was conducted up to pressures of 17 GPa at room temperature. Twenty-five phonons were identified below 600 cm^{-1} for the initial monoclinic low-pressure polymorph of $Zn(IO_3)_2$. The pressure response of the modes with wavenumbers above 150 cm^{-1} has been characterized, with modes exhibiting non-linear responses and frequency discontinuities that have been proposed to be related to the existence of phase transitions. Analysis of the high-pressure spectra acquired on compression indicates that $Zn(IO_3)_2$ undergoes subtle phase transitions around 3 and 8 GPa, followed by a more drastic transition around 13 GPa.

Keywords: iodate; infrared spectroscopy; high pressure; phase transitions

Citation: Liang, A.; Turnbull, R.; Bandiello, E.; Yousef, I.; Popescu, C.; Hebboul, Z.; Errandonea, D. High-Pressure Spectroscopy Study of $Zn(IO_3)_2$ Using Far-Infrared Synchrotron Radiation. *Crystals* **2021**, *11*, 34. https://doi.org/10.3390/cryst11010034

Received: 17 December 2020
Accepted: 28 December 2020
Published: 30 December 2020

Publisher's Note: MDPI stays neutral with regard to jurisdictional claims in published maps and institutional affiliations.

1. Introduction

Metal iodates $[M(IO_3)_x]$ receive great attention because of their non-linear optical properties. In particular, they are promising materials for second-harmonic generation (SHG) [1–4]. They have been also proposed for applications as dielectric materials, in non-linear optics, and for desalination and water treatment [1–4]. Amongst the iodates, the quasi-two-dimensional zinc iodate, $Zn(IO_3)_2$, has recently been studied by several research groups [5–8]. The crystal structure, thermal stability, and other physical properties of this compound have been characterized at ambient pressure [9–11]. However, nothing is known of its structural and vibrational behavior under high-pressure (HP) conditions. At ambient conditions, $Zn(IO_3)_2$ crystallizes into a monoclinic structure, which is represented in Figure 1. The monoclinic structure consists of ZnO_6 octahedral units connected by IO_3 triangular pyramidal units. In these units, the pentavalent iodine atom forms three covalent bonds with oxygen atoms, leaving two 5s electrons free to act as lone electron pairs (LEP). The presence of the LEP has been shown to induce an interesting HP behavior in iodates related to $Zn(IO_3)_2$ like $Fe(IO_3)_3$ [12,13], with two isostructural phase transitions taking place at pressures below 10 GPa in this compound. These transitions have been detected from changes in the pressure dependences of phonon frequencies, which show a nonlinear behavior, as well as kinks and slope changes in the frequency-versus-pressure plot at the identified transition pressures [12,13]. Such unusual behavior of phonons has been proposed to be connected with gradual modifications of the iodine coordination, which is favored by the LEP of iodine and has been connected with the isostructural phase transitions discovered in $Fe(IO_3)_3$ [12,13]. Since the LEP of iodine is a typical feature of many $M(IO_3)_x$ iodates, it would be interesting to explore if low-pressure phase transitions are a typical feature of this family of compounds. Given the structural similarities between

Fe(IO$_3$)$_3$ and Zn(IO$_3$)$_2$ (both structures consist of MO$_6$ octahedral units connected by iodine atoms coordinated by three oxygen atoms with non-bonding LEP orbitals), the possibility of isostructural phase transitions at low pressures in the unexplored Zn(IO$_3$)$_2$ motivated the present work, in which we studied the pressure dependence of phonons by means of HP infrared (IR) spectroscopy. This is a technique useful for understanding the changes induced by compression in the physical–chemical properties of iodates. Additionally, the discovery of the existence of phase transition in Zn(IO$_3$)$_2$ could be very useful for the development of solid-state cooling technologies, which take advantage of the pressure-induced barocaloric effect [14].

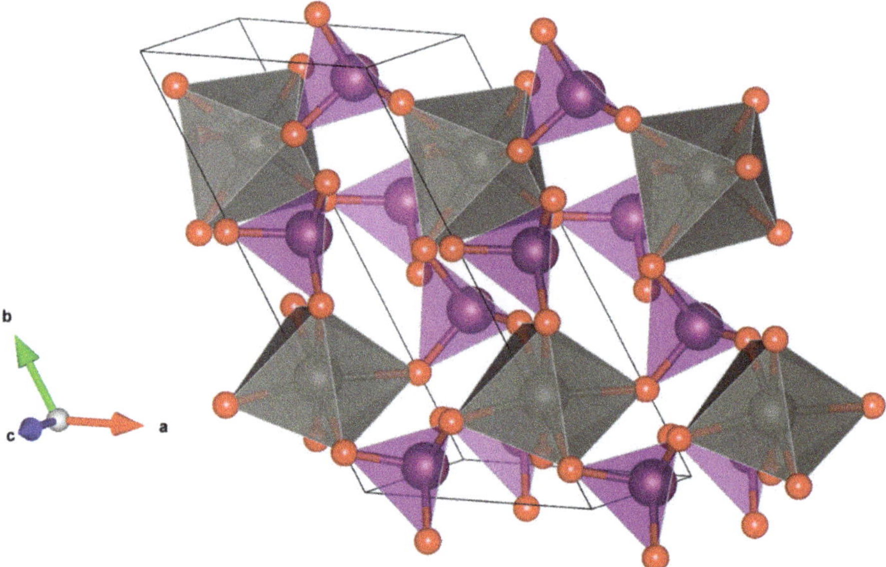

Figure 1. Crystal structure of Zn(IO$_3$)$_2$. ZnO$_6$ octahedral units are shown in grey and IO$_3$ trigonal pyramidal units are shown in purple. Small red circles are oxygen atoms. The monoclinic unit cell is shown with black solid lines.

Infrared spectroscopy has been shown to be an effective diagnostic for the detection of pressure-induced phase transitions in solids [15]. In the present work, IR spectroscopy provides information on the vibrational properties of Zn(IO$_3$)$_2$, which, until the present work, have remained relatively underexplored. In fact, less than one-third of the 51 phonons predicted by group theory analysis have been reported experimentally for Zn(IO$_3$)$_2$ [5–9,16]. Therefore, the present far-infrared spectroscopy study of Zn(IO$_3$)$_2$ under high-pressure conditions makes a timely and valuable contribution. Zn(IO$_3$)$_2$ has not previously been studied under high-pressure conditions. The information obtained from this study is relevant not only to improve the knowledge on the aforementioned issues but also to provide an accurate determination of the phonon frequencies, which are needed to properly model other physical properties such as heat capacity and thermal expansion. The use of high-brilliance infrared synchrotron radiation facilitated the measurement of tiny samples loaded in diamond-anvil cells (DACs) with a very good signal-to-noise ratio. Using this methodology, we have been able to identify 25 phonons between 98 and 600 cm^{-1}. We then follow the pressure-induced evolution of these phonons up to 17 GPa. The analysis of the results leads us to propose the existence of three phase transitions, as will be discussed in the manuscript.

2. Materials and Methods

Experiments were performed on $Zn(IO_3)_2$ powders synthesized from aqueous solution according to the synthesis method and sample characterization found in Ref. [6]. The crystal structure was confirmed by powder XRD measurements (X'Pert Pro diffractometer, Panalytical, Almelo, The Netherlands) using Cu $K_{\alpha1}$ radiation, which corroborated the crystal structure reported by Liang et al. [10] (space group $P2_1$) with unit-cell parameters: $a = 5.465(4)$, $b = 10.952(8)$, $c = 5.129(4)$ Å, and $\gamma = 120.37(8)°$. HP Fourier transform infrared (FTIR) measurements (Vertex 70 spectrometer, Bruker Optik GmbH, Ettlingen, Germany) were conducted at MIRAS beamline of the ALBA synchrotron. $Zn(IO_3)_2$ samples were loaded in a DACs designed for IR spectroscopy, using IIAC-diamonds with culets of 300 µm. Stainless-steel gaskets were pre-indented to a thickness of 40 µm and drilled with a hole in the center of 150 µm in diameter. Cesium iodide (CsI) was used as the pressure-transmitting medium (PTM) [17]. The CsI PTM is not quasi-hydrostatic beyond 3 GPa; however, radial pressure gradients were smaller than 1 GPa in the pressure range of this study [18]. CsI was chosen because it has the widest IR transmission window amongst the possible PTMs [18]. Pressure was determined using the ruby scale [19]. Synchrotron-based FTIR-micro-spectroscopy experiments were performed in the transmission mode of operation. We used a masking aperture size of 50×50 µm^2 and a beam current inside the synchrotron ring of 250 mA. The measurements were performed by employing a 3000 Hyperion microscope coupled to a Vertex 70 spectrometer (Bruker Optik GmbH, Ettlingen, Germany). The microscope was equipped with a helium-cooled bolometer detector optimized for operation in the range covering the far-infrared spectral region. A Mylar beam splitter was used in the spectrometer. Spectra were measured using a $15\times$ Schwarzschild magnification objective (NA = 0.52) coupled to a $15\times$ Schwarzschild magnification condenser. Measurements at selected pressures were collected using the OPUS 8.2 software (Bruker Optik GmbH, Ettlingen, Germany) in the 90–600 cm^{-1} range with a spectral resolution of 4 cm^{-1} and 256 co-added scans per spectrum. The analysis of FTIR results was carried using the Multiple Peak Fit Tool of the OriginPro software (OriginLab Corporation, Northampton, MA, United States) and employing Gaussian functions to model the peaks.

3. Results and Discussion

The results at the lowest measured pressure (0.9 GPa) are shown in Figure 2 together with the multiple-peak fit used to identify phonons. We have identified 25 modes, as can be seen in the figure, thereby greatly extending the number of modes previously observed in zinc iodate. The frequencies of all 25 of these modes are summarized in Table 1 and compared with those reported in the literature. $Zn(IO_3)_2$ exhibits 51 modes (26A + 25B) according to group theory. All of these 51 modes are both Raman-active and IR-active. They can be described as internal vibrations of the of the IO_3 units and external vibrations (commonly known as lattice modes) involving the relative movements of IO_3 (behaving as rigid units) and Zn, which are observed in the low-frequency region [7,20]. Symmetric (v_1) and asymmetric (v_3) stretching vibrations of the pyramidal IO_3 units are in the 780−630 cm^{-1} and 820−730 cm^{-1} wavenumber regions, respectively [7,21,22]. The symmetric (v_2) and asymmetric bending vibrations (v_4) are in the 400−320 cm^{-1} wavenumber region and around 450−400 cm^{-1}, respectively [7,21,22]. In our experiments, we detected four v_4 modes. The frequencies of the modes we found at 425, 440, and 452 cm^{-1} are less than 9 cm^{-1} larger than those measured in previous ambient-pressure Raman and IR experiments [6,7,16]. This is consistent with the fact that our experiment was performed at a pressure of 0.9 GPa. The expected phonon hardening [23], or increase in the phonon frequency, is due to the shortening bond distance caused by increasing pressure in our experiment. The mode reported at 405 cm^{-1} in Ref. [5] was not detected in previous works from the literature, but in our case could be probably assigned to a shoulder of the peak with wavenumber 388 cm^{-1}, which can be identified at 402 cm^{-1}. Additionally, the mode previously reported at 524 cm^{-1} in Ref. [8] is not consistent either with the present or previous experiments [5–7,16]. Indeed, there are no phonons in $Zn(IO_3)_2$, or in any divalent metal iodate [7], in the $600-500\ cm^{-1}$ range. The origin of such a phonon could be related with twinning of the crystal structure when large single crystals are grown [9], being a possible overtone, since $524\ cm^{-1}$ is approximately double of the frequency of the lattice modes.

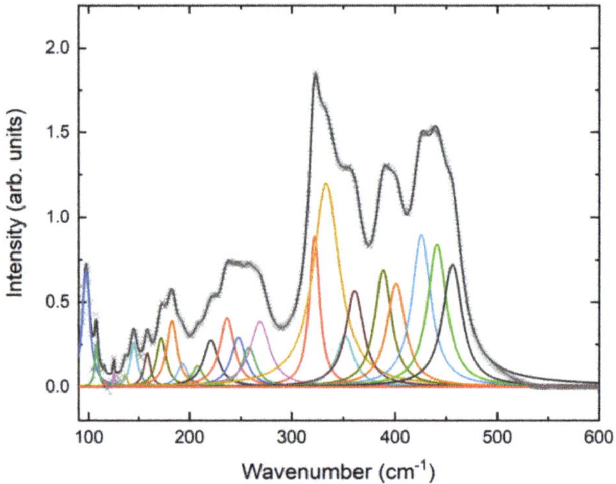

Figure 2. Far-IR spectrum of $Zn(IO_3)_2$ measured at 0.9 GPa and room temperature. The Gaussians used for the multiple peak fit are also shown in different colors. Each of them corresponds to the phonons summarized in Table 1.

Table 1. A comparison of the phonon frequencies determined by IR- and Raman-spectroscopy from this work (at 0.9 GPa) and from previous studies (ambient pressure). For the present study, frequencies are given with errors. The horizontal dashed line indicates the cut-off of the IR set-up, below which it was not possible to detect signal due to experimental constraints.

Assignment	$\omega\ (cm^{-1})$ This Work IR	$\omega\ (cm^{-1})$ [5] IR	$\omega\ (cm^{-1})$ [6] IR	$\omega\ (cm^{-1})$ [7] IR	$\omega\ (cm^{-1})$ [7] Raman	$\omega\ (cm^{-1})$ [8] IR>	$\omega\ (cm^{-1})$ [16] Raman
							61
							67
				73			
					80		80
	98(2)						
	107(2)			101			100
	116(2)				113		111
	125(2)						
	135(2)				132		139
Lattice	145(2)			141			148
modes	158(2)				152		155
	172(2)				173		173
	183(2)			180			
	193(2)				189		187
	208(2)						
	220(2)						
	236(2)						
	247(2)						
	258(2)			255			
	269(2)				267		265

Table 1. *Cont.*

Assignment	ω (cm^{-1}) This Work IR	ω (cm^{-1}) [5] IR	ω (cm^{-1}) [6] IR	ω (cm^{-1}) [7] IR	ω (cm^{-1}) [7] Raman	ω (cm^{-1}) [8] IR>	ω (cm^{-1}) [16] Raman
	322(2)			327	327		327
	336(2)						
	348(2)						
ν_2	353(2)			354	354		351
		366					
	388(2)				391		
	402(2)	405					
	425(2)		418	418	424		422
ν_4	440(2)						432
	452(2)		444				
						524	

Regarding the other bending vibrations, five ν_2 modes were detected in our experiments. The frequencies of three of them agree well with previous studies [6,7,16], with ours observed at 3–5 cm^{-1} higher frequencies due to the higher pressure (0.9 GPa) in the sample. The brilliance and resolution of the experimental set-up enabled the identification of two modes that were previously undetected. Additionally, the mode reported at 366 cm^{-1} in Ref. [5] was not observed in our measurements, and it was not reported in the rest of the works in the literature [6–8,16]. We note that the frequency of this mode is approximately double the frequency of the 183 cm^{-1} mode detected in the present work and elsewhere in the literature [6–8,16] and that it is therefore likely to be an overtone.

Regarding the lattice modes, sixteen in total were observed. Ten of these modes have been detected in previous studies [7,16], thereby providing good agreement with the results of the present work (especially considering the frequency increase due to the 0.9 GPa experimental sample pressure). In the literature, four modes have been detected by Raman and IR spectroscopy [7,8,16], which could be detected in our experiment because it is below the cut-off frequency of our setup. Lattice modes are mainly related to translation and libration movements of IO_3, and the coupling of these movements with movements of Zn atoms. In particular, the 145 cm^{-1} phonon has been assigned to a O–Zn–O deformation of the ZnO_6 octahedron [8]. This is consistent with the fact that a mode involving the same deformation of ZnO_6 has the same frequency in $ZnWO_4$ [24] and $ZnMoO_4$ [25]. Additionally, the modes below 120 cm^{-1} are likely to be due to pure translational or librational movements of IO_3, because of the large mass of the iodine atom. This hypothesis is consistent with the observation that these modes have nearly the same frequency in $Mn(IO_3)_2$, $Ni(IO_3)_2$, $Co(IO_3)_2$, and $Zn(IO_3)_2$ [8] as well as in $Fe(IO_3)_3$ [10].

We will now comment on the data acquired at higher pressures. For the sake of accuracy, we will concentrate on wavenumbers higher than 150 cm^{-1} because the signals at lower energy became noisy with increasing pressure. Figure 3 displays IR spectra measured at different pressures (indicated in the figure). It is clear in Figure 3 that there are qualitative changes in the spectra at 3.6 GPa, in particular for wavelengths smaller than 300 cm^{-1}. In particular, three modes (marked by asterisks) increase in intensity. The transition pressure is close to the loss hydrostaticity of the pressure medium (3 GPa) [18]. The influence of non-hydrostaticity in phase transitions induced by pressure in $Zn(IO_3)_2$ is beyond the scope of this study. Additional changes occur in the IR spectra at 8.8 GPa. In particular, several modes broaden, and three low-frequency modes (also marked with asterisks) become enhanced. Finally, there is a considerable broadening of phonon bands at 13 GPa and higher pressures. Similar changes have been assigned to phase transitions in the case of $Fe(IO_3)_3$ [12,13]. Thus, phase transitions could be the cause of the changes observed in the IR spectra. Further evidence supporting the existence of phase transitions in $Zn(IO_3)_2$ is obtained via the analysis of the pressure dependence of the mode frequencies described immediately below.

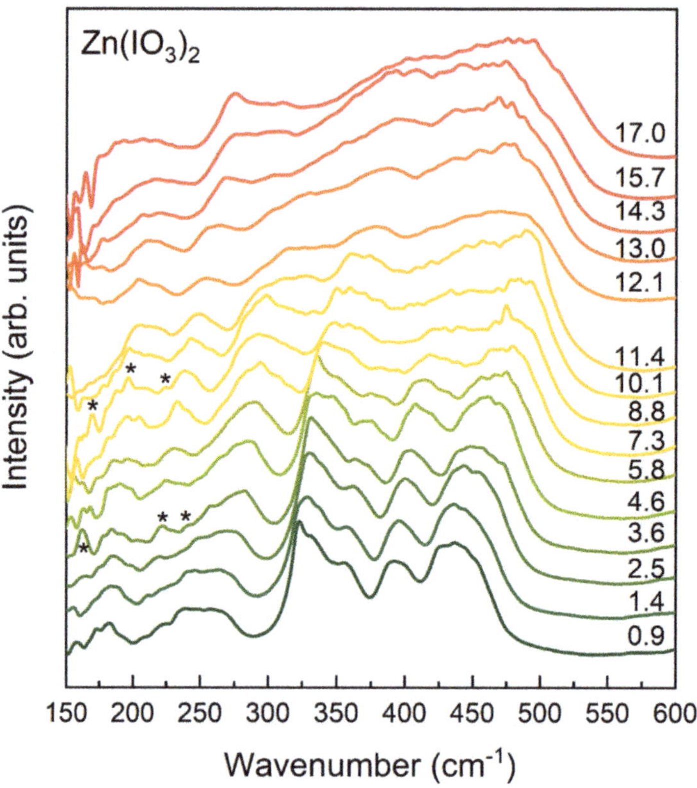

Figure 3. IR spectra acquired at increasing pressures (indicated in GPa in the figure). The asterisks indicate changes in the phonons described in the text.

From the analysis of the HP experiments, we determined the pressure dependence of eight internal bending modes and ten lattice modes. The results are shown in Figures 4 and 5. In Figure 4, it can be clearly seen that there are changes, at 3.6 and 8.8 GPa, in the pressure dependence of several modes; in particular, the bending modes with frequencies of 360 and 440 cm^{-1} among others. There are also clear changes in the modes with frequency 353 and 425 cm^{-1} at 8.8 GPa. On top of this, at 13 GPa, there are noticeable changes in the eight bending modes shown in Figure 4. In particular, there are discontinuities in at least two of the frequencies, indicating the occurrence of more important structural changes in the third phase transition than in the other two transitions.

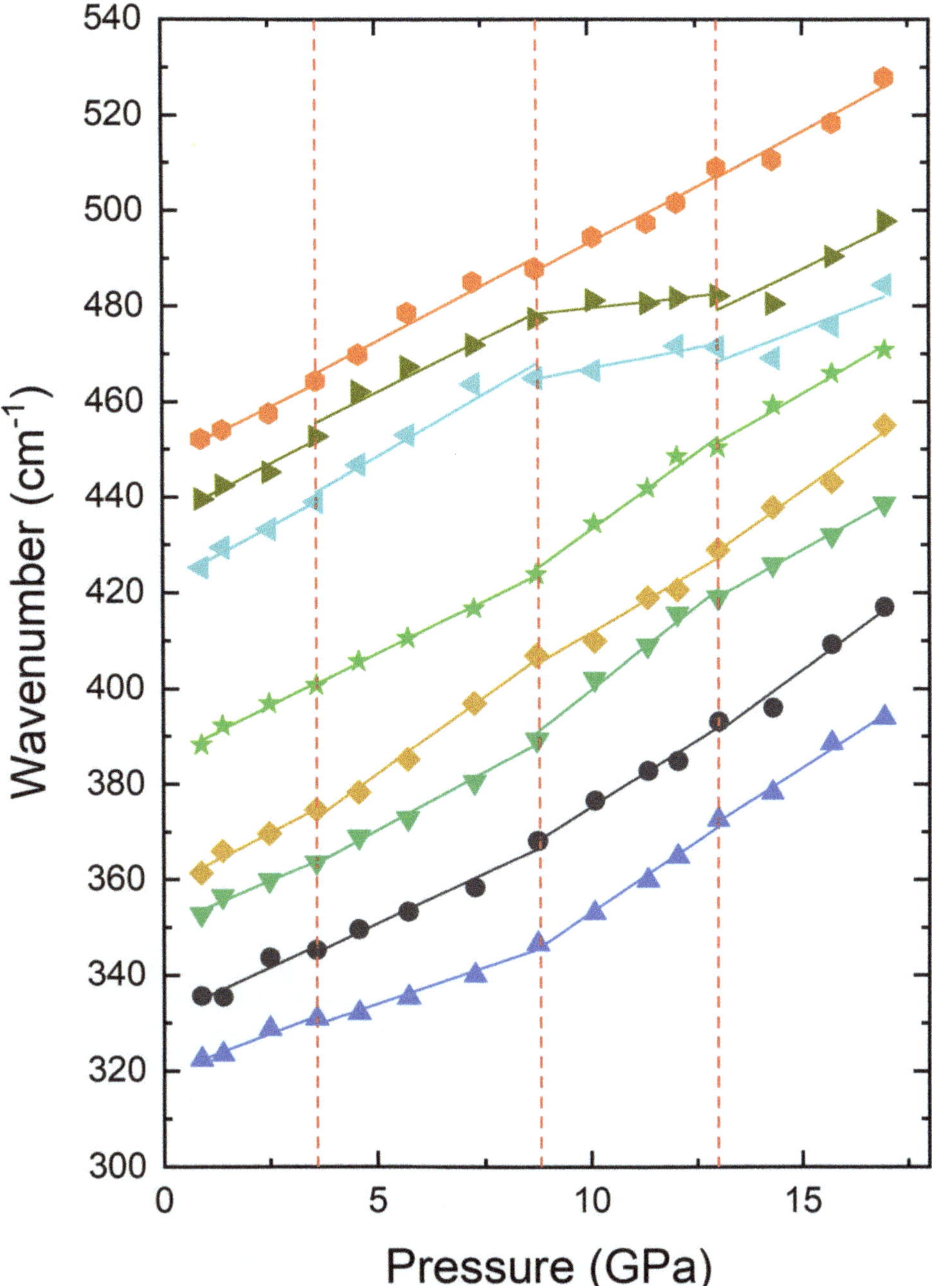

Figure 4. Pressure dependence of IR modes obtained from experiments. Only IO_3 internal bending modes are shown. Vertical lines indicate suggested transition pressures.

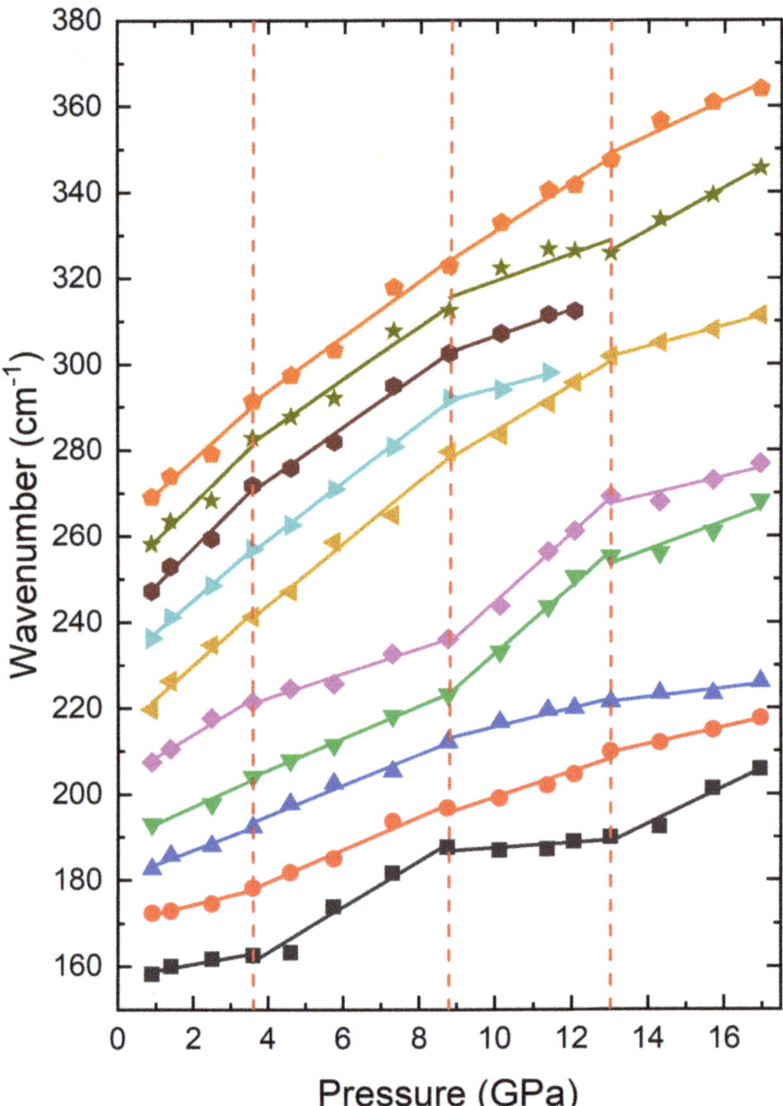

Figure 5. Pressure dependence of IR modes obtained from experiments. Only lattice modes are shown. Vertical lines indicate suggested transition pressures.

We will comment now on the pressure dependence of the ten lattice modes that were observed under sample compression. For these modes, the changes in the frequency pressure dependence, at phase transitions, are more evident than in the bending modes, as can be seen in Figure 5. In particular, slope changes can be seen at the transition pressures (3.6, 8.8, and 13 GPa), which are represented by vertical lines in the figure. Such changes are very noticeable in the lowest-frequency mode shown in Figure 5. They are also quite evident for the mode at 208 cm^{-1} at 0.9 GPa. As happened in the bending modes, the changes at 13 GPa are more noticeable than changes at the other transitions. In particular, there are discontinuities in the frequencies of at least two modes. The changes in the frequency pressure dependences at the phase transitions detected at 3.6 and 8.8 GPa, and the strongly non-linear pressure dependences, are qualitatively similar to those observed for phonons in $Fe(IO_3)_3$ at 2 and 6 GPa [12,13]. In $Fe(IO_3)_3$, these changes have been related to isostructural phase transitions occurring in Ref. [12,13], which are themselves related to gradual pressure-induced changes in the coordination sphere of iodine, which are affected by the presence of lone electron pairs. The similarities between $Zn(IO_3)_2$ and $Fe(IO_3)_3$ therefore provide further support to the observation of phase transitions in the $Zn(IO_3)_2$ investigated here. Future studies using other characterization techniques, including X-ray diffraction and Raman spectroscopy should be performed to confirm the present interpretation of results and to determine if the transitions at 3.6 and 8.8 GPa are isostructural or not. We hope our work will trigger such studies as well as computer simulation studies.

Before concluding, we would like to add a comment on the transition at 13 GPa. The changes observed around 13 GPa are more apparent than those observed around 3.6 and 8.8 GPa. This suggests that the changes around 13 GPa can be linked to the occurrence of a first-order structural transition. The fact that the changes are accompanied by a marked broadening of phonon bands suggests a disorder of the crystal structure [26], which could be related to the presence of non-hydrostatic stresses at 13 GPa and higher pressures [27].

To conclude, we note that all observed modes harden under compression. In the pressure range up to 3.6 GPa (before the first phase transition), the pressure coefficients are all within the range of 1–9 cm^{-1}/GPa. This can be seen in Table 2, where we represent phonons frequencies (ω) and pressure coefficients ($d\omega/dP$) for different modes in the different phases. We have named the phases as follows: phase I (low-pressure phase), II, III, and IV (successive HP phases). Table 2 shows clearly the change in the pressure dependences at 3.6 GPa and 8.8 GPa as well as the changes in frequencies and pressure dependencies at 13 GPa, supporting the existence of the proposed transitions. The pressure coefficients in the low-pressure phase are comparable to pressure coefficients in $Fe(IO_3)_3$ [13], $LiIO_3$ [28], and KIO_3 [29] in the same frequency region, indicating that $Zn(IO_3)_2$ is extremely compressible, similar to these other iodates. Additionally, the pressure coefficients for the different HP phases are of the same order of magnitude as in the low-pressure phase, suggesting that the changes in the coordination polyhedra are not drastic and that the transition is probably related to gradual changes in cation coordination number [30].

Table 2. Phonon frequencies (ω) of the modes observed under compression for phases I (at 0.9 GPa), II (at 3.6 GPa), III (at 8.8 GPa, and IV (at 13 GPa). The pressure coefficients ($d\omega/dP$) obtained from linear fits are also given.

ω (cm^{-1}) Phase I 0.9 GP	$d\omega/dP$ (cm^{-1}/GPa)	ω (cm^{-1}) Phase II 3.6 GPa	$d\omega/dP$ (cm^{-1}/GPa)	ω (cm^{-1}) Phase III 8.8 GPa	$d\omega/dP$ (cm^{-1}/GPa)	ω (cm^{-1}) Phase IV 13 GPa	$d\omega/dP$ (cm^{-1}/GPa)
158(2)	1.5(1)	163(2)	5.3(1)	188(2)	0.6(1)	190(2)	4.2(1)
172(2)	2.1(1)	178(2)	3.8(1)	197(2)	3.0(1)	210(2)	2.0(1)
183(2)	3.3(1)	192(2)	3.6(1)	212(2)	2.2(1)	222(2)	1.1(1)
193(2)	4.0(1)	204(2)	3.7(1)	223(2)	7.8(1)	255(2)	3.3(1)
207(2)	5.3(1)	221(2)	2.9(1)	236(2)	8.0(1)	269(2)	2.1(1)
220(2)	7.8(1)	241(2)	7.2(1)	280(2)	5.3(1)	302(2)	2.4(1)
236(2)	7.5(1)	257(2)	6.8(1)	292(2)	2.3(1)	–	–
247(2)	8.7(1)	272(2)	6.2(1)	302(2)	3.1(1)	–	–
258(2)	8.6(1)	283(2)	6.1(1)	312(2)	3.2(1)	326(2)	4.9(1)
269(2)	7.9(1)	291(2)	6.4(1)	323(2)	5.7(1)	348(2)	4.1(1)
322(2)	4.1(1)	331(2)	4.2(1)	347(2)	5.6(1)	373(2)	6.4(1)
336(2)	3.4(1)	345(2)	3.1(1)	368(2)	6.0(1)	393(2)	5.6(1)
353(2)	3.8(1)	364(2)	4.9(1)	389(2)	7.1(1)	419(2)	4.9(1)
361(2)	4.6(1)	375(2)	6.4(1)	407(2)	5.2(1)	429(2)	6.3(1)
388(2)	4.8(1)	401(2)	5.2(1)	424(2)	1.8(1)	450(2)	3.4(1)
425(2)	4.6(1)	439(2)	4.4(1)	465(2)	0.5(1)	471(2)	4.3(1)
440(2)	4.4(1)	453(2)	4.7(1)	477(2)	0.7(1)	482(2)	4.8(1)
452(2)	4.5(1)	464(2)	4.4(1)	488(2)	6.5(1)	509(2)	5.1(1)

4. Conclusions

Synchrotron far-infrared spectroscopy measurements have allowed us to determine that $Zn(IO_3)_2$ undergoes three phase transitions at 3.6, 8.8, and 13 GPa. The phase transitions are identified from changes in the infrared spectra. The first two transitions resemble those previously observed in $Fe(IO_3)_3$ at similar pressures and are probably isostructural transitions favored by the presence of lone electron pairs in $Zn(IO_3)_2$. The third phase transition appears to be a first-order transition that is connected to the occurrence of more important structural changes. Assignment of phonon modes has been discussed and their pressure dependence reported. We found that the lattice modes are more sensitive than the bending modes of IO_3 to pressure-induced structural changes.

Author Contributions: Conceptualization, D.E.; IR experiments, A.L., R.T., E.B., I.Y., and C.P.; formal analysis, A.L. and D.E.; sample preparation, Z.H.; writing, review, and editing, all the authors. All authors have read and agreed to the published version of the manuscript.

Funding: This work was supported by the Spanish Ministry of Science, Innovation, and Universities under grants PID2019-106383GB-C41 and RED2018-102612-T (MALTA Consolider-Team Network) and by Generalitat Valenciana under grant Prometeo/2018/123 (EFIMAT). R.T. acknowledges funding from the Spanish MINECO via the Juan de la Cierva Formación program (FJC2018-036185-I). A.L. and D.E. would like to thank the Generalitat Valenciana for the Ph.D. fellowship GRISO-LIAP/2019/025.

Institutional Review Board Statement: Not applicable.

Informed Consent Statement: Not applicable.

Data Availability Statement: Data is contained within the article.

Acknowledgments: The authors thank ALBA synchrotron for providing beamtime at MIRAS beamline for performing synchrotron far-infrared experiments.

Conflicts of Interest: The authors declare no conflict of interest.

References

1. Bonacina, L.; Mugnier, Y.; Courvoisier, F.; Le Dantec, R.; Extermann, J.; Lambert, Y.; Boutou, V.; Galez, C.; Wolf, J.P. Polar $Fe(IO_3)_3$ nanocrystals as local probes for nonlinear microscopy. *Appl. Phys. B Lasers Opt.* **2007**, *87*, 399–403. [CrossRef]

2. Hebboul, Z.; Galez, C.; Benbertal, D.; Beauquis, S.; Mugnier, Y.; Benmakhlouf, A.; Bouchenafa, M.; Errandonea, D. Synthesis, characterization, and crystal structure determination of a new lithium zinc iodate polymorph $LiZn(IO_3)_3$. *Crystals* **2019**, *9*, 464. [CrossRef]

3. Jia, Y.J.; Chen, Y.G.; Guo, Y.; Guan, X.F.; Li, C.; Li, B.; Liu, M.M.; Zhang, X.M. $LiM_{II}(IO_3)_3$ (M_{II} = Zn and Cd): Two Promising Nonlinear Optical Crystals Derived from a Tunable Structure Model of α-$LiIO_3$. *Angew. Chemie Int. Ed.* **2019**, *58*, 17194–17198. [CrossRef] [PubMed]

4. Hebboul, Z.; Ghozlane, A.; Turnbull, R.; Benghia, A.; Allaoui, S. Simple new method for the preparation of $La(IO_3)_3$ nanoparticles. *Nanomaterials* **2020**, *10*, 2400. [CrossRef] [PubMed]

5. Patil, A.B. FTIR study of second group iodate crystals grown by gel method. *Int. J. Grid Nad. Distrib. Comput.* **2020**, *13*, 227–235.

6. Benghia, A.; Hebboul, Z.; Chikhaoui, R.; Khaldoun Lefkaier, I.; Chouireb, A.; Goumri-Said, S. Effect of iodic acid concentration in preparation of zinc iodate: Experimental characterization of $Zn(IO_3)_2$, and its physical properties from density functional theory. *Vacuum* **2020**, *181*, 109660. [CrossRef]

7. Kochuthresia, T.C.; Gautier-Luneau, I.; Vaidyan, V.K.; Bushiri, M.J. Raman and Ftir Spectral Investigations of Twinned $M(IO_3)_2$ (M = Mn, Ni, Co, AND Zn) Crystals. *J. Appl. Spectrosc.* **2016**, *82*, 941–946. [CrossRef]

8. Shanmuga Sundar, G.J.; Kumar, S.M.R.; Packiya raj, M.; Selvakumar, S. Synthesis, growth, optical, mechanical and dielectric studies on NLO active monometallic zinc iodate [$Zn(IO_3)_2$] crystal for frequency conversion. *Mater. Res. Bull.* **2019**, *112*, 22–27. [CrossRef]

9. Phanon, D.; Bentria, B.; Jeanneau, E.; Benbertal, D.; Mosset, A.; Gautier-Luneau, I. Crystal structure of $M(IO_3)_2$ metal iodates, twinned by pseudo-merohedry, with MII: MgII, MnII, CoII, NiII and ZnII. *Z. Krist.* **2006**, *221*, 635–642.

10. Liang, J.K.; Wang, C.G. The structure of $Zn(IO_3)_2$ Crystal. *Acta Chim. Sin.* **1982**, *40*, 985–993.

11. Mougel, F.; Kahn-Harari, A.; Aka, G.; Pelenc, D. Structural and thermal stability of Czochralski grown GdCOB oxoborate single crystals. *J. Mater. Chem.* **1998**, *8*, 1619–1623. [CrossRef]

12. Liang, A.; Rahman, S.; Saqib, H.; Rodriguez-Hernandez, P.; Munoz, A.; Nenert, G.; Yousef, I.; Popescu, C.; Errandonea, D. First-Order Isostructural Phase Transition Induced by High-Pressure in $Fe(IO_3)_3$. *J. Phys. Chem. C* **2020**, *124*, 8669–8679. [CrossRef]

13. Liang, A.; Rahman, S.; Rodriguez-Hernandez, P.; Muñoz, A.; Manjón, F.J.; Nenert, G.; Errandonea, D. High-pressure Raman study of Fe(IO$_3$)$_3$: Soft-mode behavior driven by coordination changes of iodine atoms. *J. Phys. Chem. C* **2020**, *124*, 21329–21337. [CrossRef]

14. Sagotra, A.K.; Errandonea, D.; Cazorla, C. Mechanocaloric effects in superionic thin films from atomistic simulations. *Nat. Commun.* **2017**, *8*, 963. [CrossRef] [PubMed]

15. Ross, N.L.; Detrie, T.A.; Liu, Z. High-pressure raman and infrared spectroscopic study of prehnite. *Minerals* **2020**, *10*, 312. [CrossRef]

16. Peter, S.; Pracht, G.; Lange, N.; Lutz, H.D. Zinkiodate ± Schwingungsspektren (IR, Raman) und Kristallstruktur von Zn(IO$_3$)$_2$·2H$_2$O Zinc Iodates ± Infrared and Raman Spectra, Crystal Structure. *Z. Anorg. Allg. Chem.* **2000**, *626*, 208–215. [CrossRef]

17. Asaumi, K.; Kondo, Y. Effect of very high pressure on the optical absorption spectra in CsI. *Solid State Commun.* **1981**, *40*, 715–718. [CrossRef]

18. Celeste, A.; Borondics, F.; Capitani, F. Hydrostaticity of pressure-transmitting media for high pressure infrared spectroscopy. *High Press. Res.* **2019**, *39*, 608–618. [CrossRef]

19. Mao, H.K.; Xu, J.; Bell, P.M. Calibration of the ruby pressure gauge to 800 kbar under quasi-hydrostatic conditions. *J. Geophys. Res.* **1986**, *91*, 4673–4676. [CrossRef]

20. Bushiri, M.J.; Kochuthresia, T.C.; Vaidyan, V.K.; Gautier-Luneau, I. Raman scattering structural studies of nonlinear optical M(IO$_3$)$_3$ (M = Fe, Ga, and In) and linear optical β-In(IO$_3$)$_3$. *J. Nonlinear Opt. Phys. Mater.* **2014**, *23*, 1450039. [CrossRef]

21. Crettez, J.M.; Gard, R.; Remoissenet, M. Near and far infrared investigations from α and β lithium iodate crystals. *Solid State Commun.* **1972**, *11*, 951–954. [CrossRef]

22. Pimenta, M.A.; Oliveira, M.A.S.; Bourson, P.; Crettez, J.M. Raman study of Li$_{1-x}$H$_x$IO$_3$ crystals. *J. Phys. Condens. Matter* **1997**, *9*, 7903–7912. [CrossRef]

23. Errandonea, D.; Muñoz, A.; Rodríguez-Hernández, P.; Gomis, O.; Achary, S.N.; Popescu, C.; Patwe, S.J.; Tyagi, A.K. High-Pressure Crystal Structure, Lattice Vibrations, and Band Structure of BiSbO. *Inorg. Chem.* **2016**, *55*, 4958–4969. [CrossRef] [PubMed]

24. Errandonea, D.; Manjón, F.J.; Garro, N.; Rodríguez-Hernández, P.; Radescu, S.; Mujica, A.; Muñoz, A.; Tu, C.Y. Combined Raman scattering and ab initio investigation of pressure-induced structural phase transitions in the scintillator ZnWO$_4$. *Phys. Rev. B Condens. Matter Mater. Phys.* **2008**, *78*, 054116. [CrossRef]

25. Cavalcante, L.S.; Moraes, E.; Almeida, M.A.P.; Dalmaschio, C.J.; Batista, N.C.; Varela, J.A.; Longo, E.; Siu Li, M.; Andrés, J.; Beltrán, A. A combined theoretical and experimental study of electronic structure and optical properties of β-ZnMoO$_4$ microcrystals. *Polyhedron* **2013**, *54*, 13–25. [CrossRef]

26. Tschauner, O.; Errandonea, D.; Serghiou, G. Possible superlattice formation in high-temperature treated carbonaceous MgB2 at elevated pressure. *Physica B* **2006**, *371*, 88–94. [CrossRef]

27. Errandonea, D.; Meng, Y.; Somayazulu, M.; Häusermann, D. Pressure-induced → ω transition in titanium metal: A systematic study of the effects of uniaxial stress. *Physica B* **2005**, *355*, 116–125. [CrossRef]

28. Mendes Fílho, J.; Lemos, V.; Cerdeira, F.; Katiyar, R.S. Raman and x-ray studies of a high-pressure phase transition in β-LiIO$_3$ and the study of anharmonic effects. *Phys. Rev. B* **1984**, *30*, 7212–7218. [CrossRef]

29. Shen, Z.X.; Wang, X.B.; Tang, S.H.; Li, H.P.; Zhou, F. High pressure raman study and phase transitions of KIO$_3$ non-linear optical single crystals. *Rev. High Press. Sci. Technol. No Kagaku To Gijutsu* **1998**, *7*, 751–753. [CrossRef]

30. Sans, J.A.; Vilaplana, R.; Lora da Silva, E.; Popescu, C.; Cuenca-Gotor, V.P.; Andrada-Chacoón, A.; Munñoz, A.; Sánchez-Benitez, J.; Gomis, O.; Pereira, A.L.J.; et al. Characterization and decomposition of the natural van der Waals SnSb$_2$Te$_4$ under compression. *Inorg. Chem.* **2020**, *59*, 9900–9918. [CrossRef]

Article

High-Pressure Structural Behavior and Equation of State of Kagome Staircase Compound, Ni$_3$V$_2$O$_8$

Daniel Diaz-Anichtchenko [1], **Robin Turnbull** [1], **Enrico Bandiello** [1], **Simone Anzellini** [2] and **Daniel Errandonea** [1],*

[1] Departamento de Física Aplicada-ICMUV, Universitat de València, Dr. Moliner 50, 46100 Burjassot, Spain; daniel.diaz@uv.es (D.D.-A.); robin.turnbull@uv.es (R.T.); enrico.bandiello@uv.es (E.B.)
[2] Diamond Light Source Ltd., Diamond House, Harwell Science and Innovation Campus, Didcot, Oxfordshire OX11 0DE, UK; simone.anzellini@diamond.ac.uk
* Correspondence: daniel.errandonea@uv.es

Received: 17 September 2020; Accepted: 5 October 2020; Published: 8 October 2020

Abstract: We report on high-pressure synchrotron X-ray diffraction measurements on Ni$_3$V$_2$O$_8$ at room-temperature up to 23 GPa. According to this study, the ambient-pressure orthorhombic structure remains stable up to the highest pressure reached in the experiments. We have also obtained the pressure dependence of the unit-cell parameters, which reveals an anisotropic compression behavior. In addition, a room-temperature pressure–volume third-order Birch–Murnaghan equation of state has been obtained with parameters: V_0 = 555.7(2) Å^3, K_0 = 139(3) GPa, and K_0' = 4.4(3). According to this result, Ni$_3$V$_2$O$_8$ is the least compressible kagome-type vanadate. The changes of the crystal structure under compression have been related to the presence of a chain of edge-sharing NiO$_6$ octahedral units forming kagome staircases interconnected by VO$_4$ rigid tetrahedral units. The reported results are discussed in comparison with high-pressure X-ray diffraction results from isostructural Zn$_3$V$_2$O$_8$ and density-functional theory calculations on several isostructural vanadates.

Keywords: vanadate; kagome compound; high pressure; X-ray diffraction; equation of state

---

1. Introduction

Metal orthovanadates with formula M$_3$V$_2$O$_8$, where M is a divalent metal, have been the focus of research in recent years, mainly because of their optical, dielectric, and magnetic properties [1–4]. These properties make the M$_3$V$_2$O$_8$ family of compounds useful in several technological applications, from photocatalytic water splitting [3] to light-emitting diodes [5] and ion batteries [6]. In addition, compounds such as Co$_3$V$_2$O$_8$ and Ni$_3$V$_2$O$_8$ have very interesting magnetic and ferroelastic properties [2], which are intimately related to the presence of kagome staircase two-dimensional (2D) magnetic layers [2].

M$_3$V$_2$O$_8$ compounds are also intriguing from a crystallographic perspective [7,8]. They share a particular type of crystal structure, which is shown schematically in Figure 1. The structure is orthorhombic (space group *Cmca*, No. 64) with eight formula units per unit cell. In the crystal structure, the M atoms are octahedrally coordinated by oxygen atoms (see Figure 1a). In particular, the MO$_6$ octahedral units exhibit an edge-sharing pattern, forming a quasi-planar kagome staircase (see Figure 1b), which, in the case of Ni$_3$V$_2$O$_8$, results in inequivalent super-exchange interactions within the magnetic lattice [9]. Consequently, an anisotropic magnetic coupling develops in the stair-like kagome layers, leading to the emergence of multiple temperature-induced magnetic phase transitions [9]. The kagome staircases are separated by VO$_4$ tetrahedral units, giving the M$_3$V$_2$O$_8$ family of compounds a pseudo-two-dimensional layered characteristic, which, as discussed below, leads to an anisotropic compressional behavior when an external pressure is applied.

AVO_4 orthovanadates, where A is a trivalent atom, have been extensively studied under high-pressure (HP) conditions [10,11]. As a result, first-order phase transitions, involving large volume changes, have been discovered at pressures below 10 GPa (for instance, the zircon-scheelite and zircon-monazite transitions in alkaline earth vanadates [10,11]). Many of these transitions lead to interesting phenomena such as a collapse of the electronic band gap [12]. In contrast, very little effort has been dedicated to study kagome staircase orthovanadates under HP. Indeed, only two works can be found in the literature [7,9]. One of them reports powder X-ray diffraction (XRD) data on $Zn_3V_2O_8$ up to 15 GPa [7]. No phase transition was detected and an anisotropic response to compression was determined. The second one reports the influence of pressure on the magnetic properties of $Ni_3V_2O_8$ up to 2 Gpa [9]. Therefore, despite the interesting physics involved, the HP behavior of $M_3V_2O_8$ orthovanadates is an underexplored research area.

Here, we report a study of the HP behavior of the structural properties of $Ni_3V_2O_8$. The information obtained from such studies can form a foundation for future studies on the influence of pressure in physical properties, including magnetic properties. We have performed synchrotron powder XRD experiments in $Ni_3V_2O_8$ up to 23 Gpa at ambient temperature, thereby obtaining information on the structural stability of compressibility of the compound. The rest of the paper is organized as follows. Section 2 is devoted to the description the experimental methods. In Section 3, the results are reported and discussed in comparison with $Zn_3V_2O_8$ and other isostructural $M_3V_2O_8$ vanadates. Concluding remarks are given in Section 4.

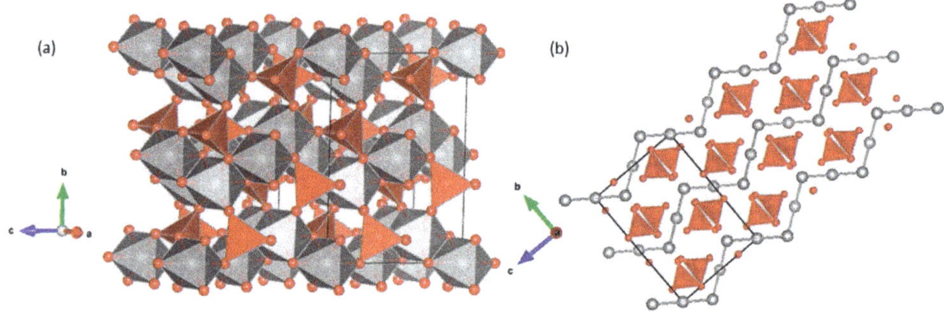

Figure 1. (**a**) Crystal structure of $Ni_3V_2O_8$. NiO_6 octahedra are shown in grey and VO_4 tetrahedra are in red. Small red circles are oxygen atoms. (**b**) Projection of the crystal structure showing VO_4 tetrahedra (in red) and Ni atoms (in gray) connected to make more evident the kagome staircase framework. Three unit cells are included in (**b**). The unit cell is shown is black solid lines.

2. Materials and Methods

Powder samples of $Ni_3V_2O_8$ were synthesized via a solid-state reaction starting from NiO (99.995% purity) and V_2O_5 (99.9% purity). The starting reagents were obtained from Alfa Aesar (Tewksbury, MA, United States). They were mixed meticulously and subsequently heated in an Al_2O_3 crucible in air at 800 °C for 16 h. The product was then ground and pressed into a pellet, to be afterward sintered at 900 °C for an additional 16 h. The obtained sample was finally ground manually in an agate mortar to obtain a micron size powder. In order to characterize it, we performed powder XRD measurements using a X'Pert Pro diffractometer from Panalytical (Almelo, The Netherlands) with Cu $K_{\alpha 1}$ radiation ($\lambda = 1.5406$ Å). The measurements were performed in the angular range $10° < 2\theta < 70°$, by continuous scanning with a step size of 0.02° and total step time of 200 s.

Angle-dispersive X-ray diffraction experiments at room-temperature and high pressure were performed at beamline I15 of the Diamond Light Source using a membrane-type diamond-anvil cell (DAC) with diamond-culet sizes of 450 μm in diameter and monochromatic X-rays of $\lambda = 0.42466$ Å. The X-ray beam was focused down to 10×10 μm^2. A rocking (±10°) of the DAC was used to improve the homogeneity of the Debye rings. A fine powder of $Ni_3V_2O_8$ was loaded in a 150 μm hole of a rhenium gasket pre-indented to a 40 μm thickness. One grain of Cu was loaded together with the sample and used as internal standard for pressure determination. For this purpose, we used the equation of state (EOS) determined by Dewaele et al. under hydrostatic conditions [13]. A 16:3:1 methanol–ethanol–water (MEW) mixture was used as a pressure-transmitting medium [14,15]. The powder XRD patterns were collected using a Pilatus 2M detector (DECTRIS, Baden, Switzerland) and transformed into one-dimensional patterns using the DIOPTAS suite [16]. The sample-to-detector distance was measured following standard procedure from the diffraction rings of LaB_6.

3. Results and Discussion

The results of the ambient-conditions XRD measurements are shown in Figure 2. All of the observed reflections can be accounted for by the orthorhombic crystal structure reported in the literature (space group *Cmca*, No. 64) [17], leading to small residuals (see Figure 2) and converging to small R-factors: R_p = 2.39% and R_{WP} = 3.79%. The resulting unit-cell parameters were a = 5.928(4) Å, b = 11.384(6) Å, and c = 8.241(5) Å, which agree to within 0.1% of literature values [17]. The obtained atomic positions are reported in Table 1, being also in good agreement with the literature [17].

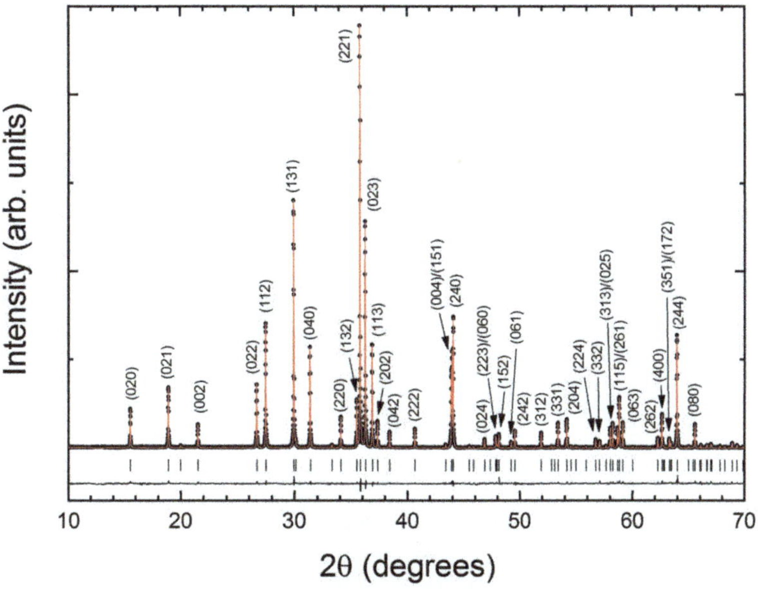

Figure 2. Background subtracted ambient-conditions powder X-ray diffraction (XRD) pattern measured in $Ni_3V_2O_8$ using Cu $K_{\alpha1}$. Experiment: black dots, Rietveld refinement: red line, residual: black line. Ticks indicate the position of peaks. The most intense peaks have been labeled with the corresponding Bragg planes.

Table 1. Fractional coordinates of atoms in the crystal structure of $Ni_3V_2O_8$.

Atom	Wyckoff Position	x	y	z
Ni_1	4a	0	0	0
Ni_2	8e	0.25	0.1319(4)	0.25
V	8f	0	0.3766(5)	0.1191(5)
O_1	8f	0	0.2491(9)	0.2302(9)
O_2	8f	0	0.0011(9)	0.2445(9)
O_3	16g	0.2663(9)	0.1193(9)	0.0009(9)

Figures 3–5 show sequential integrated XRD patterns acquired up to 23 GPa on compression of $Ni_3V_2O_8$. In Figure 3, it can be seen that over the complete pressure range of the experiments, all the reflections can be assigned to the known orthorhombic structure of $Ni_3V_2O_8$ and to Cu (the pressure marker), which is supported by a profile matching analysis using the Le Bail method. Therefore, no phase transition is observed in $Ni_3V_2O_8$ up to 23 GPa. More details of the HP X-ray patterns can be seen in Figures 4 and 5. Figure 4 shows results for P ≤ 7 GPa and Figure 5 shows results from 7.6 to 23 GPa. Beyond 20 GPa, we detected the presence of one extra reflection originating from the Re gasket, which appears because of the reduction in the size of the sample chamber. The data in Figures 3 and 4 show a change of relative peak intensities under increasing compression. This is related to the development of preferred crystallite orientation relative to the X-ray beam. Under compression, we also observe that the position of peaks indexed as $0k0$ is less sensitive to pressure, and that they move less towards higher angles than the rest of the peaks. This can be observed in Figure 4 by comparing the pressure evolution of peaks identified with planes (131) and (040), since the (131) peak approaches the stationary (040) peak. As we will discuss below, this observation is rooted in the non-isotropic behavior of $Ni_3V_2O_8$. Another consequence of the anisotropic behavior is the gradual merging of peaks; for instance, those identified with planes (221) and (023). Their merging under compression can be seen in Figure 3, Figure 4, and Figure 5. In addition, beyond 7.6 GPa, we have observed a gradual broadening of the peaks (Figure 5), which is related to the gradual loss of quasi-hydrostaticity [15,18].

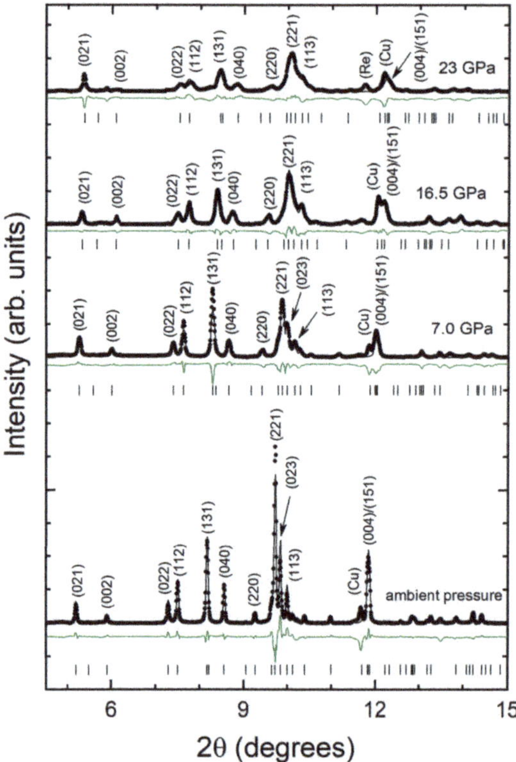

Figure 3. Integrated XRD patterns of $Ni_3V_2O_8$ from 0 to 23 GPa. Pressures are indicated on each pattern and the Bragg planes corresponding to the most intense reflections have been labeled. The experiments are shown with symbols. The refinements, assuming the ambient-pressure orthorhombic structure, are shown with black lines. The difference between the observed data and Le Bail refinement is shown in green. Ticks indicate the positions of $Ni_3V_2O_8$ peaks.

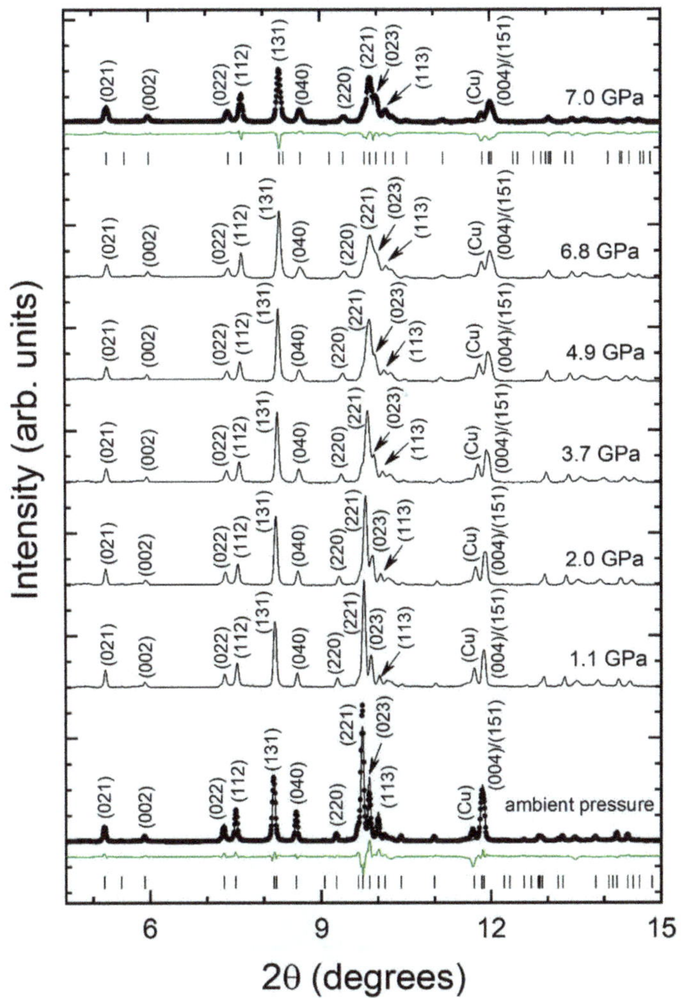

Figure 4. Integrated XRD patterns of $Ni_3V_2O_8$ measured from 0 to 7 GPa. Pressures are indicated on each pattern and Bragg planes corresponding to the most intense reflections have been labeled. At the bottom and top, traces observed experimental data are shown with black symbols. The refinements are shown with black lines. The difference between the observed data and Le Bail refinement is shown in green.

Through performing Le Bail refinements of the powder XRD patterns, we have obtained the pressure dependence of the unit-cell parameters. The XRD patterns at all pressures can be identified with the ambient-pressure orthorhombic structure of $Ni_3V_2O_8$; however, the XRD patterns measured at pressures higher than 15.1 GPa have been excluded from this analysis due to the partial overlap of the Cu peak used to determine pressure with a diffraction peak from the sample (see Figure 5), because this could lead to inaccuracies in the pressure determination larger than 0.2 GPa. The unit-cell parameters as a function of pressure are shown in Figure 6. This information is important for modeling the suppression of ferroelectricity by pressure in $Ni_3V_2O_8$ [9]. The response of the crystal structure to compression is clearly non-isotropic, with the *b* and *a* axes, respectively, having the smallest and largest compressibility. The anisotropic compressibility observed $Ni_3V_2O_8$ is fully compatible with the anisotropic thermal expansion of the same compound [9]. Both results can be rationalized based on the layered characteristic of the kagome staircase in the crystal structure and the relative compressibilities of the constituent polyhedra, as discussed below.

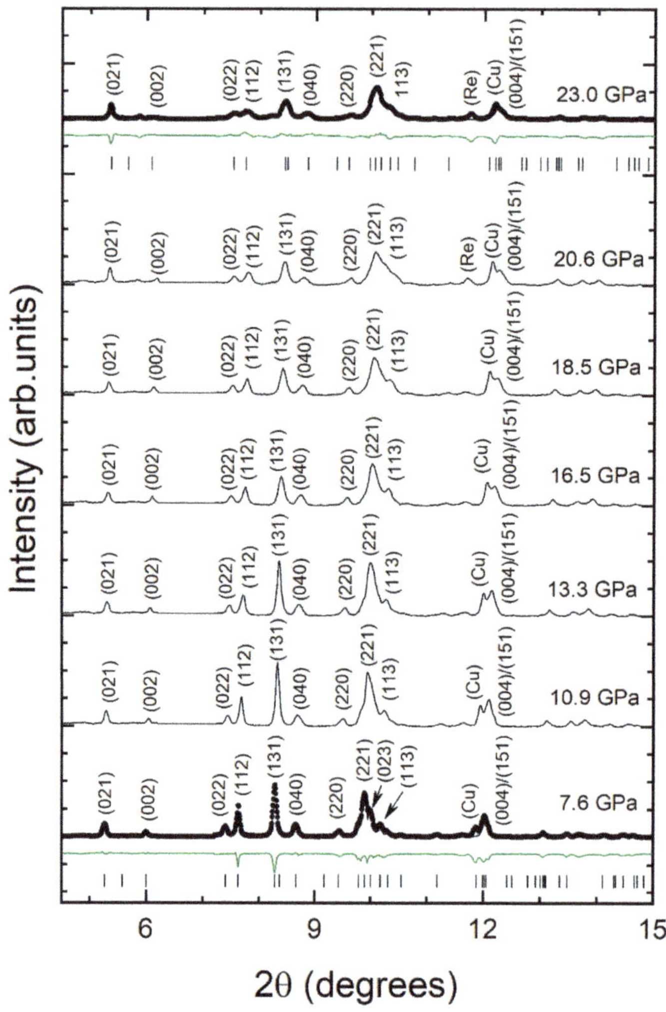

Figure 5. Integrated XRD patterns of $Ni_3V_2O_8$ measured from 7.6 to 23 GPa. Pressures are indicated on each pattern, and Bragg planes corresponding to the most intense reflections have been labeled. At the bottom and top, traces observed experimental data are shown with black symbols. The refinements are shown with black lines. The difference between the observed data and Le Bail refinement is shown in green. In the figure, it can be clearly seen the merging and broadening of peaks described in the text.

From the hydrostatic pressure range of our experiments ($P \leq 7.6$ GPa), we have determined the linear compressibilities at ambient pressure, $\kappa_x = -\frac{1}{x}\frac{\partial x}{\partial P}$, where x represents the unit cell lengths a, b, or c. The linear compressibilities are summarized in Table 2, where it can be seen that they increase following the sequence $\kappa_b < \kappa_c < \kappa_a$. The a and c-axes have larger compressibilities than the b-axis by approximate factors of 1.5 and 1.3, respectively. A similar behavior has previously been observed in $Zn_3V_2O_8$.

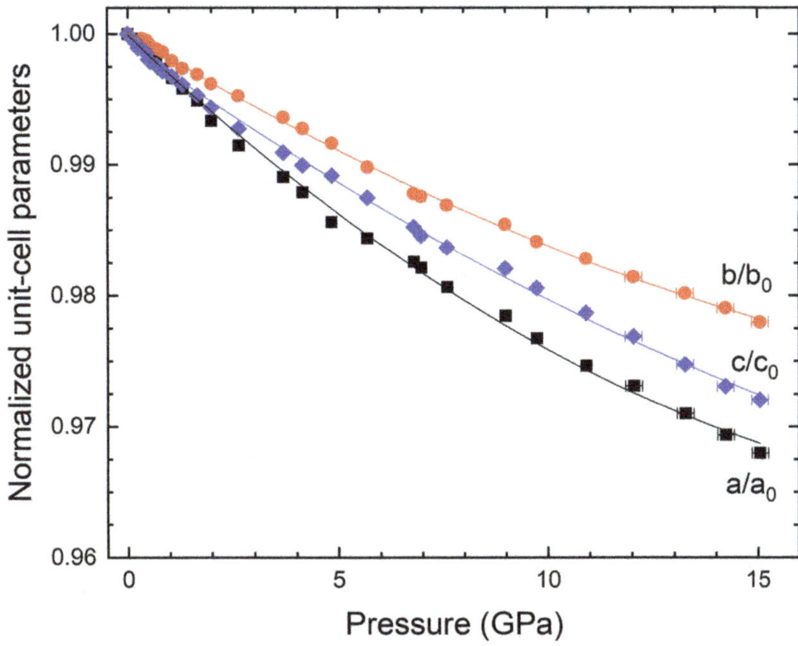

Figure 6. Normalized unit-cell parameters of $Ni_3V_2O_8$ as a function of pressure. Symbols are the results from Le Bail refinements. The lines describe unidirectional equations of state (EOSs) described in the text. Where not shown error bars are comparable to symbols' size.

The reason for the anisotropic behavior of $Ni_3V_2O_8$ and $Zn_3V_2O_8$ (and probably all isomorphic vanadates) might be related to the kagome layered characteristic of the crystal structure. It is well-known that in other ternary oxides, such as $NiWO_4$ and $ZnWO_4$ [19], the pressure-induced changes in crystal structure are largely determined by the NiO_6 and ZnO_6 octahedral units because of their large compressibility relative to the VO_4 tetrahedron. The much smaller VO_4 tetrahedron has been determined from the study of many different vanadates to be an essentially rigid unit, which, due to V 3d - O 2p hybridization [20], changes little under compression [10]. As we have already described (see Figure 1), the crystal structure of $M_3V_2O_8$ compounds is composed of 2D kagome staircases of compressible MO_6 octahedra, which therefore constitute compressible layers that lie perpendicular to the b-axis. In between these layers of NiO_6 and ZnO_6 octahedra, there are located the less compressible VO_4 tetrahedra, which are arranged to form pillars between the kagome layers, reducing the compressibility along the b-axis. In fact, the in-plane effective bulk modulus [21] for kagome layers is 132 GPa, and the effective bulk modulus in the perpendicular direction is 186 GPa. Thus, the layered characteristic of the crystal structure of kagome-type vanadates provides a rational explanation to their anisotropic behavior under compression. The present results suggest that it is not unreasonable to speculate that anisotropic compressibility would be a finger print of the broad family of kagome-type compounds. Such anisotropic behavior is expected to affect super-exchange interactions between magnetic atoms, thereby modifying the temperature of different magnetic transitions. This would modify the Néel temperature as reported for $Ni_3V_2O_8$ [9]. By extrapolating the result reported by Chaudhury et al. [9], it can be speculated that under compression the Néel temperature can be increased from 9.8 K at ambient pressure to a temperature close to 15 K at 20 GPa.

Table 2. Linear compressibilities at zero pressure (left) and room-temperature equation of state (EOS) parameters of $Ni_3V_2O_8$. The center (right) column shows the converged fitting parameters for the third-order Birch–Murnaghan EOS for $P \leq 7.6$ (15.1) GPa. V_0 is the ambient-pressure volume. K_0 is the bulk modulus. K_0' is the pressure derivative of the bulk modulus.

$\kappa_a = 2.7(1)\ 10^{-3}\ \mathrm{GPa}^{-1}$	$V_0 = 555.\,7(2)\ \text{Å}^3$	$V_0 = 555.5(2)\ \text{Å}^3$
$\kappa_b = 1.79(6)\ 10^{-3}\ \mathrm{GPa}^{-1}$	$K_0 = 139(3)\ \mathrm{GPa}$	$K_0 = 118(1)\ \mathrm{GPa}$
$\kappa_c = 2.33(5)\ 10^{-3}\ \mathrm{GPa}^{-1}$	$K_0' = 4.4(3)$	$K_0' = 11.1(9)$

The unit-cell volume of $Ni_3V_2O_8$ as a function of pressure is shown in Figure 7. The results show an apparent change in compressibility at 7.6 GPa, which cannot be related to a phase transition, since the XRD data provide no evidence for it up to the maximum pressure investigated (23 GPa). A similar behavior has been recently reported in other ternary oxides near this pressure [22]. The apparent change in compressibility is in fact related to the known hydrostatic limit for MEW [15]. Fitting a third-order Birch–Murnaghan (BM) EOS [23] (using EosFit [24]) in the hydrostatic regime ($P \leq 7.6$ GPa) gives the parameters given in the central column of Table 2. To illustrate the importance of constraining EOS fits to hydrostatic data, for comparison, we also fitted the EOS with all data up to 15.1 GPa, leading to the values given in the right column of Table 2. The inclusion of non-hydrostatic data leads to an underestimation of the ambient pressure bulk modulus (K_0) and to an unusual large pressure derivative ($K_0' = 11.1(9)$).

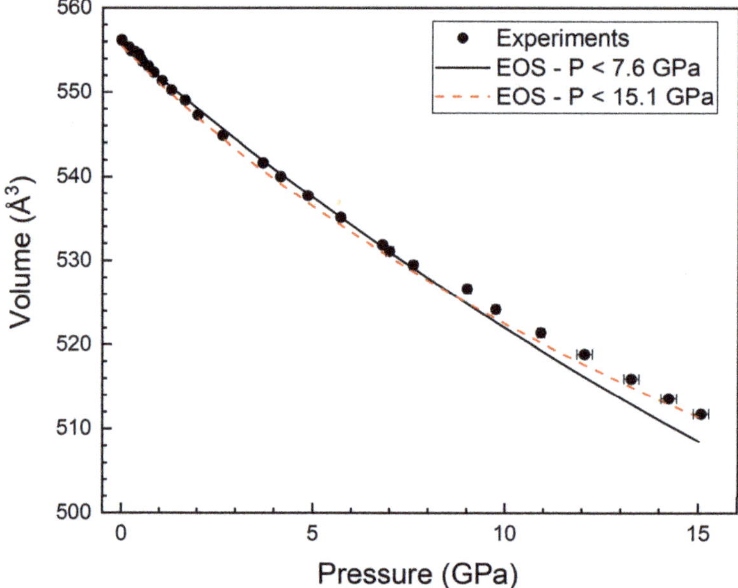

Figure 7. Pressure dependence of the unit-cell volume of $Ni_3V_2O_8$. Solid circles are obtained from experiments. Where not shown, error bars are comparable to symbols' size. The black solid (red dashed) line is the equation of state obtained from data for $P \leq 7.6$ (15.1) GPa. As stated in the text, results deviate from the quasi-hydrostatic EOS for $P \geq 7.6$ GPa likely due to the influence of non-hydrostaticity.

We will now comment on the bulk modulus of $Ni_3V_2O_8$ within the context of other $M_3V_2O_8$ vanadates [25,26] using the bulk moduli summarized in Table 3. In Table 3, it can be seen that $Ni_3V_2O_8$ is the most incompressible compound within $M_3V_2O_8$ family of isomorphic vanadates. For the discussion, based on the fact that NiO_6 units are more compressible than rigid VO_4 units [10,19], we will assume that the bulk modulus of $Ni_3V_2O_8$ is mainly determined by the average Ni-O distance (d_{Ni-o}) at ambient pressure and the formal charge of Ni (Z_{Ni}). Such an empirical approximation works very well for ternary oxides [27,28], whereby the estimated bulk modulus is given by $B_E = \frac{610\,Z_{Ni}}{d_{Ni-O}^3}$. Using this approximation, a bulk modulus of 140 GPa is obtained for $Ni_3V_2O_8$, which agrees within errors with the experimental value obtained from the hydrostatic-pressure regime (139(3) GPa). This suggests that the bulk modulus obtained from density-functional theory (DFT) calculations (130.4–131.2 GPa, given in Table 3) has been underestimated by ~7% of the experimental value determined here. This fact is probably related to the overestimation of the ambient-pressure volume by DFT calculations [29], which is a typical feature of the general-gradient approximation used in previous computer simulations [25,26].

By using the same empirical approximation, we have estimated the bulk modulus for other $M_3V_2O_8$ vanadates, which are summarized in Table 3 for comparison with the literature. For example, in the case of $Zn_3V_2O_8$, the estimated bulk modulus is consistent with the fact that this compound has an experimentally determined bulk modulus smaller than that for $Ni_3V_2O_8$, which indicates that the approximation can be also applied to other kagome-type vanadates. In performing the empirical approximations summarized in Table 3, we found that DFT calculations have also underestimated the bulk modulus of $Cu_3V_2O_8$ and $Mn_3V_2O_8$. Using the empirical approximation, which works well for the studied vanadates, we can predict bulk moduli of 132 and 131 GPa for $Mg_3V_2O_8$ and $Co_3V_2O_8$, respectively, which have not yet been investigated under compression. The conclusions reported in this work may be also applicable to isomorphic arsenates [30], to $(Ni,Mg)_3(VO_4)_2$ solid solutions [31], and to hydrated kagome-structured vanadates, such as $Cu_3V_2O_7(OH)_2 \cdot 2H_2O$ volborthite [32]. For the particular case of $Ni_3As_2O_8$, a bulk modulus of 205 GPa is predicted from the empirical model used in this work.

Table 3. Unit-cell volume, V_0, and bulk modulus, K_0, of different $M_3V_2O_8$ compounds.

Compound	V_0 ($Å^3$)	Experimental K_0 (GPa)	Empirical K_0 (GPa)	DFT K_0 (GPa)
$Ni_3V_2O_8$	555.5(3)	139(3) [a]	140 [c]	
$Ni_3V_2O_8$	562.64			130.4–131.2 [b]
$Mg_3V_2O_8$	576.92		132 [c]	
$Co_3V_2O_8$	578.96		131 [c]	
$Zn_3V_2O_8$	585.1(1)	120(2) [d]	128 [c]	
$Cu_3V_2O_8$	586.95		127 [c]	95–110 [e]
$Mn_3V_2O_8$	622.1		115 [c]	89.7–93 [b]

[a] Calculated from X-ray diffraction (XRD) measurements here reported. [b] Density-functional theory (DFT) calculations [25]. [c] Obtained using the empirical model described in the text. [d] From previous XRD experiments [7]. [e] DFT calculations [26].

4. Conclusions

Through synchrotron powder X-ray diffraction measurements, we have determined that at ambient temperature the orthorhombic kagome staircase crystal structure of $Ni_3V_2O_8$ is stable at least up to 23 GPa, the maximum pressure achieved in present experiments. We have also found that the response to compression is anisotropic. The compressibility of the *b*-axis is approximately 40% lower than the compressibility of the other axes. This special direction is perpendicular to the kagome layers, and the compression along the *b*-axis is the smallest due to the presence of incompressible VO_4 tetrahedral units, which interconnect the kagome staircases of NiO_6 octahedra. From the pressure dependence of the unit-cell volume we fitted a third-order Birch–Murnaghan equation of state, giving a bulk modulus of 139(3) GPa, which is the largest among isostructural $M_3V_2O_8$ vanadates. A systematic comparison of bulk moduli in isomorphic $M_3V_2O_8$ vanadates is made based on an empirical method for predicting bulk moduli in ternary oxides. The results are consistent with those determined from experimental data here for $Ni_3V_2O_8$ and previously for $Zn_3V_2O_8$ [7]. Therefore, we also present predicted bulk moduli of $M_3V_2O_8$ compounds not studied yet under high pressure, $Co_3V_2O_8$ and $Mg_3V_2O_8$.

Author Contributions: Formal analysis, D.D.-A. and D.E.; Investigation, R.T., E.B. and S.A.; Methodology, D.E.; Supervision, D.E.; Writing—review & editing, D.D.-A., R.T., E.B., S.A. and D.E. All authors have read and agreed to the published version of the manuscript.

Funding: This work was supported by the Spanish Ministry of Science, Innovation, and Universities under grants PID2019-106383GB-C41 and RED2018-102612-T (MALTA Consolider-Team network) and by Generalitat Valenciana under grant Prometeo/2018/123 (EFIMAT). R.T. acknowledges funding from the Spanish Ministry of Economy and Competitiveness MINECO) via the Juan de la Cierva Formación program (FJC2018-036185-I).

Acknowledgments: The authors thank S. N. Achary (Bhabha Atomic Research Center, India) for enlightening discussions. The authors would like to thank Diamond Light Source for beamtime provision (proposal CY21610).

Conflicts of Interest: The authors declare no conflict of interest.

References

1. Vijayakumar, S.; Lee, S.-H.; Ryu, K.-S. Synthesis of $Zn_3V_2O_8$ nanoplatelets for lithium-ion battery and supercapacitor applications. *RSC Adv.* **2015**, *5*, 91822–91828. [CrossRef]

2. Cabrera, I.; Kenzelmann, M.; Lawes, G.; Chen, Y.; Chen, W.; Erwin, R.; Gentile, T.R.; Leao, J.B.; Lynn, J.W.; Rogado, N.; et al. Coupled Magnetic and Ferroelectric Domains in Multiferroic $Ni_3V_2O_8$. *Phys. Rev. Lett.* **2009**, *103*, 087201. [CrossRef]

3. Bîrdeanu, M.-I.; Vaida, M.; Ursu, D.; Fagadar-Cosma, E. Obtaining and characterization of $Zn_3V_2O_8$ and $Mg_3V_2O_8$ pseudo binary oxide nanomaterials by hydrothermal method. In *HIGH ENERGY GAMMA-RAY ASTRONOMY: 6th International Meeting on High. Energy Gamma-Ray Astronomy*; AIP Publishing LLC: Melville, NY, USA, 2017; p. 030006. [CrossRef]

4. Mazloom, F.; Masjedi-Arani, M.; Salavati-Niasari, M. Novel size-controlled fabrication of pure $Zn_3V_2O_8$ nanostructures via a simple precipitation approach. *J. Mater. Sci. Mater. Electron.* **2015**, *27*, 1974–1982. [CrossRef]

5. Qian, T.; Fan, B.; Wang, H.; Zhu, S. Structure and luminescence properties of $Zn_3V_2O_8$ yellow phosphor for white light emitting diodes. *Chem. Phys. Lett.* **2019**, *715*, 34–39. [CrossRef]

6. Liu, Y.; Li, Q.; Ma, K.; Yang, G.; Wang, C.; Qian, L. Graphene Oxide Wrapped CuV_2O_6 Nanobelts as High-Capacity and Long-Life Cathode Materials of Aqueous Zinc-Ion Batteries. *ACS Nano* **2019**, *13*, 12081–12089. [CrossRef] [PubMed]

7. Díaz-Anichtchenko, D.; Santamaria-Perez, D.; Marqueño, T.; Pellicer-Porres, J.; Ruiz-Fuertes, J.; Ribes, R.; Ibañez, J.; Achary, S.N.; Popescu, C.; Errandonea, D. Comparative study of the high-pressure behavior of ZnV_2O_6, $Zn_2V_2O_7$, and $Zn_3V_2O_8$. *J. Alloy. Compd.* **2020**, *837*, 155505. [CrossRef]

8. Rogado, N.; Lawes, G.; Huse, D.; Ramirez, A.; Cava, R. The Kagomé-staircase lattice: Magnetic ordering in $Ni_3V_2O_8$ and $Co_3V_2O_8$. *Solid State Commun.* **2002**, *124*, 229–233. [CrossRef]

9. Chaudhury, R.P.; Yen, F.; Cruz, C.D.; Lorenz, B.; Wang, Y.Q.; Sun, Y.Y.; Chu, C.W. Pressure-temperature phase diagram of multiferroic $Ni_3V_2O_8$. *Phys. Rev. B* **2007**, *75*, 012407. [CrossRef]

10. Errandonea, D. High pressure crystal structures of orthovanadates and their properties. *J. Appl. Phys.* **2020**, *128*, 040903. [CrossRef]

11. Errandonea, D.; Garg, A.B. Recent progress on the characterization of the high-pressure behaviour of AVO_4 orthovanadates. *Prog. Mater. Sci.* **2018**, *97*, 123–169. [CrossRef]

12. Bandiello, E.; Sánchez-Martín, J.; Errandonea, D.; Bettinelli, M. Pressure Effects on the Optical Properties of $NdVO_4$. *Crystals* **2019**, *9*, 237. [CrossRef]

13. Dewaele, A.; Loubeyre, P.; Mezouar, M. Equations of state of six metals above 94GPa. *Phys. Rev. B* **2004**, *70*, 094112. [CrossRef]

14. Klotz, S.; Chervin, J.-C.; Munsch, P.; Le Marchand, G. Hydrostatic limits of 11 pressure transmitting media. *J. Phys. D Appl. Phys.* **2009**, *42*, 075413. [CrossRef]

15. Errandonea, D.; Meng, Y.; Somayazulu, M.; Hausermann, D. Pressure-induced alpha-to-omega transition in titanium metal: A systematic study of the effects of uniaxial stress. *Physica B* **2005**, *355*, 116–125. [CrossRef]

16. Prescher, C.; Prakapenka, V.B. DIOPTAS: A program for reduction of two-dimensional X-ray diffraction data and data exploration. *High. Press. Res.* **2015**, *35*, 223–230. [CrossRef]

17. Sauerbrei, E.E.; Faggiani, R.; Calvo, C. Refinement of the crystal structure of $Co_3V_2O_8$ and $Ni_3V_2O_8$. *Acta Crystallogr. Sect. B Struct. Crystallogr. Cryst. Chem.* **1973**, *29*, 2304–2306. [CrossRef]

18. Errandonea, D.; Muñoz, A.; Gonzalez-Platas, J. Comment on "High-pressure x-ray diffraction study of YBO_3/Eu^{3+}, $GdBO_3$, and $EuBO_3$: Pressure-induced amorphization in $GdBO_3$". *J. Appl. Phys.* **2014**, *115*, 216101. [CrossRef]

19. Errandonea, D.; Ruiz-Fuertes, J. A Brief Review of the Effects of Pressure on Wolframite-Type Oxides. *Crystals* **2018**, *8*, 71. [CrossRef]

20. Laverock, J.; Piper, L.F.J.; Preston, A.R.H.; Chen, B.; McNulty, J.; Smith, K.E.; Kittiwatanakul, S.; Lu, J.W.; Wolf, S.A.; Glans, P.-A.; et al. Strain dependence of bonding and hybridization across the metal-insulator transition of VO_2. *Phys. Rev. B* **2012**, *85*, 081104. [CrossRef]

21. Zhang, Y.; Qiu, X.; Fang, D. Mechanical Properties of two novel planar lattice structures. *Int. J. Solids Struct.* **2008**, *45*, 3751–3768. [CrossRef]

22. Turnbull, R.; Errandonea, D.; Cuenca-Gotor, V.P.; Sans, J.Á.; Gomis, O.; Gonzalez, A.; Rodríguez-Hernandez, P.; Popescu, C.; Bettinelli, M.; Mishra, K.K.; et al. Experimental and theoretical study of dense YBO_3 and the influence of non-hydrostaticity. *J. Alloy. Compd.* **2021**, *850*, 156562. [CrossRef]

23. Birch, F. Finite elastic strain of cubic crystals. *Phys. Rev.* **1947**, *71*, 809–824. [CrossRef]

24. Gonzalez-Platas, J.; Alvaro, M.; Nestola, F.; Angel, R.J. EosFit7-GUI: A new GUI tool for equation of state calculations, analyses, and teaching. *J. Appl. Crystallogr.* **2016**, *49*, 1377–1382. [CrossRef]

25. Koc, H.; Palaz, S.; Mamedov, A.M.; Ozbay, E. Electronic and elastic properties of the multiferroic crystals with the Kagome type lattices -$Mn_3V_2O_8$ and $Ni_3V_2O_8$: First principle calculations. *Ferroelectrics* **2019**, *544*, 11–19. [CrossRef]

26. Jezierski, A.; Kaczkowski, J. Electronic structure and thermodynamic properties of $Cu_3V_2O_8$ compound. *Phase Transit.* **2015**, *88*, 1–9. [CrossRef]

27. Errandonea, D.; Manjón, F.J. Pressure effects on the structural and electronic properties of ABX_4 scintillating crystals. *Prog. Mater. Sci.* **2008**, *53*, 711–773. [CrossRef]

28. Errandonea, D.; Muñoz, A.; Rodríguez-Hernández, P.; Gomis, O.; Achary, S.N.; Popescu, C.; Patwe, S.J.; Tyagi, A.K. High-Pressure Crystal Structure, Lattice Vibrations, and Band Structure of $BiSbO_4$. *Inorg. Chem.* **2016**, *55*, 4958–4969.

29. Zhang, G.-X.; Reilly, A.M.; Tkatchenko, A.; Scheffler, M. Performance of various density-functional approximations for cohesive properties of 64 bulk solids. *New J. Phys.* **2018**, *20*, 063020. [CrossRef]

30. Barbier, J.; Frampton, C. Structures of orthorhombic and monoclinic $Ni_3(AsO_4)_2$. *Acta Crystallogr. Sect. B Struct. Sci.* **1991**, *47*, 457–462. [CrossRef]

31. Nord, A.G.; Werner, P.-E. Cation distribution studies of three (Ni, Mg) orthovanadates. *Zeitschrift für Kristallographie - Crystalline Materials* **1991**, *194*, 49. [CrossRef]

32. Wang, P.; Yang, H.; Wang, D.; Chen, Y.; Dai, W.L.; Zhao, X.; Yang, J.; Wang, X. Activation of kagome lattice-structured $Cu_3V_2O_7(OH)_2 \cdot 2H_2O$ volborthite via hydrothermal crystallization for boosting visible light-driven water oxidation. *Phys. Chem. Chem. Phys.* **2018**, *20*, 24561–24569. [CrossRef] [PubMed]

Review

Phase Relations of Earth's Core-Forming Materials

Tetsuya Komabayashi

School of GeoSciences and Centre for Science at Extreme Conditions, University of Edinburgh, Edinburgh EH9 3FE, UK; Tetsuya.komabayashi@ed.ac.uk; Tel.: +44-131-650-8518

Abstract: Recent updates on phase relations of Earth's core-forming materials, Fe alloys, as a function of pressure (P), temperature (T), and composition (X) are reviewed for the Fe, Fe-Ni, Fe-O, Fe-Si, Fe-S, Fe-C, Fe-H, Fe-Ni-Si, and Fe-Si-O systems. Thermodynamic models for these systems are highlighted where available, starting with 1 bar to high-P-T conditions. For the Fe and binary systems, the longitudinal wave velocity and density of liquid alloys are discussed and compared with the seismological observations on Earth's outer core. This review may serve as a guide for future research on the planetary cores.

Keywords: phase relation; Earth's core; iron alloys; high-pressure; high-temperature; thermodynamics; extreme conditions

1. Introduction

The phase relations of Fe-alloys have been of primary importance in many research disciplines. For the engineering and metallurgy fields, the stability diagram of materials as a function of temperature (T) and composition (X) serves as a map to their target phase and its properties. In pursuit of this, phase relations at 1 bar were extensively studied for most of the binary Fe-alloy systems by experiment and theory. The stability diagrams of Fe-alloys are also important for Earth science as the major component of Earth's central core is iron. Because Earth's core is under high pressure (P) and temperature conditions, the stability diagrams need to be established for such extreme conditions.

High-pressure study of phase relations of Fe-alloys started around 1950 with shock loading experiments [1]. In the early days, the research was led by laboratory experiment which included static and dynamic compression techniques. Since around 2000, the theoretical approach with first-principles calculations became competitive with the experimental thanks to the enhanced computer performance and development of appropriate simplification of calculation. Currently, the experimental and theoretical approaches are considered to be complementary to each other. However, different approaches may give different results as is often the case, and therefore it is imperative to understand what properties are agreed on and what are not between the different/same techniques.

In this review, I will summarise recent updates on the phase relations of Earth's core-forming materials, Fe alloys, as a function of P-T-X for the Fe, Fe-Ni, Fe-O, Fe-Si, Fe-S, Fe-C, Fe-H, Fe-Ni-Si, and Fe-Si-O systems. I highlight thermodynamic models based on experimental data for these systems, where available, starting with 1 bar to high-P-T conditions. An experiment-based thermodynamic model, which includes thermodynamic potential functions of phases (e.g., the Gibbs free energy as a function of P-T-X), may provide information that is difficult to constrain by experiment such as the entropy of fusion and velocity of liquid under extreme conditions. Therefore, thus-calculated properties could serve as "experimental constraints" which can be compared with those from the first-principles calculations. For the Fe and binary systems, the longitudinal wave velocity and density of liquid alloys are discussed and compared with the seismological observations on Earth's outer core. This review may serve as a guide for the future research on the cores of terrestrial planets.

Citation: Komabayashi, T. Phase Relations of Earth's Core-Forming Materials. *Crystals* **2021**, *11*, 581. https://doi.org/10.3390/cryst11060581

Academic Editors: Daniel Errandonea and Simone Anzellini

Received: 31 March 2021
Accepted: 11 May 2021
Published: 22 May 2021

Publisher's Note: MDPI stays neutral with regard to jurisdictional claims in published maps and institutional affiliations.

2. Theoretical Background

Here I briefly introduce thermodynamic functions that are needed for discussions in this paper.

2.1. Equation of State

The equation of state (EoS) is needed for calculating the pressure effect on the Gibbs free energy, density, and elasticity of a phase. The detailed theoretical background of EoS can be found elsewhere [2,3].

The room-temperature molar volume of a phase upon compression can often be expressed with the third-order Birch–Murnaghan (BM) or Vinet EoS as:

$$P_{300} = \frac{3K_0}{2}\left[\left(\frac{V_0}{V}\right)^{\frac{7}{3}} - \left(\frac{V_0}{V}\right)^{\frac{5}{3}}\right]\left\{1 - \frac{3}{4}(4 - K\prime)\left[\left(\frac{V_0}{V}\right)^{\frac{2}{3}} - 1\right]\right\} \dots \text{BM} \tag{1}$$

$$P_{300} = 3K_0 x^{-2}(1 - x)\exp\left[\frac{3}{2}(K\prime - 1)(1 - x)\right] \dots \text{Vinet} \tag{2}$$

where $x \equiv (V/V_0)^{1/3}$ and, P_{300}, K_0, K', and V_0 are the pressure at $T = 300$ K, the isothermal bulk modulus, its pressure derivative, and the molar volume at $P = 1$ bar and $T = 300$ K, respectively.

In order to extrapolate an EoS to high temperature, either (i) the thermal pressure model or (ii) thermal expansivity model can be used.

(i) The thermal pressure model is described as:

$$P(V,T) = P(V,300\text{ K}) + Pth\ (V,T) \tag{3}$$

where $P(V,T)$, $P(V,300\text{ K})$, and $Pth(V,T)$ are the total pressure, pressure at 300 K for a given sample volume, and thermal pressure at a given temperature. The thermal pressure part may often be approximated as:

$$Pth = \alpha K_T \times (T - 300) \tag{4}$$

where α is the thermal expansion coefficient and K_T is the isothermal bulk modulus. The αK_T value can be assumed to be constant near or above the Debye temperature. Therefore, a simple relation of $\alpha K_T = \alpha_0 K_0$ holds for materials with low Debye temperatures such as metals, where the subscript 0 indicates at 1 bar and 300 K.

Alternatively, the Mie–Grüneisen–Debye model can be used for the thermal pressure part in Equation (3) [3]:

$$Pth = \gamma/V\ \Delta Eth\ (\theta,T) \tag{5}$$

where γ is the Grüneisen parameter, ΔEth is the change in thermal energy, and θ is the Debye temperature. The thermal energy can often be calculated from the Debye approximation:

$$Eth = \frac{9nRT}{(\theta/T)^3}\int_0^{\theta/T} \frac{\xi^3}{e^\xi - 1}d\xi \tag{6}$$

where n, θ, and R are the number of atoms per formula unit, Debye temperature, and gas constant, respectively. The Debye temperature and Grüneisen parameter can be assumed to be functions of volume as:

$$\theta = \theta_0 \exp[(\gamma_0 - \gamma)/q] \tag{7}$$

and

$$\gamma = \gamma_0\ (V/V_0)^q \tag{8}$$

where θ_0, γ_0, and q are the Debye temperature, Grüneisen parameter at 1 bar and 300 K, and a dimensionless parameter, respectively. As mentioned above, θ_0 for a metal is likely close to or below room temperature.

In the thermal pressure model (Equation (3)), there may be additional terms for anharmonic and electronic contributions. For example, hcp Fe shows large contributions from these terms [4]. The presence or absence of these additional contributions can be resolved when the αK_T term shows temperature dependence (Equation (4)) [5].

(ii) The thermal expansion coefficient can be modelled for a phase under high pressure using the Anderson–Grüneisen parameter, δ_T [6]:

$$\frac{\partial \ln \alpha}{\partial \ln V} = \delta_T = \delta_0 \eta^\kappa \tag{9}$$

where $\eta \equiv V/V_0$, δ_0 is the value of δ_T at $P = 1$ bar and κ is a dimensionless parameter. This equation yields:

$$\frac{\alpha}{\alpha_0} = \exp\left[-\frac{\delta_0}{\kappa}(1 - \eta^\kappa)\right] \tag{10}$$

2.2. Thermodynamics

Once a thermodynamic potential has been assessed, one can derive any equilibrium properties of the system [7]. The Gibbs free energy of a phase at a given *P-T* condition ($G_{P,T}$) is expressed as:

$$G_{P,T} = G_{1\text{bar},T} + \int_{1\text{bar}}^{P} V_T dP \tag{11}$$

where $G_{1\text{bar},T}$ is the Gibbs free energy at $P = 1$ bar and T of interest and V_T is the molar volume at T. $G_{1\text{bar},T}$ is often available from the metallurgy database for Fe-alloys (e.g., [8]). The volumetric term in Equation (11) is calculated from the *P-V-T* EoS of the phase. Komabayashi [9,10] adopted the thermal expansion model (Equation (10)) in the free energy calculations. When a phase is a solution between end-members, such as Fe-alloys, its mixing property needs to be considered to obtain $G_{P,T}$. Many of the Fe-alloy liquids show nonideal mixing at 1 bar and its evolution with increasing pressure is of interest.

As mentioned above, one of the major advantages of thermodynamic modelling is that one can derive the properties that cannot be directly or easily constrained by experiment, such as the EoS of liquid phases. Komabayashi [9,10] evaluated EoS parameters for Fe-alloy liquid phases by calculating their melting curves which reproduce experimental data.

3. Phase Relations

Throughout the paper, I describe the composition of materials, for example, Fe-5 wt% Ni-4 wt% Si, as Fe-5Ni-4Si.

3.1. Major Components

3.1.1. Fe

Major issues in the phase relations of the main component of Earth's core, Fe(-Ni), include the *P-T* location of the transition boundary between face-centred cubic (fcc) and hexagonal close-packed (hcp) structures, the stable structure in the inner core, and the melting point at the inner core-outer core boundary (ICB) corresponding to a pressure of 330 GPa.

- **The fcc-hcp Transition Boundary**

An important phase relation in iron is the fcc-hcp transition boundary. This boundary determines the *P-T* location of an invariant point where the fcc, hcp, and liquid phases are stable [11–15] (Figure 1). Results of Shen et al. [12] and Boehler [13] highlighted a major inconsistency among experimental studies. The temperature values of the invariant point by both data are nearly the same of 2800 K, but the pressure values were very different: 60 GPa (Shen et al. [12]) and 100 GPa (Boehler [13]) (Figure 1), as a consequence of different *P-T* slopes of the fcc-hcp boundary. Komabayashi et al. [14] revisited the boundary with internally resistive-heated diamond anvil cells (DAC) and precisely determined its slope. They also pointed out that the different slopes in the previous works could be partly resolved by taking into account the difference in estimates of thermal pressure during the experiments. Figure 1 summarises the results of experimental determinations of the fcc-hcp transition [11–14,16–18]. Among them, the *P-T* data where the pressures at high temperatures were calculated from internal pressure standards are fairly consistent [11,14,17,18]. In contrast, the high-temperature DAC studies in [12,13,16] did not consider thermal pressure effects. Therefore, it may be a coincident that Boehler's [13] result is rather consistent with Komabayashi et al.'s [14] data (Figure 1). The results of [12,16] would be relatively consistent with Komabayashi et al. [14] if the thermal pressure effects are considered which is about a 20 GPa increase upon heating to 2800 K based on the EoS of Fe [4]. As discussed later, Komabayashi [9] constructed a thermodynamic model of Fe including the fcc-hcp transition [14].

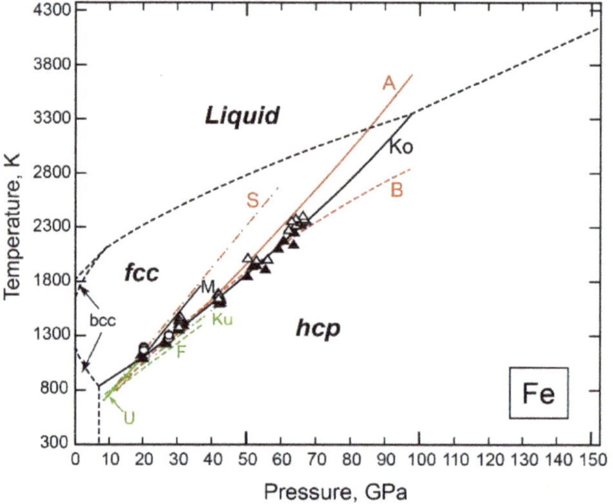

Figure 1. Pressure-temperature locations of the boundary of iron between face-centred cubic (fcc) and hexagonal close-packed (hcp) structures (S, Shen et al. [12]; B, Boehler [13]; M, Mao et al. [16]; U, Uchida et al. [11]; F, Funamori et al. [17]; Ku, Kubo et al. [18]; Ko, Komabayashi et al. [14]; and A, Anzellini et al. [15]). The other reaction boundaries (shown by dashed lines) are from Komabayashi [9].

More recently, Anzellini et al. [15] also determined the fcc-hcp transition boundary as part of their determination of the melting points of Fe to 200 GPa in laser-heated DAC. Their data show slightly higher, but still consistent, transition temperatures with Komabayashi et al.'s [14] data (Figure 1).

- **The Stable Structure in the Inner Core**

The stable iron structure in the inner core has been a literally "central" issue about the Earth as its physical properties, such as the melting/crystallising point and element partitioning upon crystallization, control core dynamics. Furthermore, the constituent phase structure would govern inner core dynamics, for example, the seismic anisotropy observed there, namely, seismic waves travel faster along the Earth's polar axis by 3–4% compared with equatorial directions [19–22], which will not be discussed in this paper as the focus is on the phase relations.

A growing number of data has accumulated supporting the conventional view that the hcp phase is the stable phase in the inner core [15,23–26]. Tateno et al. [23] reached the *P-T* conditions relevant to the centre of the Earth using a static device (i.e., DAC) (Figure 2a). Although samples in their experiments showed contamination by carbon from diamond anvils, their conclusion that the hcp phase is the stable iron phase in the inner core is widely accepted. Other structures however have also been proposed by experiment and theory. Shock compression data by [27] showed two discontinuities in sound velocity. Assuming the second discontinuity was due to melting, the first one might have been from a solid-solid transition, and therefore, new phases were considered in the following studies (e.g., [28,29]).

A mixture of fcc and hcp phase was identified in X-ray diffraction (XRD) patterns in DAC samples quenched from high temperatures greater than 3700 K at above 160 GPa [30]. From their own first-principles calculations, Mikhaylushkin et al. [30] showed that the fcc-hcp transition boundary would be more temperature dependent at higher pressures due to the change in magnetic property [30] (Figure 2a). Stixrude [24] confirmed this temperature-dependent fcc-hcp transition slope of [30] by his own first-principles calculations within the quasiharmonic approximation, while he concluded that the hcp phase should be stable under Earth's core conditions. The *P-T* conditions of the experiments by [30] were close to the melting curve (Figure 2) and the Gibbs energies of hcp and fcc are similar at these *P-T* conditions in the vicinity of the fcc-hcp transition boundary, and therefore the possibility that the Fe sample once melted and the mixture of fcc and hcp phase crystallized from the melt upon quenching in [30] cannot be ruled out.

Possible stability of a bcc phase in the inner core was also proposed from classical molecular dynamics simulations [29] and later supported by several groups with the ab initio molecular or lattice dynamics simulations [31,32] (Figure 2a). Those studies claimed that the bcc phase at core pressures would not be dynamically stable at low temperatures, but should become stable at core temperatures due to the entropy effect. Luo et al. [31] discussed that the stability of the bcc phase cannot be properly addressed within the quasiharmonic approximation. There were no reports from experiment which observed the presence of the high-pressure bcc phase in pure iron. Since the expected *P-T* conditions are extremely high for a DAC study, there were not many XRD data relevant to the presence/absence of the high-pressure bcc phase [15,25]. In addition, so-called "fast crystallization" in these *P-T* conditions might have hindered those studies from obtaining a clear diffraction line.

- **The Melting Point at the ICB Pressure, 330 GPa**

The melting temperature of iron under core pressure has been extensively argued since 1986. Extrapolated shock wave data [27] indicated that the melting point of iron at 330 GPa was 6500 ± 300 K, while DAC studies reported significantly lower temperature of 4850 ± 200 K when extrapolated from 200 GPa [13] (Figure 2b). The melting points on the Hugoniots of different groups' shock compression data were generally consistent [33,34]. The latest shock compression data in which XRD patterns of liquid iron were obtained as a melting criterion reinforced the validity of the previous shock wave data [35] (Figure 2b). As such, the shock compression data are well consistent among different groups. First-principles calculations predicted melting points of hcp iron which are consistent with the shock compression data [36–38] although Laio et al.'s [39] calculations were consistent with the DAC data by [13] (Figure 2b). If the high-pressure bcc phase is stabilized, the melting point of Fe at the ICB pressure would be increased by about 300 K [29]. As such, there was a consensus that the melting point of iron at the ICB is in the order of 6300–6500 K between the studies with shock compression experiments and first-principles calculations.

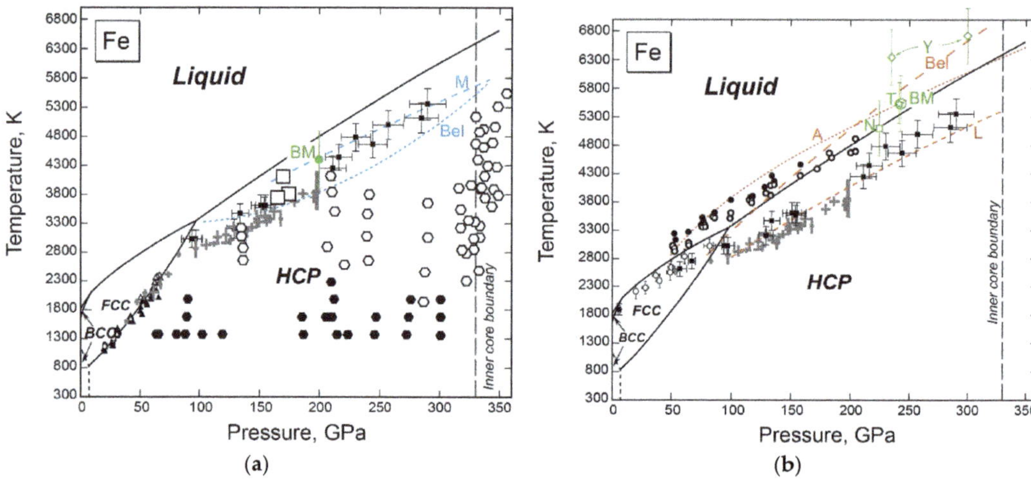

Figure 2. A phase diagram for iron [9] for discussions of (**a**) subsolidus and (**b**) melting relations. In (**a**), the plotted data are, the fcc–hcp transition boundaries in diamond anvil cells (DAC), Komabayashi et al. [14] (open triangle, fcc; solid triangle, hcp) and Boehler [13] (cross); stability of the hcp phase in DAC, Tateno et al. [23] (open hexagon) and Kuwayama et al. [26] (solid hexagon); stability of the fcc phase in DAC, Mikhayulskin et al. [30] (square box); melting of the hcp phase in DAC, Boehler [13] (cross) and Sinmyo et al. [25] (small solid square with error bar); a phase transition in shock compression, Brown and McQueen [27] (BM, dot); the fcc-hcp transition by theory, Mikhayulskin et al. [30] (M, blue dashed line); the hcp-bcc transition by theory, Belonoshko et al. [29] (Bel, blue dotted curve). In (**b**), the plotted data in addition to those in (**a**) are, melting in DAC, Jackson et al. [40] (open hexagon with error bar) and Anzellini et al. [15] (open circle, subsolidus; solid circle, melted); melting of the hcp phase by theory, Alfè et al. [38] (A, red dotted line), Belonoshko et al. [37] (Bel, red long dashed line), and Laio et al. [39] (L, red short dashed line); melting in shock compression, Brown and McQuee [27] (BM, green open circle), Yoo et al. [34] (Y, green open diamond), Nguyen et al. [33] (N, green open square), and Turneaure et al. [35] (T, green open star).

Newly improved static experimental measurements were reported by Anzellini et al. [15] in which melting was detected by the presence of diffuse scatterings from Fe liquid in laser-heated DAC. The previous melting experiments were based on the loss of diffractions lines from solids [12] or the detection of movement on the sample surface [13] as melting criteria. The results of [15] are well consistent with the shock wave and first-principles calculations, and the melting point of Fe at the ICB pressure should be 6230 ± 500 K [15]. Since the report by [15], the presence of a diffuse scattering has become a standard criterion for melting in the experimental mineral physics community. Published in the same year, Jackson et al. [40] also determined the melting points of fcc iron under pressure to 82 GPa using synchrotron Mössbauer spectroscopy which enabled them to monitor the dynamics of the iron atoms (Figure 2b). Their data showed somewhat lower melting temperatures than those of [15], but consistent with earlier [12,13] and recent measurements [25]. On the other hand, Morard et al. [41] discussed that Jackson et al.'s [40] data can be consistent with Anzellini et al.'s [15] when their pressure values were recalculated with revised thermal pressure values. As discussed later, the data of Jackson et al. [40] can be integrated into a thermodynamic model which is consistent with multi-anvil phase equilibrium data in the Fe-O system.

The quest for the accurate melting curve of iron in DAC is still on-going. Sinmyo et al. [25] used an internally resistive-heating system in DAC to place constraints on the melting point under high pressure to 290 GPa (Figure 2b). Their experiments had an advantage of stable and precise high-temperature generation as was the case for the fcc-hcp transition by [14]. The melting criteria in Sinmyo et al.'s [25] experiments did not include the presence of diffuse scatterings in XRD as was taken in [15] because the observation of a diffuse scattering requires a sufficient amount of sample liquid, which Sinmyo et al. [25] claimed could be a source of overestimation of the experimental temperature on laser heating in [15]. Sinmyo et al.'s [25] melting points of hcp iron are systematically lower than Anzellini et al.'s [15] and higher than Boehler's [13] data and the extrapolated melting point at 330 GPa is 5500 ± 220 K.

- **An Integrated Thermodynamic Model**

Having reviewed the key phase relations of iron above, I here discuss the holistic picture of iron phase relations. Different pieces of experimental information can be integrated into a self-consistent thermodynamic model [9,42,43]. Among the latest models, the Fe database constructed by [9] has been widely used as a model based on latest high-*P-T* experimental measurements [4,14,15,44] together with 1 bar metallurgy data [43], although the model does not include the high-pressure bcc phase. Komabayashi [9] started with examining the thermodynamics of subsolidus phase relations including the fcc-hcp transition and extended it to include the liquid phase. The liquid EoS was assessed so that thermodynamic calculations would reproduce experimental data. As discussed later in the Fe-O system (Section 3.2.1), the melting points of Fe need to be lower than those of FeO at pressures to ~20 GPa, and therefore Fe melting points in [9] are somewhat lower than in Anzellini et al. [15] and more consistent with Jackson et al.'s [40] data (Figure 2b). Note that considering the experimental uncertainties, Komabayashi's [9] model is still consistent with both [15,40]. At higher pressures, the calculated melting curve becomes more consistent with Anzellini et al.'s [15] data, and reproduces the results of shock wave experiments and first-principles calculations (Figure 2b). In the following discussions, the iron database constructed by [9] will serve as a reference system.

- **Liquid Properties under Core Conditions**

Here I discuss some of the resulting properties of the thermodynamic model [9]. The entropy of fusion at 330 GPa calculated from the model is 7.17 J/K/mol, which is fairly consistent with 8.73 J/K/mol by the first-principles calculation [45]. The good agreement between the two different approaches implies that the values are reliable. Indeed, these values are close to the gas constant and this would prove that the Richard's rule (a constant entropy change on melting for metals on the order of the gas constant) (e.g., [46]) is still valid under such extreme *P-T* conditions.

The longitudinal wave velocity (Vp) and density of iron liquid calculated from the thermodynamic model [9] are compared with those from shock wave experiments [47], DAC experiments [48], and first-principles calculations [49,50], together with seismological models (preliminary reference Earth model (PREM), Dziewonski and Anderson [51]) over the outer core depths between the core-mantle boundary (CMB) and ICB (Figure 3). The different types of measurements, except those by [50], show consistent Vp and density of iron liquid. The validity of the EoS for liquid iron in [9] was further confirmed by the first-principles calculation to 2 terapascals and 10,000 K [49]. As such, the Fe database of Komabayashi [9] based on static experimental data is consistent with shock compression studies and theoretical studies in phase relation and EoS of the phases. If one models the EoS for liquid iron using Sinmyo et al.'s [25] melting curve, its resulting density will be greater than the case of Komabayashi's [9] melting curve, which will have to be tested by comparing with those different types of measurements in the future.

Figure 3 shows that liquid iron shows a greater density by 7.1% and a reduced velocity by 3.7% compared with the PREM at the ICB conditions [9]. Similarly, the solid hcp Fe shows 4.5% greater density than the inner core. These values, so-called core density deficit (cdd), are the important starting point for the discussions of the kinds and amounts of light elements dissolved in the core. The cdd of the inner core has often been revised upon publication of a new EoS for the hcp phase and the most recent estimate is based on Fei et al. [52] of 3.6% at 6000 K. The cdd obviously depends on core temperature which can be constrained by the melting point at the ICB. The densities of iron liquid and solid, and the melting point of iron at the ICB can only be simultaneously constrained by internally consistent thermodynamic models (e.g., [9]).

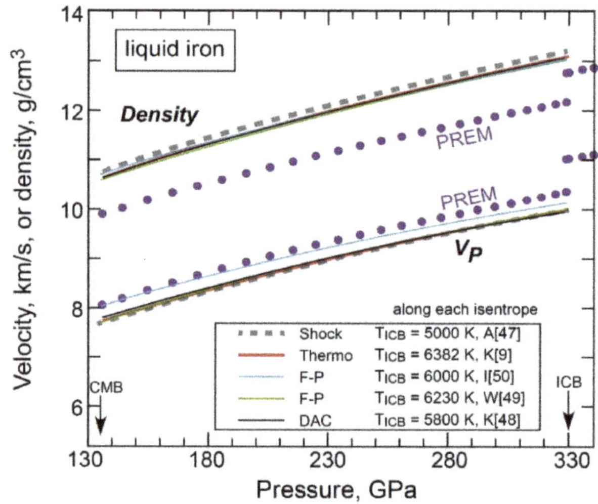

Figure 3. Density and longitudinal wave velocity (*Vp*) profiles for pure Fe over the outer core pressure range along each isentrope: shock wave study (Anderson and Ahrens [47]); thermodynamic model (Komabayashi [9]); first-principles calculation (Ichikawa et al. [50]; Wagle and Steinle-Neumann [49]); static measurements in DAC (Kuwayama et al. [48]). Data of the preliminary reference Earth model (PREM, Dziewonski and Anderson [51]) are also shown.

3.1.2. Fe-Ni

Phase relations of the Fe-Ni system are reviewed here. In particular (i) the fcc-hcp transition boundary, (ii) the stable structure in the inner core, and (iii) the melting point at the ICB, are discussed in comparison with the cases of pure Fe (Section 3.1.1). The Ni content in Earth's core is expected to be 5–15 wt% from the cosmochemical arguments (see [53]) and therefore many of the existing reports were made on the Fe-rich portion, namely Fe-5-15Ni.

- **The fcc-hcp Transition Boundary**

Figure 4a summarizes the results of experimental determinations of the fcc-hcp phase transitions in the Fe-Ni system together with the results of pure iron (Fe-10Ni, [54]; Fe-5.4Ni, [55]; Fe-10.2Ni, [56]; Fe-9.7Ni, [57]; and Fe, [12,14]). Lin et al. [54], Mao et al. [55], and Shen et al. [12] did not consider thermal pressure effects upon heating, while Dubrovinsky et al. [56] and Komabayashi et al. [57] estimated the pressures at high temperatures, and therefore the comparison should be made for each of the former and latter groups. Both groups show that the addition of Ni to Fe reduces the transition temperature, expanding the stability field of the fcc phase. Komabayashi et al. [57] reported a narrower two-phase region where the fcc and hcp phases coexist than the other three works (Figure 4a). This may be a direct result of the use of the internal resistive-heating system with improved precision in temperature [57].

- **The Stable Structure in the Inner Core**

Since the fcc-hcp transition boundary in Fe-10Ni is placed at only slightly lower temperatures than in pure Fe, the stable phase in Fe-Ni Earth's inner core is expected to be hcp as well (e.g., [57]). However, Dubrovinsky et al. [56] reported the formation of a bcc phase in Fe-10Ni at above 225 GPa and 3400 K (Figure 4b). They also conducted first-principles calculations and found that the high-pressure bcc phase was dynamically unstable within the quasiharmonic approximation as is the case for pure Fe [31]. Their calculations demonstrated that Fe and Fe-10Ni bcc phases showed similar phonon spectra which implied that the effect of Ni is minor, and therefore they expected that high temperature should stabilize the bcc phase due to the entropic effect, same as for pure Fe [56]. Then, Sakai et al. [58] and Tateno et al. [59] conducted laser-heated DAC experiments on the same composition Fe-10Ni, but did not observe any phases other than hcp in their DAC experiments (Figure 4b).

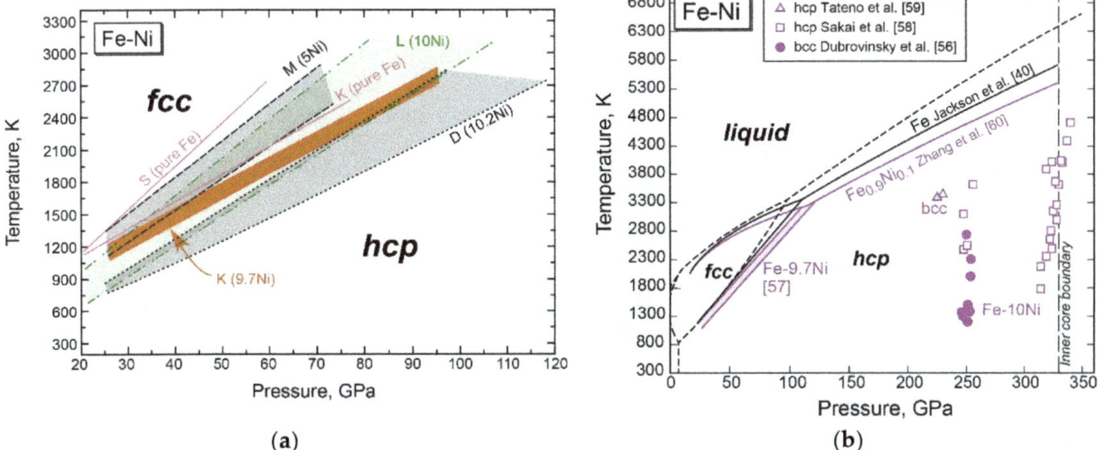

| (a) | (b) |

Figure 4. (**a**) Comparison of the *P-T* location of the fcc-hcp transition boundaries (L, Lin et al. [54]; M, Mao et al. [55]; D, Dubrovinsky et al. [56]; K, Komabayashi et al. [57]). The Ni content in wt% in each study is shown. For reference, the fcc-hcp transition boundaries in pure iron are shown (S, Shen et al. [12]; K, Komabayashi et al. [14]). Lin et al. [54], Mao et al. [55], and Shen et al. [12] did not take the thermal pressure upon laser heating into consideration, while Dubrovinsky et al. [56] and Komabayashi et al. [14,57] evaluated the pressures at high temperatures. (**b**) A phase diagram of Fe-10Ni (purple): the fcc-hcp transition boundaries [57]; melting of the fcc phase [60]. For comparison, iron phase relations are shown (black dashed line [9]; black solid line [40]). Note that Zhang et al. [60] took the same method as in Jackson et al. [40] to determine the fcc melting temperatures, and therefore, the effects of Ni can be seen from comparison of these two datasets. The stability of the hcp phase and high-pressure bcc phase was examined by Sakai et al. [58], Tateno et al. [59], and Dubrovinsky et al. [56].

- **The Melting Point at the ICB**

The melting points of the fcc and hcp phases in Fe-10Ni were measured to 125 GPa using laser-heated DAC and synchrotron Mössbauer spectroscopy by Zhang et al. [60], which was the same methodology as in [40] for pure Fe (Figure 4b). The results indicate that the addition of Ni negligibly changes the melting points of the fcc phase. However, the triple point where the hcp, fcc, and liquid phases are stable is moved to a higher pressure by the addition of Ni because of the steeper d*P*/d*T* slope of the fcc-hcp transition boundary [57], which results in a depression of the melting point of the hcp phase by about 200 K for Fe-10Ni (Figure 4b). Similar discussions were recently made by Torchio et al. [61] from their new experimental measurements of melting point in more Ni-enriched systems to Fe-36Ni. As discussed later in the Fe-Ni-Si system, Ni-bearing systems may be more complicated than expected from relevant binary systems.

3.2. Binary Systems with Light Elements

Here I review phase relations of binary systems with light elements. The phase relations of iron by Komabayashi [9] are compared as a reference system.

3.2.1. Fe-O

The Fe-O system shows many intermediate compounds between Fe and O. The relevant compositional range for the Earth's core falls between Fe and FeO and the subsystem Fe-FeO has been extensively studied.

- **Solid FeO**

FeO shows rich polymorphism from 1 bar to core *P-T* conditions including magnetic and metal-insulator transitions [62–70]. A metallic B8 structure was previously considered to be relevant to Earth's core due to an inferred strongly pressure-dependent phase boundary between B1 and B8 phases at about 70 GPa [63,65,66,71]. However, the slope of the boundary turned out to be less pressure-dependent [68] and the B8 phase was found to transition to a B2 phase under Earth's core *P-T* conditions [69]. On the other hand, the B1 phase is stable as a liquidus phase over a wide *P-T* range. This phase was considered to be insulating, but Fischer et al. [72] and Ohta et al. [70] demonstrated from either experiment, theory, or in combination, that it undergoes an insulator-metal transition at 30–90 GPa depending on temperature (Figure 5).

- **Melting of FeO**

Komabayashi [9] modelled the thermodynamics of FeO melting, while assessing the EoS of FeO liquid. The key phase relation is that B1 FeO is stable over a wide *P-T* range from 1 bar to 240 GPa and 5000 K as a liquidus phase (Figure 5), which enabled constraining the liquid EoS parameters using the EoS of the counterpart of solid B1 phase (e.g., [73]). The melting temperatures of B1 FeO were experimentally constrained to 77 GPa [74–77] and Ozawa et al. [69] constrained the stability of solid FeO phases (B1 and B2) up to about T = 5000 K at P = 240 GPa. Komabayashi [9] obtained EoS parameters of liquid FeO that give a calculated melting curve consistent with selected experimental measurements. The calculated melting curve is shown in Figure 5. Komabayashi [9] pointed out that consistency between the melting points of Fe and FeO should be considered (Figure 6a,b); the melting temperature of FeO needs to be higher than that of pure iron at about 15 GPa in order to reproduce binary phase assemblages determined in multi-anvil experiments (e.g., [77]) (Figure 6b). The melting temperatures of FeO determined by [74] are much lower than those of pure iron [15] at about 15 GPa and hence were not integrated in the thermodynamic model by [9] which is consistent with [15].

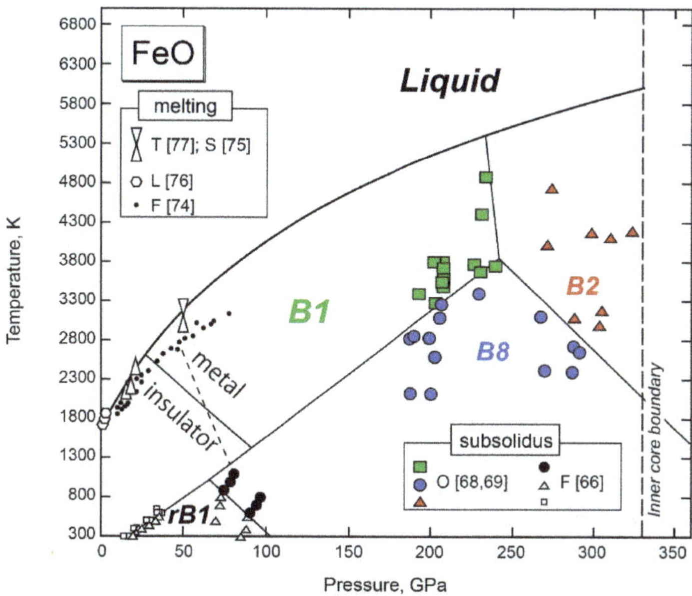

Figure 5. Phase relations of FeO. The solid black curve is the melting curve of the B1 structure extended to the inner core-outer core boundary, which was calculated with the thermodynamic model [9]. Thin solid straight lines are solid-solid reactions which divide stability fields of the phases. The B1 structure undergoes insulator-metal transition (dashed line [72]; solid line [70]). Melting experimental data to constrain the equation of state (EoS) of the liquid are plotted (Lindsley [76]; Tsuno et al. [77]; Seagle et al. [75]). High-pressure stability of solid FeO is indicated (Ozawa et al. [68,69]). The stability of rhombohedral B1 (rB1) phase is by Fei and Mao [66]. The small filled circle is the melting point of FeO (Fischer and Campbell [74]) which was not used for the liquid EoS determination in [9].

- **The Fe-FeO System**

The Fe-FeO system is characterized by the presence of a large miscibility gap above the solidus between Fe-rich metallic liquid and FeO-rich ionic liquid [77–79] (Figure 6a). The mixing property of liquid Fe and FeO was well examined at 1 bar [80,81]. The pressure dependence of the mixing interaction parameters was assessed by Frost et al. [82] based on two-liquid immiscibility data obtained in their own experiments up to 25 GPa. At greater pressures, the system behaves as a simple binary eutectic system [75] (Figure 6c,d). Komabayashi [9] calculated the eutectic relations to 50 GPa (Figure 6b,c) using the mixing parameters by [82] and at the core pressures (Figure 6d) with ideal mixing employed. Although Frost et al.'s [82] model was aimed at calculating the oxygen partitioning between the mantle and core, their mixing parameters reproduced the experimental data to 50 GPa well (Figure 6c). Komabayashi [9] also calculated the pressure dependence of eutectic composition, which is compared with experimental measurements in Figure 7 [75,77,78,83–86]. The calculated oxygen content in eutectic melt shows a steep rise at 30–60 GPa [9] which is consistent with earlier experimental data in laser-heated DAC [75]. Recent experiments further reinforced this pressure effect [85,86] (Figure 7), although Morard et al. [85] reported the steep rise at some greater pressures (~80 GPa).

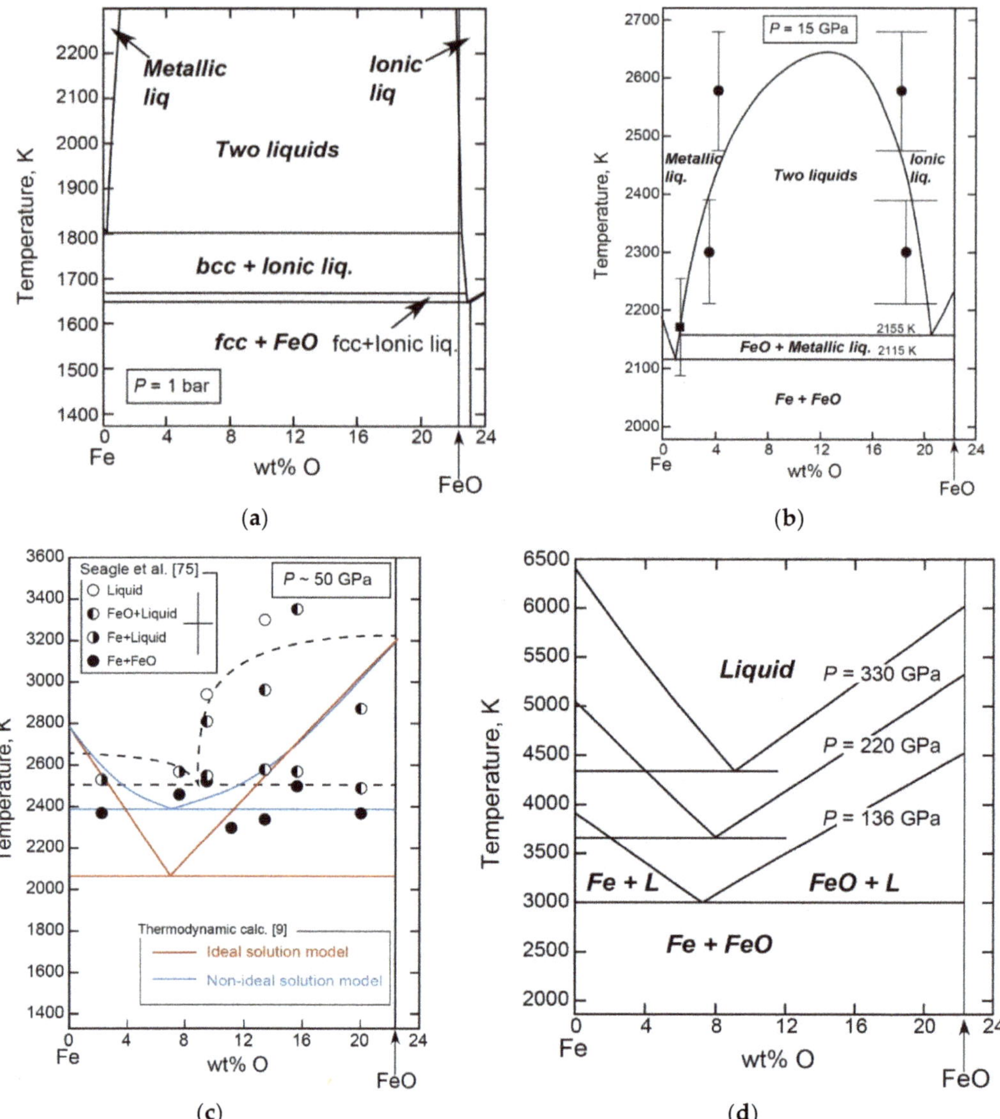

Figure 6. Eutectic relations in the system Fe-FeO at (**a**) 1 bar, (**b**) 15 GPa, (**c**) 50 GPa, and (**d**) 136–330 GPa. (**a**) is from [79] and (**b**–**d**) are from [9]. In (**b**), calculated phase relations [9] and experimental data of the compositions of liquids [77] are plotted: circle, coexisting two liquids; square, a liquid coexisting with solid FeO. In (**c**), calculated eutectic relations [9] are compared with laser-heated DAC experimental data [75]. The experimental error bar is shown in the inset, and the dashed line is the liquidus curve that Seagle et al. [75] drew. The calculations were made with the ideal solution and nonideal solution models [82] assumed. In (**d**), eutectic relations to the inner core boundary pressure were calculated with the ideal solution employed for the liquids. L, liquid.

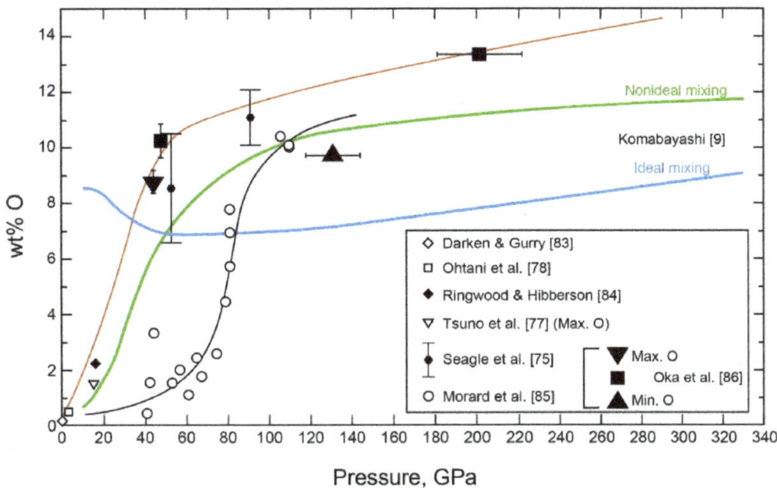

Figure 7. Change in eutectic composition of the Fe-FeO system as a function of pressure. Calculated eutectic compositions are shown as green (nonideal) and blue (ideal) lines [9]. Experimental estimates are plotted (Darken and Gurry [83]; Ohtani et al. [78]; Ringwood and Hibberson [84]; Tsuno et al. [77]; Seagle et al. [75]; Morard et al. [85]; Oka et al. [86]). The red and black curves were fitted to [86] and [85], respectively.

I here calculated a *G-X* relation for liquids and solids in the Fe-FeO system from the thermodynamic database [9,82] (Figure 8) to make detailed analyses of the increasing oxygen content in eutectic melt with increasing pressure. At a low pressure of 15 GPa, the nonideal mixing of Fe and FeO liquids forms two minima in the free energy curve, which accounts for the liquid immiscibility for intermediate bulk compositions. The eutectic point where solid fcc Fe and liquid coexist is very close to Fe (Figure 8, upper panel). With increasing pressure, the pressure effect stabilizes the liquids in the intermediate compositional range, and therefore the free energy curve becomes a simple downward convex curve where the two-liquid field diminishes (<50 GPa) while the eutectic point is moving towards the FeO side (Figure 8, middle panel). At a greater pressure of 90 GPa, the eutectic point is at about 9 wt% O (Figure 8, lower panel) and there is no further move with further increasing pressure because there is no longer drastic change in the shape of the free energy curve.

A liquid immiscibility occurs when the mixing enthalpy is a large positive value and disappearance of the two-liquid field often occurs with increasing temperature at a fixed pressure, for example, at 1 bar, due to the effect of configurational entropy. However, the Fe-FeO system shows disappearance of the liquid immiscibility with increasing pressure at a fixed temperature as well [82]. This indicates that the mixing enthalpy becomes smaller with pressure; in other words, the end-member liquids which are metallic Fe-rich liquid and ionic FeO-rich liquid become compatible at high pressures. Komabayashi [9] discussed that a potential source for the reduction in nonideality with increasing pressure is metallization of FeO liquid. Metallization of subsolidus B1 FeO was observed by experiment and theory [70,72], which occurred at 30–90 GPa depending on temperature. The structure of a liquid may change according to that of the counterpart crystalline phase (e.g., [87]). Hence FeO liquid might become a metal at pressures in the vicinity of the transition in the solid, which might make it compatible with Fe liquid. As such, the rapid increase in the oxygen content in eutectic melt at 30–60 GPa could be as a consequence of metallization of liquid FeO [9,85]. Although there is a slight difference in the pressure where the oxygen content starts rising between the studies [9,75,85,86], the important point is that the pressure dependence of the nonideal mixing parameters established for another geological reaction [82] reproduces the change in eutectic composition with pressure [9] and this is also consistent with the pressure range of metallization of B1 FeO [70,72].

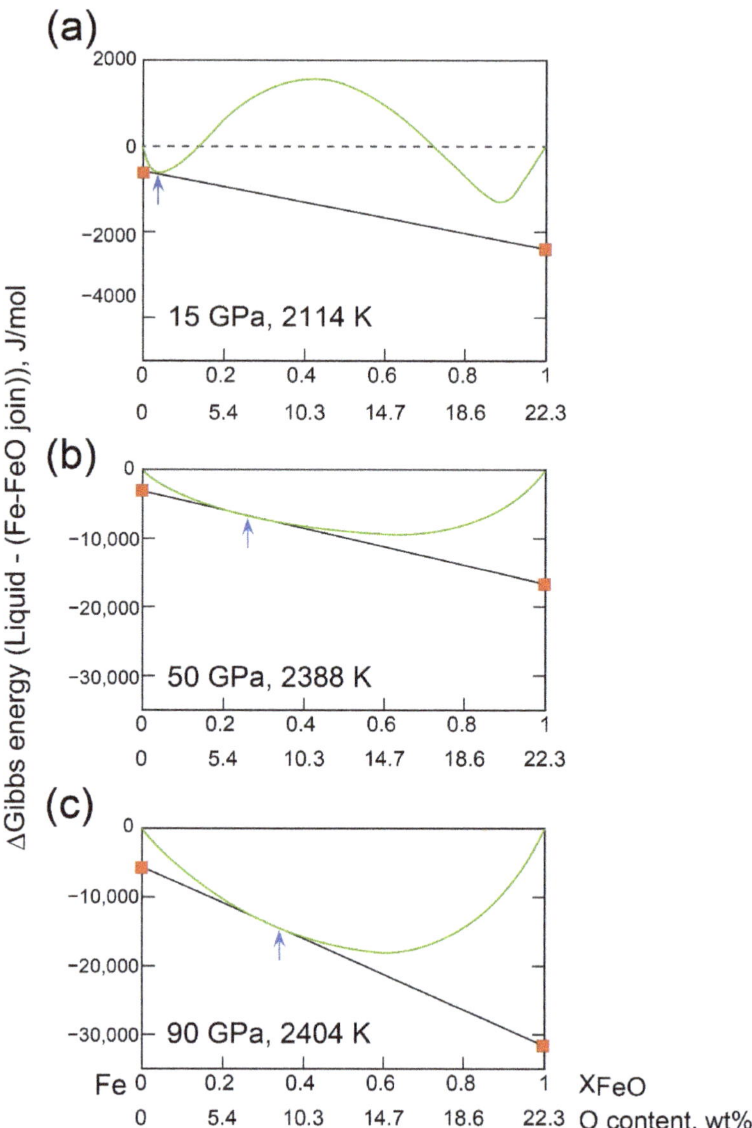

Figure 8. Gibbs energy-composition (*G-X*) diagrams for the Fe-FeO system at the eutectic temperatures for (**a**) 15 GPa, (**b**) 50 GPa, and (**c**) 90 GPa, which were calculated with the thermodynamic model [9]. The green curve represents the free energy curve for the liquids and the red squares are the free energies of solid Fe and FeO. The eutectic points are indicated by arrow. From low to high pressures, the nonideality of the Fe-FeO liquids decreases, which moves the eutectic point from oxygen-depleted side to -enriched side. The reduction of nonideality should be associated with metallization of B1 FeO [70,72].

As for the eutectic temperature, Komabayashi [9] discussed that the nonideal mixing model reproduced DAC experimental data to 50 GPa and the ideal mixing model worked for the data of [69] to pressures above 200 GPa (Figure 9). Recent laser-heated DAC data [85,86] are also consistent with the ideal mixing model. This agreement between the thermodynamic calculations and experiments indicates that the mixing property indeed changes from the highly nonideal at low pressures to ideal under core pressures. The calculated eutectic compositions at the core pressures are slightly depleted in oxygen than the experimental measurements (Figure 7). This could be resolved if the melting points of FeO would be elevated, which can be tested in the future.

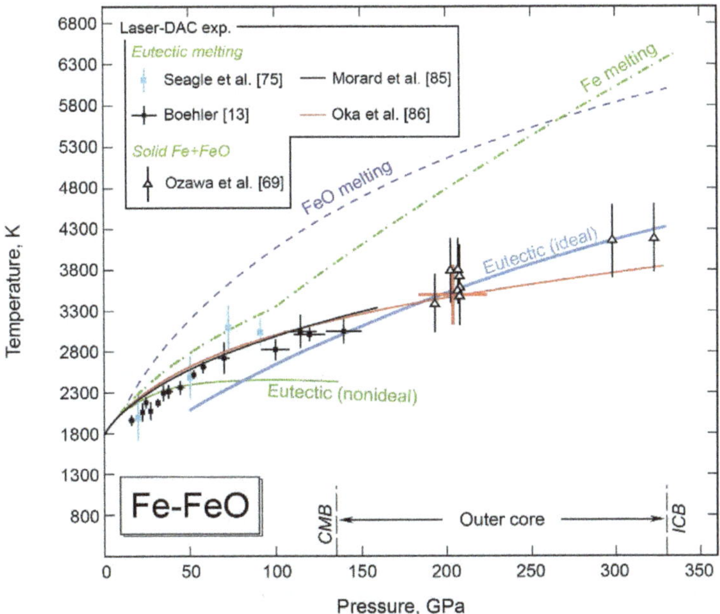

Figure 9. Eutectic melting temperatures in the Fe-FeO system together with the melting curve for each end-member composition. The green and blue solid lines give the calculated eutectic temperatures assuming the nonideal and ideal solution for liquids, respectively [9]. The squares with error bars are experimentally determined eutectic temperatures (blue, Seagle et al. [75]; black, Boehler [13]). The triangles are points of stability of solid Fe + FeO observed in DAC experiments [69]. The solid black and red curves are experimentally constrained eutectic curves by Morard et al. [85] and Oka et al. [86], respectively. The red cross denotes the error bar for [86]. CMB, core-mantle boundary; ICB, inner core boundary.

- **The Liquid Properties**

The Vp and density of Fe-FeO liquids were calculated from the thermodynamic model [9]. The results are shown in Figure 10 together with first-principles calculations [88,89]. The calculated results by [9] show that the addition of oxygen to Fe liquid reduces both the density and Vp, while the theoretical calculations show a reduction in density but an increase in Vp. Figure 10b shows the effects of oxygen on the liquid iron properties: density, adiabatic bulk modulus (Ks), and Vp. Komabayashi's [9] model predicted a negative effect (i.e., decrease) on the velocity by the addition of oxygen, while the first-principles calculations by Badro et al. [88] and Ichikawa and Tsuchiya [89] predicted positive effects (Figure 10b). However, this does not necessarily imply that the experiment-based and theory-based results are complete opposites. The result depends on the magnitude of reduction of the bulk modulus of iron liquid due to oxygen. All of [9,88,89] show that the addition of oxygen reduces both the bulk modulus and density (Figure 10b). The *Vp* of iron liquid is reduced in Komabayashi [9] because the bulk modulus decreases more than the density by mixing an oxygen component. As such, the *Vp* may increase or decrease due to oxygen depending on the difference in the bulk modulus between the end-members (i.e., Fe and FeO in [9]). In his modelling of the EoS of liquid phases, Komabayashi [9] demonstrated that the compressibility of a liquid should be close to that of the counterpart crystalline phase under core pressure, and the small bulk modulus of liquid FeO was a consequence of the small bulk modulus of solid B1 FeO. The EoS of solid B1 FeO was established based on experimental data to 207 GPa and 3800 K [69,73] which are relevant to core conditions and therefore its elastic behaviour can be reliably applied. Komabayashi [9] listed a set of EoS for the solids and liquids in the Fe-O system while the first-principles calculations did not provide such a self-consistent set of EoS for the solids and liquids. Further comparison is needed to resolve the difference between experiment and theory.

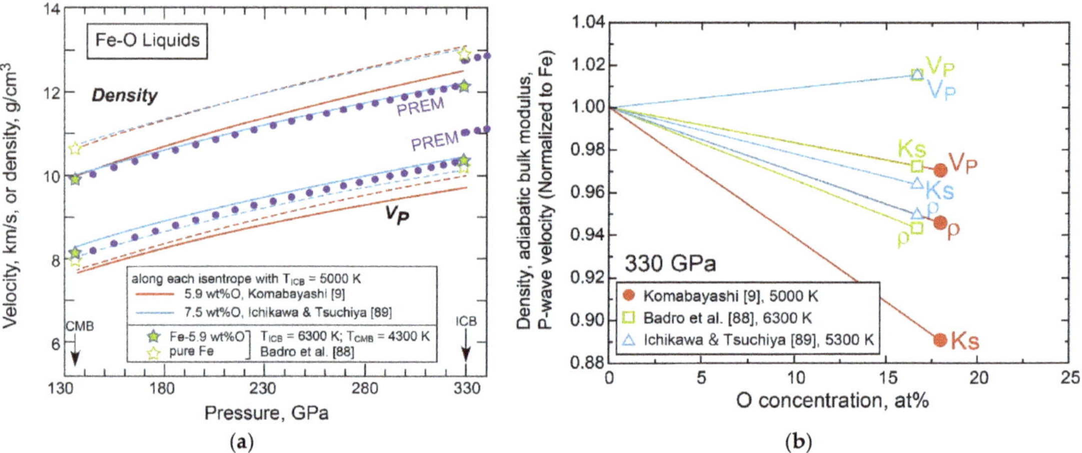

Figure 10. (a) Density and *Vp* profiles for Fe-O outer core models along each isentrope from thermodynamic calculations (Fe-5.9O, [9]) and first-principles calculations (Fe-7.5O, [89]). Furthermore, results of another theoretical calculation at the CMB and ICB are shown [88]. Note that the plots for Fe-5.9O needed a slight extrapolation in composition for [88]. For comparison, pure iron data are plotted (red dashed line with T_{ICB} = 6382 K [9]; blue dashed line with T_{ICB} = 6000 K [50]). (b) Effects of oxygen on the *Vp*, adiabatic bulk modulus (Ks), and density (ρ) of liquid iron at 330 GPa. Data from different approaches are compared (Komabayashi [9]; Badro et al. [88]; Ichikawa and Tsuchiya [89]). Note that the parameters are normalized to those for pure iron. The effect of oxygen on Ks is more pronounced than on ρ, resulting in a reduction of *Vp* in [9].

The above analyses suggest that Fe-O liquids may or may not account for the outer core profiles (Figure 10a). On the other hand, hcp iron may crystallise to form the inner core, if the outer core contains 5.9–7.5 wt% oxygen [88,89] since the oxygen content in eutectic melt would be greater (Figure 7). The crystallised hcp iron would however not accommodate any amount of oxygen [68] so that it cannot account for the inner core density deficit. Therefore, the addition of oxygen alone to iron cannot fully explain the PREM profiles of the core.

3.2.2. Fe-Si

The Fe-Si system is characterised by a narrow melting loop of the bcc Fe phase at 1 bar which partitions Si into both solid and liquid. This could account for the cdd of both the solid and liquid cores if high-pressure iron phases would show similar melting relations. The phase relations and EoS of solid phases in the Fe-(Fe)Si system have been extensively studied by both experiment and theory [90–99].

- **The *T-X* Relations**

Phase relations in the Fe-Si system are complicated with many different solid phases involved at low pressures, but they become simpler with increasing pressure (Figure 11). Above 100 GPa the subsystem Fe-FeSi, which is relevant to Earth's core, only includes hcp iron, B2, and liquid phases (e.g., [95]) (Figure 11d). Fischer et al. [95,100] determined phase relations based on in situ XRD while Ozawa et al. [98] placed constraints on the width of melting loops based on chemical analyses of recovered DAC samples. The two studies are qualitatively consistent, but some details need attention. While the compositional range of fcc + B2 at 50 GPa (Figure 11b) and hcp + B2 at 125 GPa (Figure 11d) constrained by Ozawa et al. [98] are not inconsistent with the data in Fe-9Si and Fe-16Si by [95], the compositional ranges of fcc + B2 and hcp + B2 at 80 GPa (Figure 11c) which are well constrained by [95] seem too wide compared with Ozawa et al.'s [98] data at 50 and 125 GPa (Figure 11b,d). This apparent discrepancy is due to the very large compositional range of the B2 solid solution phase in [98]. This should further be examined, by an alternative approach such as first-principles calculation.

- **The fcc-hcp Transition**

Because the Fe-phases (bcc, fcc, and hcp) form extensive solid solutions with Si, as was discussed in Fe-Ni (Section 3.1.2), the fcc-hcp transition boundary is discussed here. Figure 12 summarises the experimental attempts to determine the *P-T* locations of the fcc-hcp transition boundaries in the Fe-Si system with in situ synchrotron XRD [97,99,101]. An experimental study with laser-heated DAC reported that the transition temperature was greatly reduced when 3.4 wt% Si was added to Fe [101]. In contrast, Tateno et al. [97] showed increased transition temperatures in Fe-6.5Si, which was later supported by Komabayashi et al. [99] who employed internally resistive-heated DAC experiments in Fe-4Si (Figure 12). Komabayashi et al. [99] noted that Asanuma et al. [101] assigned tiny shallow rises as peaks from the fcc phase, while the appearance of the fcc phase was clearly marked by the presence of the (200) peak in [14,57,99] (Figure 13).

- **The Stable Structure in the Inner Core**

The stable structure in a hypothetical Fe-Si inner core has been discussed in relation to a reaction boundary: hcp → hcp + B2 (Figures 11d and 14) [93,95,97,102]. Fischer et al. [95] and Tateno et al. [97] determined this boundary in Fe-9Si by synchrotron XRD in laser-heated DAC to 201 GPa and 407 GPa, respectively, and both results agree well (Figure 14). According to Tateno et al.'s [97] boundary, if the inner core temperature is below 4800 K, the inner core can be a sole hcp phase with up to 9 wt% Si.

On the other hand, theoretical studies reported a possible presence of the high-pressure bcc phase at a lower Si concentration (~5 mol% = 2.6 wt%) under the inner core conditions [103,104]. Vočadlo et al.'s [103] calculations showed that the high-pressure bcc phase is dynamically stable at core temperatures, but thermodynamically less stable than the hcp phase in pure Fe. However, the addition of a few mol% Si stabilized the bcc phase relative to the hcp phase and they concluded that the bcc phase is a strong candidate for the inner core constituent phase. This is somewhat in contrast to Belonoshko et al. [29,32] who claimed that the bcc structure is more stable than the hcp structure even in pure Fe as we have seen in Fe section (Section 3.1.1). In addition, it is uncertain whether the high-pressure bcc phase is identical to the B2 phase observed in the experiments above, although the high-pressure bcc phase should take the same atomic coordinates as the B2 phase [103].

The above discussions are based on the stability of crystal structures under subsolidus conditions (see Figure 11b–d), but the liquidus phase of the outer core that crystallises to form the inner core depends on the eutectic composition. Ozawa et al. [98] placed constraints on the eutectic composition in the Fe-Si system from textural observations made on samples recovered from DAC experiments, which is compared with existing data in Figure 15 [95,97,98,100,105]. Their results show that the Si content in eutectic melt decreases with increasing pressure and is less than 1.5 ± 0.1 wt% Si at 127 GPa (Figure 15). The expected Si content in the outer core ranges from 4.5 to 11 wt% assuming that silicon is the sole light element in the core [10,88,89,96,106], in order to account for the cdd (see also the discussions about the Vp and density of Fe-Si liquids below). Such a Si-enriched liquid core should crystallise a less dense solid, CsCl (B2)-type phase at the ICB which would not meet the inner core density. However, the crystallization of the hcp iron phase would be possible if other light elements are present in the core to account for the cdd [98]. It would therefore be concluded that silicon cannot be the sole light element in the core.

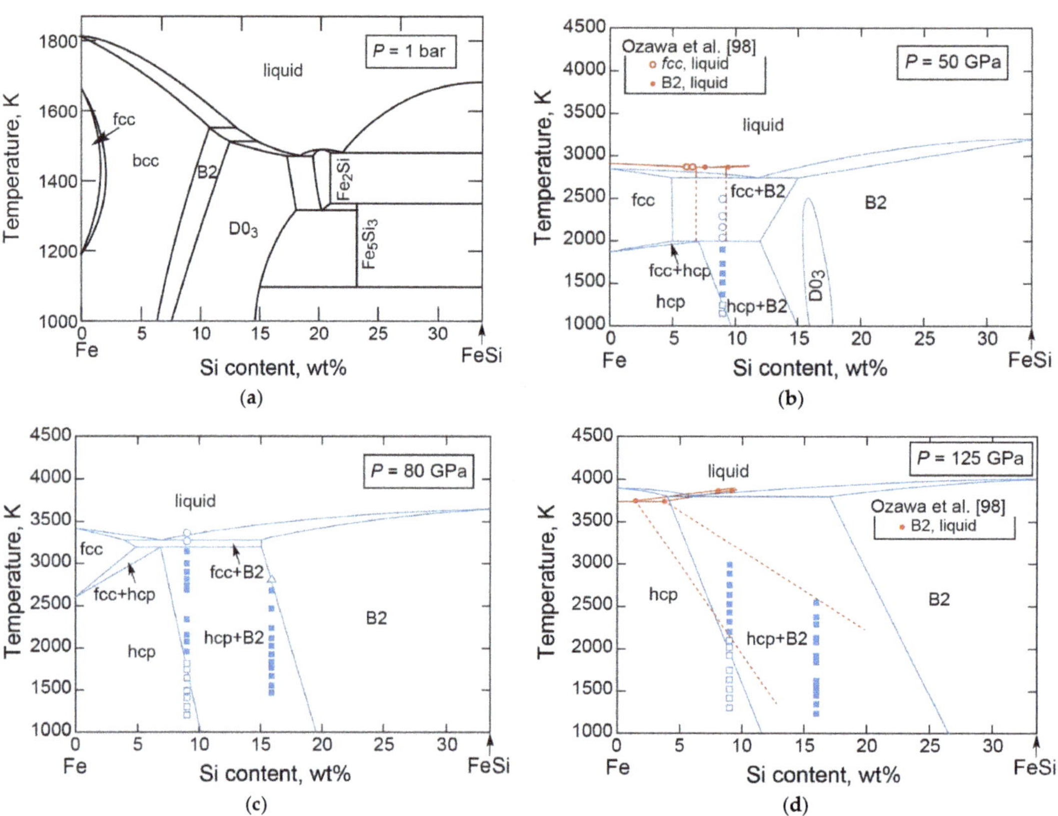

Figure 11. Eutectic relations in the Fe-Si system at (**a**) 1 bar, (**b**) 50 GPa, (**c**) 80 GPa, and (**d**) 125 GPa. (**a**) is from Ohnuma et al. [107], (**b**) from Fischer et al. [95] (blue) and Ozawa et al. [98] (red), (**c**) from Fischer et al. [95], and (**d**) from Fischer et al. [95] (blue) and Ozawa et al. [98] (red). For Ozawa et al.'s [98] data, the eutectic temperatures were constrained by [10]; solid lines are constrained by experimental data points, whereas the dashed lines are inferred boundaries which are drawn to be consistent with data in [95].

- **The Mixing Properties of Solutions**

Since many phases in the Fe-Si system show extensive mixing of Fe and Si, their mixing properties needs to be critically evaluated in order to understand the materials properties. The thermodynamic models established at 1 bar in metallurgy indicate that the low-pressure bcc, fcc, and liquid phases show strong negative nonideality, which stabilizes them more than what is expected from ideal mixing (e.g., [107]). Alfè et al. [90] discussed that the deviation from the ideal solution was weak in the hcp and liquid phases at low Si content under core conditions. More recently, Huang et al. [108] also reported that the mixing of Fe and Si in liquids at core pressures would be ideal from first-principles calculations. On the experimental side, Komabayashi [10] evaluated the mixing properties of the fcc and hcp phases based on the *P-T* locations of the fcc-hcp transition boundaries in Fe-4Si [99]. He applied the nonideal mixing model developed at 1 bar [107] and ideal mixing model to thermodynamic calculations of the transition and found that both models reproduced the experimental measurements at pressures to 40 GPa [99]. As such, it is reasonable to apply ideal mixing to the solutions in the Fe-Si system under the core *P-T* conditions.

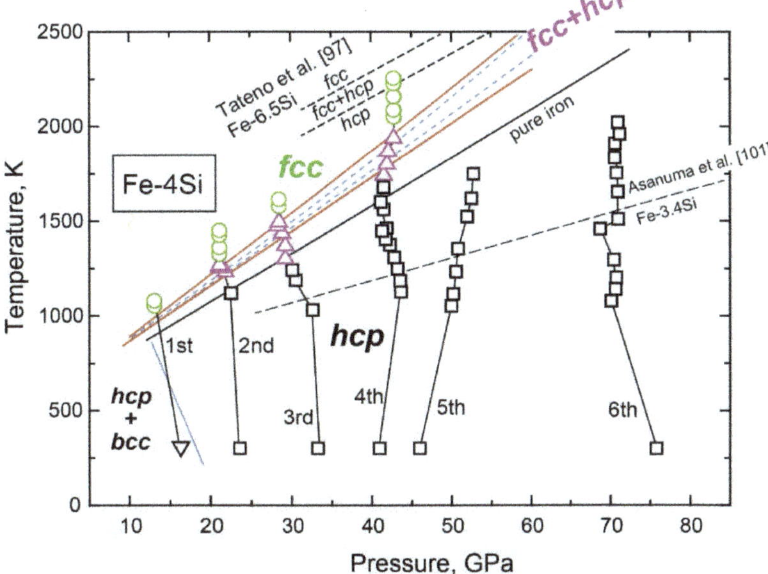

Figure 12. Calculated fcc-hcp transition boundaries for Fe-4Si with the ideal mixing model (red line) and nonideal mixing model (blue dashed line) [10]. Experimental results with in situ XRD in internally resistive-heated DAC are also plotted: inverted triangle, hcp + bcc; square, hcp; normal triangle, fcc + hcp; circle, fcc [99]. The transition boundaries in pure iron [14] (hcp-fcc), Fe-3.4Si [101] (hcp → hcp + fcc), and Fe-6.5Si [97] (hcp → hcp + fcc → fcc) are also shown.

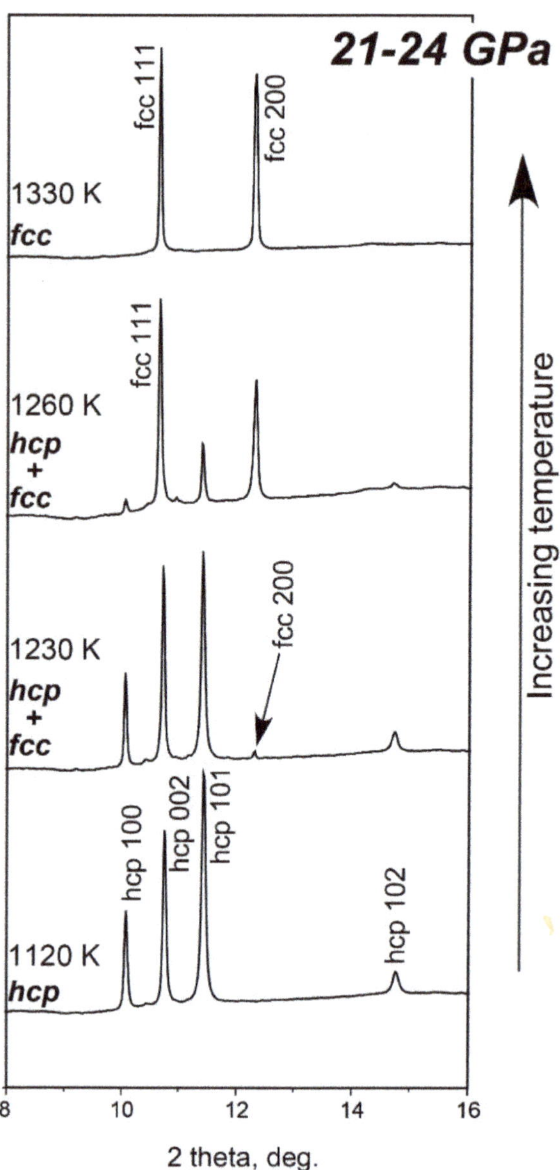

Figure 13. Series of XRD patterns collected in the run 2 in Figure 12 for increasing temperature [99]. The presence of the fcc phase was unambiguously marked by the appearance of (200) peak. The figure is taken after [99].

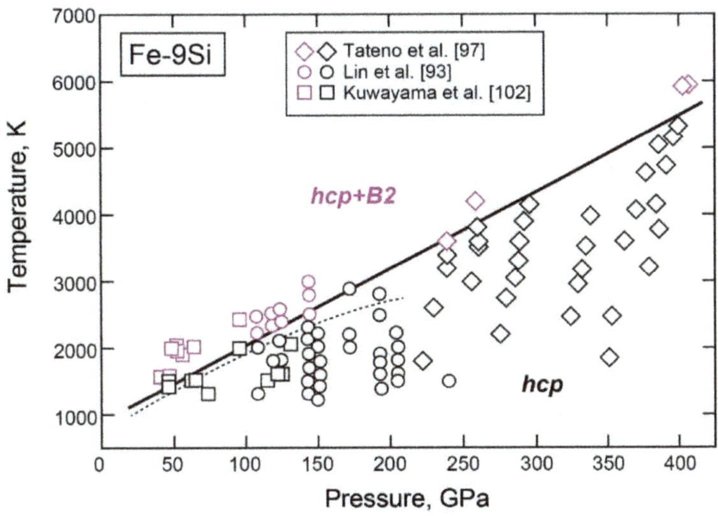

Figure 14. Phase boundary between hcp and hcp+B2 for Fe–9Si alloy [93,97,102]. The dotted curve is from [95]. The figure is taken after [97].

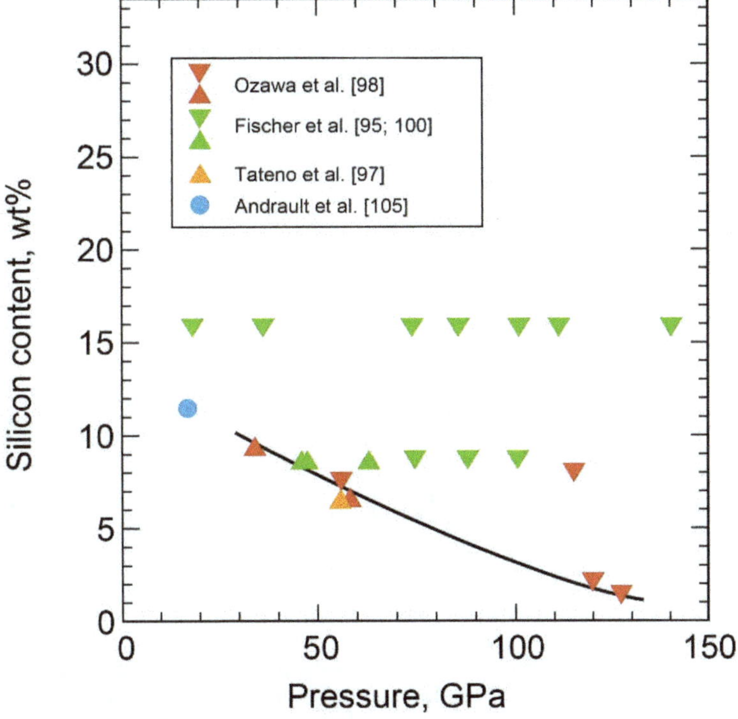

Figure 15. Silicon content in eutectic point as a function of pressure in the Fe–FeSi system [95,97,98,100,105]. The normal and inverted triangles denote the lower and upper bounds of the silicon content in eutectic melt, respectively. The thick black line shows the best-representative eutectic compositions [98].

- **The Melting Point at the ICB**

The melting temperature of Si-bearing iron phases (fcc and hcp) was not directly measured at pressures greater than 21 GPa [94], although high-pressure experiments observed melting of the B2 phase or eutectic of Fe + B2 in the Fe-(Ni)-Si system in the Si-rich portion of the system Fe-Si (Si > 10 wt%) [95,109–111]. There are no reports from theory on the melting point of hcp Fe-Si phases under core conditions, although Belonoshko et al. [104] reported the melting temperatures of bcc Fe and $Fe_{0.9375}Si_{0.0625}$ (Fe-3.2Si) at the ICB pressure of 7000 and 7200 K, respectively (Figure 16).

Another important melting property is the width of the melting loops of iron phases because it is the key to understanding the Si partitioning between the solid inner core and liquid outer core upon inner core crystallization [90,94,98]. At 1 bar the width of the melting loop for the bcc phase is as narrow as 2 wt% Si for Fe-5Si [107] (Figure 11a). Kuwayama and Hirose [94] suggested that the width of the melting loop for the fcc phase would be less than 2 wt% of Si at 21 GPa based on high pressure experiments performed in a multi-anvil apparatus. Recently Ozawa et al. [98] demonstrated from laser-heated DAC experiments that the fcc melting loop was as narrow as less than 1 wt% Si at 58 GPa. Alfè et al. [90] reported by first-principles calculation that silicon is almost equally partitioned between hcp iron and liquid at 370 GPa. As such the width of melting loop of the iron phases seems to become narrower with increasing pressure from 1 bar to over 3 Mbar.

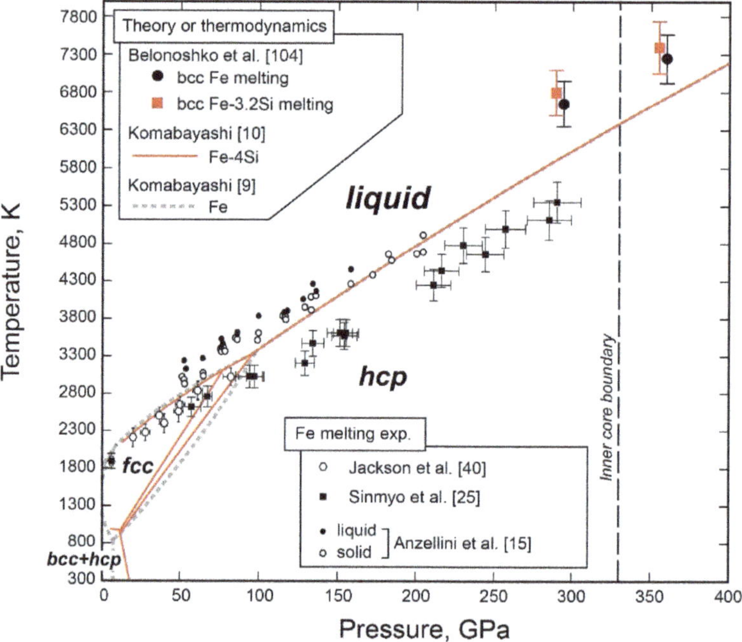

Figure 16. A phase diagram for Fe-4Si [10] in comparison to that of pure iron [9]. The fcc-hcp transition boundaries and melting curve of the hcp phase were calculated with ideal mixing of Fe and Si in the phases, while the other boundaries including the melting curve of the fcc phase were schematically drawn for Fe-4Si. The melting curves of fcc and hcp phases are compared with of pure Fe, namely, 20 K lower than pure iron at 60 GPa and the same as for iron above 90 GPa. Pure iron melting data are plotted [15,25,40]. The melting points of high-pressure bcc Fe and bcc Fe-3.2Si are also shown [104]. Note that the melting curve for Fe-4Si is to demonstrate that the effect of Si can be negligible. Ozawa et al. [98] reported that the silicon content in eutectic point is less than 4 wt% at pressures greater than 90 GPa (Figure 15).

Komabayashi [10] assessed the melting temperatures of Si-bearing fcc and hcp structures based on the width of the melting loops. The thermodynamic system that he considered was Fe-Si in which the hcp, fcc, and liquid phases were treated as solutions between Fe and hypothetical Si with the same structure, for example, hcp Fe and hcp Si. He assessed the melting temperature for the Si end-member ($T_{Si}^{melting}$) under high pressure, by reproducing the width of melting loops [90,98], adopting ideal mixing as discussed above. The key constraints came from that the entropy of fusion, $\Delta S^{melting}$, for both end-members (Fe and Si) are large of 7–10 J/K/mol (cf. $\Delta S_{Fe}^{fcc-hcp\ transition}$ = 3 J/K/mol at 60 GPa). With the ideal mixing model, when ΔS is large for both end-members, a phase transition loop would be wide, depending on the difference in transition temperature between the end-members, i.e., $\Delta T_{Fe-Si}^{melting}$ for the present discussion (e.g., Yamasaki and Banno [112]) (Figure 17). As such $\Delta T_{Fe-Si}^{melting}$ needs to be small in order to account for the reported narrow melting loops, namely, $T_{Si}^{melting}$, should be close to $T_{Fe}^{melting}$. When the melting phase loop was calculated to match the data by [98] at 60 GPa, $T_{Si}^{melting}$ needed to be 2600 K (Figure 18) which is only 300 K lower than $T_{Fe}^{melting}$ = 2910 K. Alfè et al. [90] concluded that Si is almost equally partitioned between solid and liquid to 20 mol% Si (11 wt% Si) at 7000 K and 370 GPa. Assuming an ideal solution for both liquid and solid, an equal partitioning of Si between solid and liquid implies composition-independent melting temperature. Thus-obtained melting curves for the hcp and fcc phases in Fe-4Si are shown in Figure 16. Note, that as discussed above, the Si content in the eutectic melt under the core conditions might be less than 1.5 wt% [98], and therefore Figure 16 is exclusively used for the effects of Si on the melting temperature of the hcp phases.

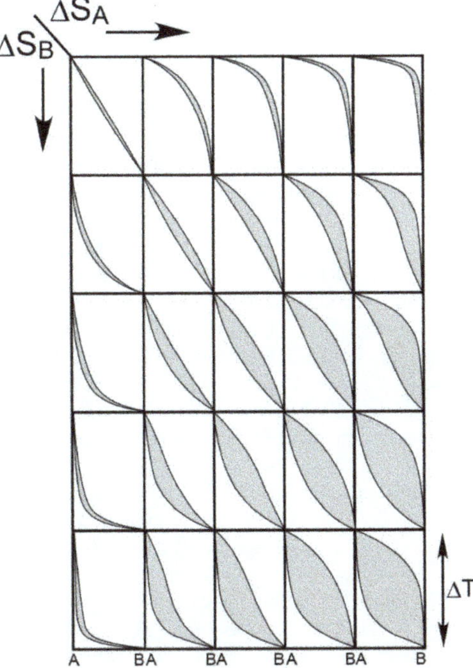

Figure 17. Variation in the shape of a two-phase loop in the binary AB system as a function of ΔS_A and ΔS_B (after [112]). Ideal mixing is assumed. When both ΔS are large, the loop is wide, unless ΔT is small. The figure is taken after [10].

For 1–8 wt% Si-bearing iron systems, which span the proposed values for the Si content in Earth's core [113–117], the expected change of liquidus temperature of Fe due to Si is about –50 K at 60 GPa and about 0 K at 330 GPa (Figure 16). This suggests that the addition of Si to Earth's core would not significantly affect the temperature at the ICB where the inner core is believed to be crystallised from the outer core.

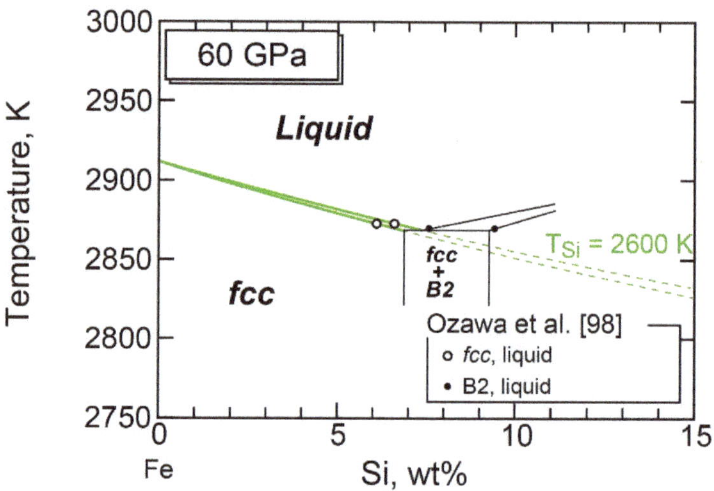

Figure 18. Calculated melting loop of the fcc phase at 60 GPa, with ideal mixing assumed between Fe and Si [10]. The green lines were obtained so that the calculations reproduce the reported Si partitioning data [98]. The Si partitioning data between the liquid and B2 phases are also plotted [98].

- **Liquid Properties**

From the above-constrained melting curve, Komabayashi [10] assessed thermal EoS of Si-bearing iron liquids. Figure 19a shows calculated density and Vp for an Fe-4Si outer core along its isentrope together with results of first-principles calculations [88,89], inelastic X-ray scattering experiments [118], and PREM [51]. Note that the Si contents in Komabayashi [10], Badro et al. [88], and Nakajima et al. [118] in Figure 19a are not meant to be best-fits to the PREM while that in Ichikawa and Tsuchiya's [89] calculations is the best-fit composition in the binary Fe-Si. All four the models show that the addition of silicon increases the velocity. However, the effects of silicon on the bulk modulus of liquid iron are in contrast between the experiment-based [10] and theory-based models [88,89] (Figure 19b). As discussed in the case of Fe-FeO (Section 3.2.1), the compressibility of a liquid should be close to that of the counterpart crystalline phase under core pressure. The EoS parameters for the Fe-Si hcp phase in Komabayashi [10] was based on compression experiments on the hcp phase by Tateno et al. [97] to 305 GPa who demonstrated that the addition of silicon surely increases the bulk modulus of hcp iron. Komabayashi [10] provided a set of EoS for the solids and liquids in the Fe-Si system while the first-principles calculations have not, the same as in the case of oxygen (see, Section 3.2.1. Fe-O).

Any of Badro et al. [88], Ichikawa and Tsuchiya [89], and Komabayashi's [10] models in Figure 19a require a Si content greater than 4 wt% to make the density and Vp profiles match the PREM. However, as discussed above, the Si content in the eutectic melt at the ICB is expected to be less than 1.5 wt% [98]. Therefore, the crystallising phase from such an outer core is likely the B2 phase, which should be less dense than the inner core. On the other hand, recent experimental measurements by Nakajima et al. [118] showed that the addition of Si to iron liquid produced an enhanced increase in the Vp and moderate reduction in density, which implies that the addition of Si alone cannot account for the observed seismic properties of the outer core. As such, all the *Vp*-density models plotted in Figure 19a suggest that silicon cannot be the sole light element in Earth's core.

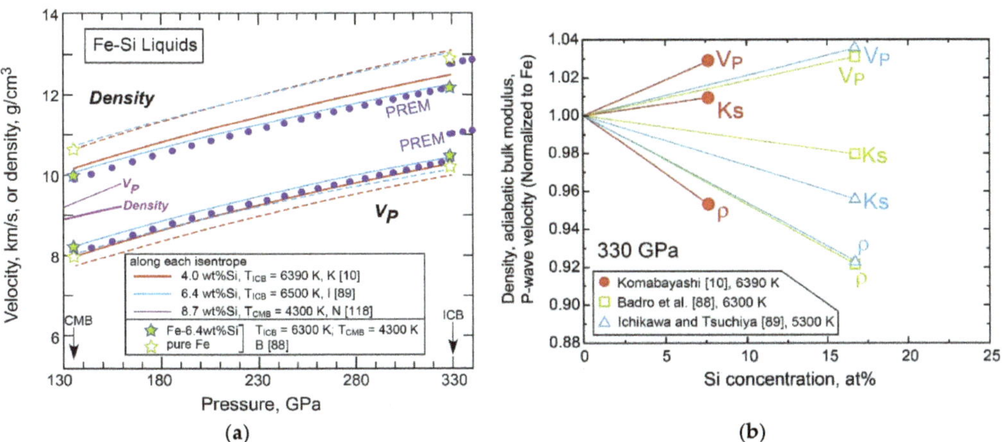

(a)   (b)

Figure 19. (**a**) *Longitudinal* wave velocities (*Vp*) and densities of Fe-Si liquids: Fe-4Si (Komabayashi [10]), Fe-6.4Si (Ichikawa and Tsuchiya [89]), Fe-6.4Si (Badro et al. [88]), and Fe-8.7Si (Nakajima et al. [118]). Each profile is along its isentrope. For comparison, pure iron data are plotted (red dashed line with T_{ICB} = 6382 K, Komabayashi [9]; blue dashed line with T_{ICB} = 6000 K, Ichikawa et al. [50]). (**b**) Effects of silicon on the *Vp*, adiabatic bulk modulus (*Ks*), and density (ρ) of liquid iron at 330 GPa [10]. Data from the first-principles studies are compared [88,89]. Note that the parameters are normalized to those for pure iron.

3.2.3. Fe-S

Of the candidate elements to account for the cdd, sulphur has been most extensively studied [119]; the Fe-S system is still currently of significant interest as new discoveries are being reported [120–125].

- **The *T-X* Relations**

    The 1-bar phase relations of the Fe-S system include many intermediate compounds (e.g., [126]) and these compounds show a series of phase changes under pressures [127,128] (Figure 20). Above 21 GPa the sulphide phase stable with Fe under subsolidus conditions is Fe₃S which takes a tetragonal system [128] forming a simple eutectic system [124,129] (Figure 20d).

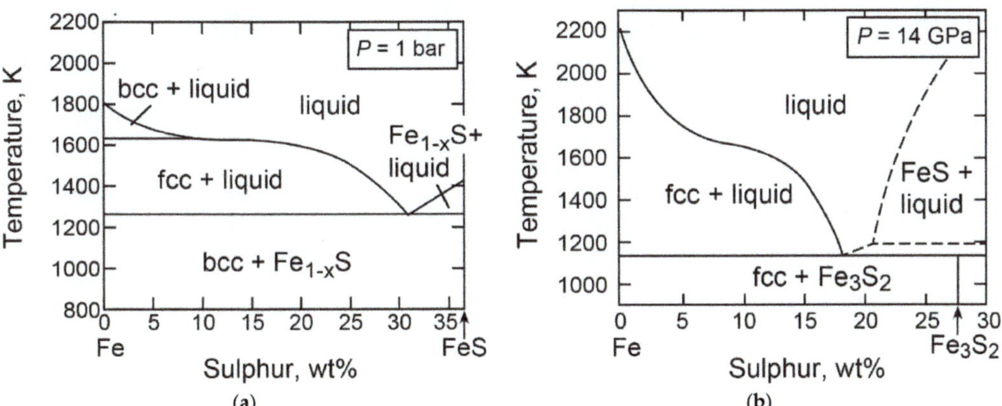

(a)   (b)

Figure 20. *Cont.*

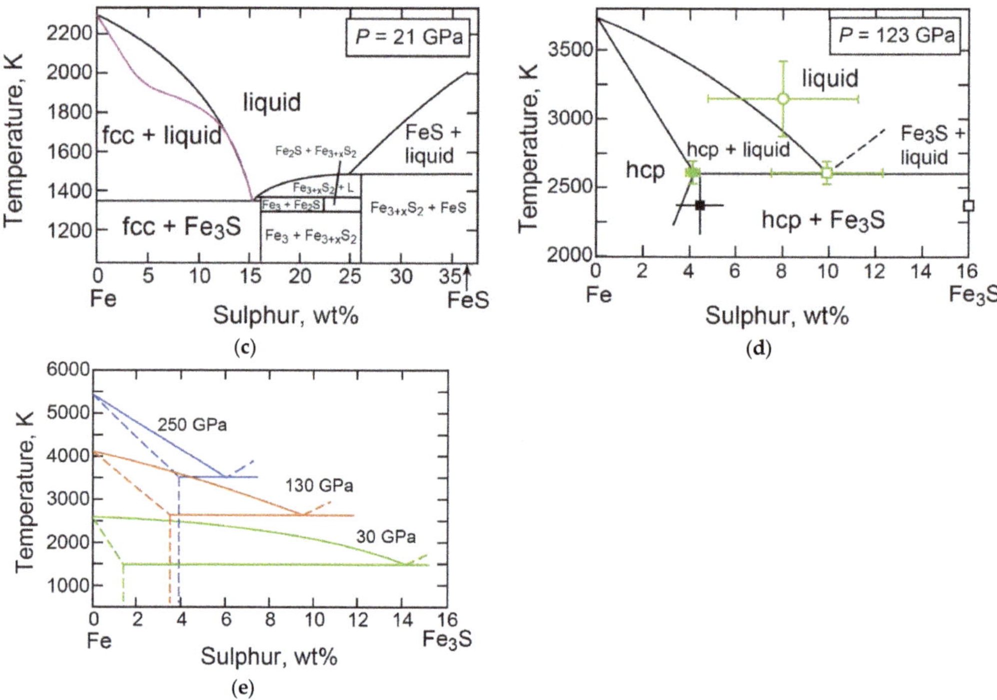

Figure 20. Experimentally constrained eutectic relations in the Fe-S system at (**a**) 1 bar, (**b**) 14 GPa, (**c**) 21 GPa, and (**d**) 123 GPa. Calculated eutectic relations are shown in (**e**) [122]. (**a**) is from Waldner and Pelton [126], (**b**) from Chen et al. [127], (**c**) from Fei et al. [128] (black line) and Pommier et al. [125] (purple line), and (**d**) from Kamada et al. [124]. In (**b**), the dashed lines were not directly constrained by [127]. In (**d**), the green open circle denotes total melting, green solid and open squares are eutectic solid and liquid, and black solid square denotes the subsolidus assemblage of hcp Fe and Fe₃S.

The eutectic composition and temperature as a function of pressure were experimentally examined [122,124,128–132] (Figure 21). Experiments also confirmed the stability of Fe₃S to 250 GPa (Figure 20e). The eutectic composition becomes Fe-richer with increasing pressure to 6 wt% S at 250 GPa (Figure 21a). The eutectic temperatures were also constrained to 250 GPa [122,124,128,130,132,133] (Figure 21b). Considering the experimental uncertainties, the eutectic temperatures do not differ much between the Fe-FeO and Fe-Fe₃S systems at the core pressures (Figure 21b).

The *T-X* liquidus curve of the Fe phases at 1 bar shows a sigmoidal shape. This is due to the nonideality of liquids and was also confirmed at 14 GPa [127] (Figure 20b). Although Fei et al. suggested a parabolic-shaped liquidus curve at 21 GPa, a recent study by Pommier et al. [125] reported a sigmoidal liquidus curve, which indicates that the nonideality still remains at this pressure (Figure 20c). There are no reports on the shape of the liquidus curve of the Fe phases at greater pressures.

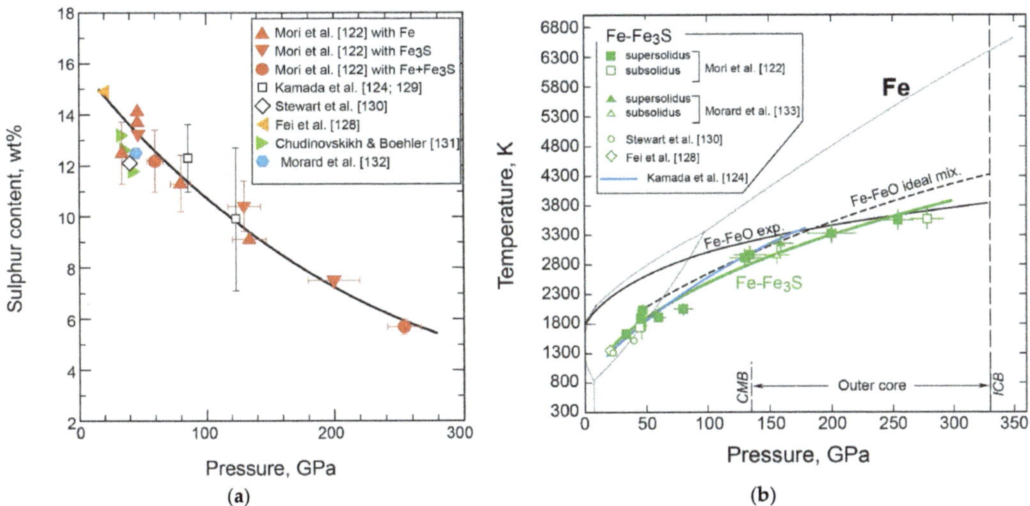

Figure 21. (**a**) Sulphur content in eutectic melt as a function of pressure in the Fe-S system [122,124,128–132]. (**b**) Eutectic temperatures of the Fe-S system [122,124,128,130,133] in comparison with Fe-FeO by experiment [85] and thermodynamic calculation with ideal mixing for liquids [9].

- **Sulphur in the hcp Phase**

  The sulphur content in hcp Fe coexisting with a liquid needs to be critically assessed because a S-rich hcp phase could be a main constituent phase of the inner core [90]. Experimental data cover a pressure range to 250 GPa, which show that the sulphur content in the hcp phase increases with pressure [122,124,129,130,134] (Figure 22a). Figure 22b shows partitioning of S between solid (hcp) and coexisting liquid in experiments by Mori et al. [122] compared with a prediction from the first-principles calculations at 330 GPa [90]. The experimental data are approaching the theoretical value for 330 GPa with increasing pressure; the data at 254 GPa is very close to the predicted curve. Thus, a significant amount of sulphur can be incorporated in the hcp phase at the ICB conditions.

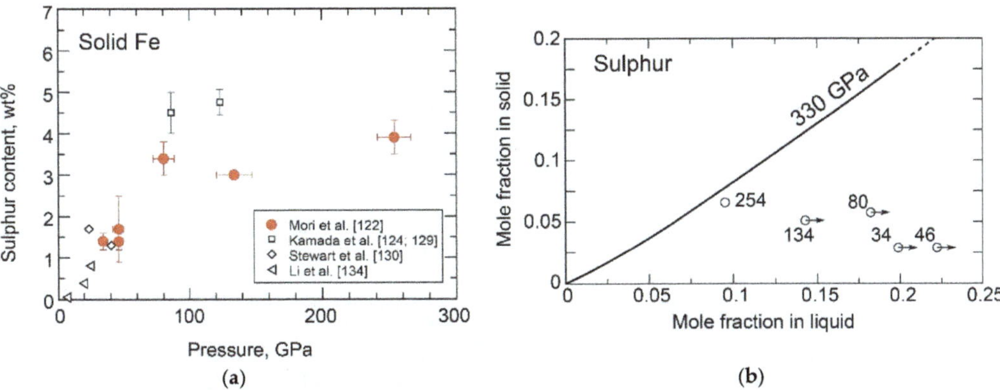

Figure 22. (**a**) Sulphur content in solid iron phases [122,124,129,130,134]. (**b**) Sulphur partitioning between solid iron and liquid. The black line is a result of the first-principles calculations for 330 GPa by Alfè et al. [90]. The circle symbols are experimental data and the numbers attached are the experimental pressures [122]. The small arrows attached to circles indicate the lower bounds of sulphur content in the iron phases.

- **The Properties of Fe$_3$S**

Since the Fe$_3$S phase is stable over 200 GPa as an end-member of the subsystem Fe-Fe$_3$S (Figure 20), constraining its physical properties provides vital information about a hypothetical sulphur-bearing core. The EoS of Fe$_3$S has been experimentally examined with in situ XRD [128,135–138]. The 300-K EoS was well constrained thanks to the wide pressure range covered up to 200 GPa [135–138] (Figure 23a). In contrast, the thermal parameters were less studied [135,137,138]. The most recent high-T data were reported by Thompson et al. [138] who proposed a thermal EoS from laser-heated DAC experiments with in situ XRD to 126 GPa and 2500 K. Figure 23b shows compression curves of Fe$_3$S calculated with the EoS by Thompson et al. [138] and Kamada et al. [136]. The EoS by [138] calculates the density of solid Fe$_3$S greater than by [136] since Kamada et al. [136] adopted the large αK_T value (0.011 GPa/K) proposed by Seagle et al. [135].

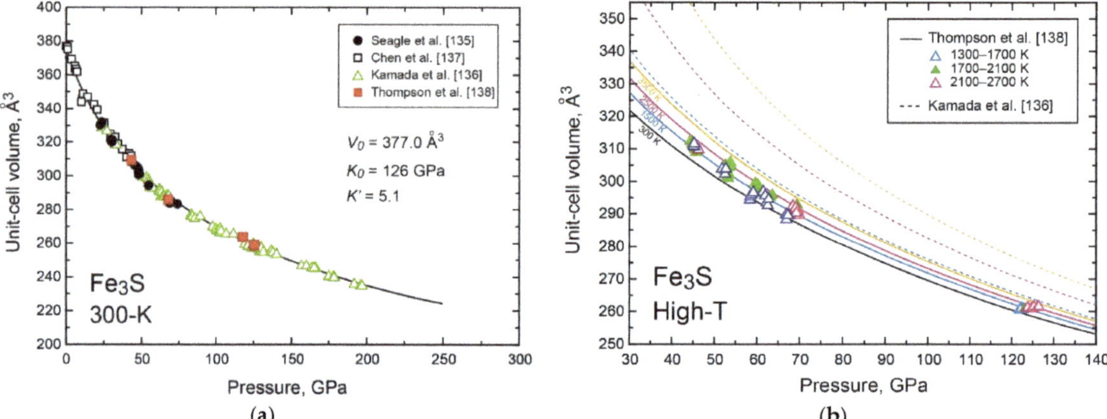

Figure 23. (a) Unit-cell volumes for Fe$_3$S at 300 K (Seagle et al. [135]; Chen et al. [137]; Kamada et al. [136]; Thompson et al. [138]). A compression curve based on the BM EoS (Equation (1)) fitted to all the data is also shown. (b) Unit-cell volumes for Fe$_3$S under high temperatures [138]. Compression curves based on the thermal EoS constructed by [138] are shown together with those by [136]. The figures are from [138].

- **Decomposition of Fe$_3$S at Core Pressures**

Solid Fe$_3$S is stable to 250 GPa and therefore its elastic parameters can be used to assess the liquid properties in the future. However, it undergoes a decomposition reaction into an Fe-rich hcp phase and a S-rich B2 phase at higher pressures, and therefore, the most Fe-rich sulphide phase stable in the inner core conditions would be the B2 phase, not Fe$_3$S [121].

The nature of the reaction,

$$\text{Fe}_3\text{S} = \text{Fe-rich hcp} + \text{S-rich B2} \tag{R1}$$

was examined by Thompson et al. [138], assuming the composition of the hcp and B2 phases to be pure Fe and Fe$_2$S, respectively [121,123]. Figure 24a plots the average atomic volumes for Fe and Fe$_3$S at 250 GPa and 300 K, based on which the volume for Fe$_2$S is estimated. The average atomic volume for the B2 phase needs to be below the Fe-Fe$_3$S line (Figure 24a), because a first-order pressure-induced transition must be accompanied with a volume reduction. Thompson et al. [138] considered two possible cases: (i) the volume change of the reaction (ΔVr) of Fe$_3$S = Fe + Fe$_2$S is zero, and (ii) $\Delta Vr = -1.5\%$. The volume for Fe$_2$S is then obtained for each case in Figure 24a. Recently, Tateno et al. [123] reported experimental data on the unit-cell volume of the Fe$_2$S phase at pressures greater than

180 GPa. Their volume of Fe$_2$S is plotted in Figure 24a, with their experimental pressure values corrected to be consistent with the pressure scale in [138]. Tateno et al.'s [123] data shows 1.0% volume reduction on reaction (R1), which is within the predicted range by [138] (Figure 24a). Thompson et al. [138] estimated compression curves for Fe$_2$S with evaluating EoS parameters (Figure 24b).

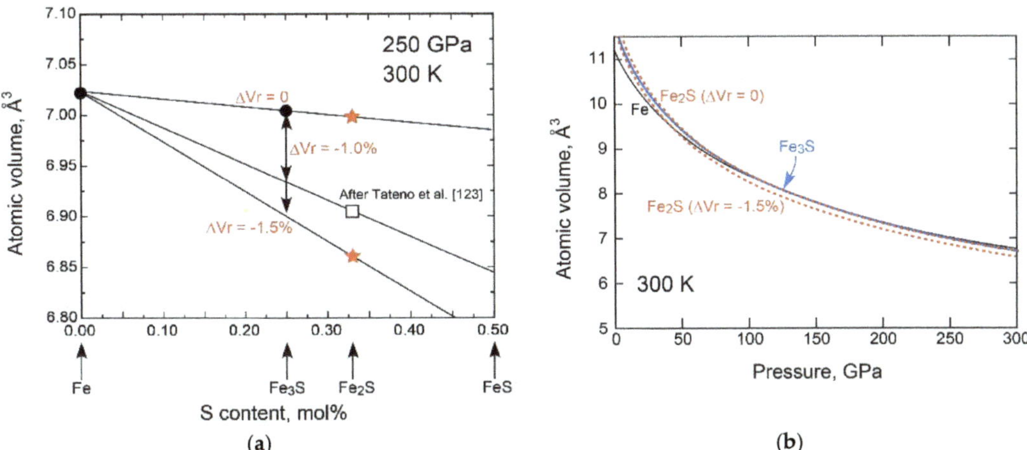

Figure 24. (a) Volume relationships for reaction (R1) [121]. The composition of the hcp phase is assumed to be pure Fe. The volumes were calculated from Dewaele et al. [4] for Fe and Thompson et al. [138] for Fe$_3$S. The star represents a hypothetical B2 phase with a composition of Fe$_2$S [121]. The volumes of the hypothetical Fe$_2$S phase were estimated for the cases of $\Delta Vr = 0$ and -1.5%, respectively. The open square indicates the volume of Fe$_2$S constrained by experiment [123]. (b) Compression curves of the hypothetical Fe$_2$S at 300 K. The figures are from [138].

- **Thermodynamic Model of the System**

The thermodynamics of the Fe-S system at 1 bar has been well studied [126,139–142]. The system is characterized by the liquid phases showing negative nonideality over the entire compositional range. The most recent study treated the liquid phases with the quasi-chemical model [126]. Solid Fe$_3$S was not included in those models as it is only stable at high pressures ($P > 20$ GPa). The nonideality under high pressure can be discussed from the shape of the liquidus curve of the Fe phases. At 1 bar to 20 GPa, experiments confirmed that the liquidus curves are sigmoidal shaped, which implies that the nonideality continues to those pressures [125–127]. There is no information about the nonideality of liquids under greater pressures.

Saxena and Eriksson [143] established a model for calculations of high-P-T phase relations including Fe$_3$S. The calculated eutectic composition at the ICB is Fe-10S, which is more S-enriched than the recent experimental determination by Mori et al. [122] (Figure 21a), but they are qualitatively consistent as for the pressure dependence of the sulphur content in eutectic melt, namely it is decreasing with increasing pressure [143].

Future models should include the recent experimental data: the updated EoS for Fe$_3$S [138], eutectic compositions to 250 GPa [122], and the breakdown reaction of Fe$_3$S into Fe(+S) hcp +Fe$_2$S [121]. Integrating these new data will make the model more robust and enable us to assess the liquid EoS to compare with the outer core PREM.

- **Density of Fe-S Liquids and Solids in the Core**

Figure 25a shows density profiles of solid Fe$_3$S from the EoS by [138] at 300, 4000, and 5500 K compared with the PREM over the core pressure range. The density of solid Fe$_3$S at 300 K is calculated to be smaller than the inner core PREM (Figure 25a). As the compression behaviour of Fe$_3$S at 300 K is well constrained to 200 GPa [135–138], the uncertainty of the calculated unit-cell volume when extrapolated to 330 GPa is $\pm0.2\%$. This ensures that solid Fe$_3$S is less dense than the inner core under core temperatures.

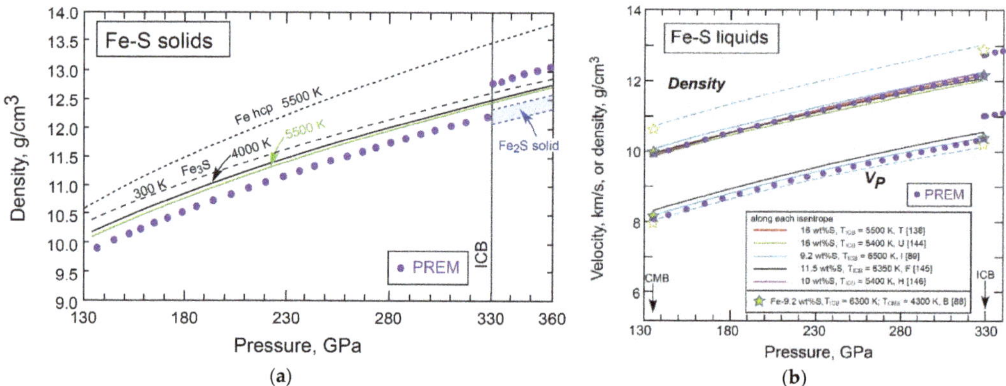

Figure 25. (**a**) Isothermal density profiles of solid Fe$_3$S based on [138] at 300 (black dashed), 4000 (black solid), and 5500 K (green solid), compared with a compression profile of hcp Fe [4] along a 5500 K isotherm and the PREM profiles [51]. An estimated density range of Fe$_2$S B2 phase at the inner core pressures and 5500 K is shown: the lower and upper bounds are for ΔVr (R1) = 0 and -1.5% respectively. (**b**) Density and Vp for Fe-S liquids over the outer core pressure range together with the PREM data. The density of liquid Fe$_3$S (Fe-16S) estimated from that of solid Fe$_3$S is from Thompson et al. [138]. Results from the first-principles calculations are also shown (Badro et al. [88]; Umemoto et al. [144]; Ichikawa and Tsuchiya [89]; Fu et al. [145]). The green open stars denote the pure Fe properties by [88]. Results for the density measurements of a shock wave compression study (Huang et al. [146]) are also plotted. Isentropic profiles of liquid Fe are also shown (dashed blue lines, Ichikawa et al. [50]) from Figure 3.

Data of Fe-S liquids are summarized in Figure 25b including estimates based on solid compression data [138], first-principles calculations [88,89,144,145], and shock compression experiments [146]. Thompson et al. [138] estimated the density profile of liquid Fe$_3$S by assuming a ΔV upon melting (ΔVm) for Fe$_3$S to be the same as for pure iron at core pressures, and assuming a temperature gradient by 1500 K through the outer core, which means 4000 K at 140 GPa and 5500 K at 330 GPa. This temperature gradient is consistent with the Grüneisen parameter of pure Fe of about 1.5 [9,147]. Thus-calculated density profile for liquid Fe$_3$S matches the outer core density within an uncertainty of 1% (Figure 25b). As such, the sulphur content in the outer core needs to be as much as 16 wt% (=Fe$_3$S) if it is the sole light element in the core [138]. This estimate can be improved by refining the values of ΔVm and the isentropic temperature gradient over the outer core.

Using first-principles calculations, Umemoto et al. [144] calculated the density of Fe$_3$S liquid along an isentrope with 5400 K at 330 GPa, which agrees with the PREM density of the outer core and the estimated Fe$_3$S liquid density by [138]. Ichikawa and Tsuchiya [89] provided the best-fit value of sulphur content in the outer core of 9.2 wt%. Badro et al.'s [88] calculations also showed that 9.2 wt% sulphur would make the liquid Vp and density match the PREM (Figure 25b). A shock compression study reported that the density of Fe-10S along an isentrope with T$_{ICB}$ = 5400 K would match the PREM [146]. As such, the estimated sulphur content in the outer core based on density ranges from 9.2 to 16 wt%, which could be narrowed down if all the density profiles were set at the same T$_{ICB}$. More data from different approaches are needed to constrain the Vp.

Liquid cores with compositions of Fe-9.2S or Fe-16S would crystallise a sulphide since the eutectic composition in the Fe-Fe$_3$S sustem was reported to become close to the Fe side with increasing pressure [122] (Figure 21a). On the other hand, as we discussed above, crystalline Fe$_3$S may not be stable under inner core conditions as it would break down to a mixture of Fe-rich hcp phase and S-rich B2 phase above 250 GPa [121]. This results in a formation of the S-rich B2 phase from a liquid with S > 6 wt% at ICB. Assuming that the B2 phase has a composition of Fe_2S and causes $\Delta Vr = 0$ to -1.5% as discussed above (Figure 24a), the density profile of B2 Fe_2S for the inner core range is calculated along a 5500 K isotherm (Figure 25a). The density of Fe_2S is likely less dense than solid Fe_3S and cannot match Earth's inner core density. As such, the liquidus phase for Fe-9.2S to Fe-16S is not the constituent phase of the inner core. In summary, while the outer core density requires as much sulphur as 9–16 wt%, the resulting liquidus phase cannot meet the density of the inner core.

3.2.4. Fe-C

The phase relations in the Fe-C system at 1 bar were extensively studied as it is a fundamental system in steelmaking. High-pressure phase diagrams were constructed to 13 GPa in the late 1960s [148,149]. In Earth science, a pioneering work on the phase relations which can be applicable to core conditions was made by Wood [150] who employed thermodynamic calculations to 330 GPa based on available experimental data including melting data to 5 GPa and thermoelastic properties of Fe-C phases. His calculations showed that the eutectic composition was very close to the Fe end-member (0.9 wt% C) at 330 GPa and solid Fe_3C may compose the inner core. Later experiments, up to 29 GPa, confirmed that Fe_7C_3 would be the first phase to crystallize out of a carbon-rich liquid core [151] which reinforced earlier results of the stability of Fe_7C_3 under high pressure [149]. Here I review *T-X* phase relations, stability relations of Fe_3C and Fe_7C_3, and liquid properties.

- **The *T-X* Relations**

Figure 26 summarises *T-X* relations in the Fe-rich portion with increasing pressure [150–154]. At 1 bar, the steelmaking industry prefers to show both stable and metastable reactions (Figure 26a) because the free energy difference between them is so small and one would encounter both reactions in an experiment [155]. Under high pressure, for example, 5 GPa (Figure 26b), Fe_3C is a stable phase and the phase diagram is unambiguous. A high-pressure phase Fe_7C_3 appears above 9 GPa, which makes Fe_3C melt incongruently [149,151] (Figure 26c). The phase relations at core pressures (Figure 26d,e) are based on thermodynamic calculations [150,154,156] and experiments [156].

The eutectic point is at Fe-4.2C at 1 bar and its carbon content may or may not change with increasing pressure (Figure 27a). At the ICB pressure, while the recent thermodynamic model predicted it to be around 2 wt% [154], experiments in laser-heated DAC combined with chemical analysis on recovered samples showed that almost constant eutectic compositions with 3.6–4.5 wt% C at 23–255 GPa [156]. On the other hand, the eutectic temperatures are consistent between different studies at least up to 70 GPa [85,154,156,157] (Figure 27b).

- **Carbon in the hcp Phase**

Carbon dissolves into the crystal structure of iron. The maximum carbon concentration in bcc iron at 1 bar is about 0.02 wt% at 1010 K [152], which increases in fcc iron to about 2.00 wt% [152], and decreases with increasing pressure [158]. Mashino et al. [156] showed that carbon content in solid hcp iron remains in the order of 1 wt% up to 255 GPa (Figure 28).

From first-principles calculations, Caracas [159] found that the density deficit of the inner core can be matched for 1–2.5 wt% carbon in hcp Fe, depending on the thermal profile. This is consistent with an experimental investigation by Yang et al. [160] who made compression experiments in DAC and showed that 1.30 and 0.43 wt% carbon may explain the inner core density deficit at 5000 K and 7000 K, respectively. As such the carbon content in hcp iron required for the 4% cdd at 6000 K is only 1–2 wt% because incorporation of carbon in the hcp structure expands the volume [160]. Such C-bearing hcp phases can directly crystallise out of an Fe-C outer core as the eutectic composition would be at 2–4 wt% C at the ICB as discussed above [154,156], depending of the carbon content in the outer core, which will be discussed later.

- **The Properties and Decomposition of Fe₃C at Core Pressures**

  Figure 26d,e show that Fe_3C is destabilised between 136 and 330 GPa according to the thermodynamic model by [154], which has an impact on the liquidus phase field adjacent to the eutectic point, namely from Fe_3C to Fe_7C_3. Since Fe_3C and Fe_7C_3 phases have different carbon contents, thermoelasticity, and physical properties, their stability relations are critically important for models of a carbon-bearing inner core. Both of these iron carbides have been shown to plausibly match some solid-state physical properties of the inner core, such as shear wave velocity and Poisson's ratio [161–165]. As Fei and Brosh's [154] model was focused on melting relations, I here review the subsolidus relations regarding stability of Fe_3C examined by experiment and first-principles calculation.

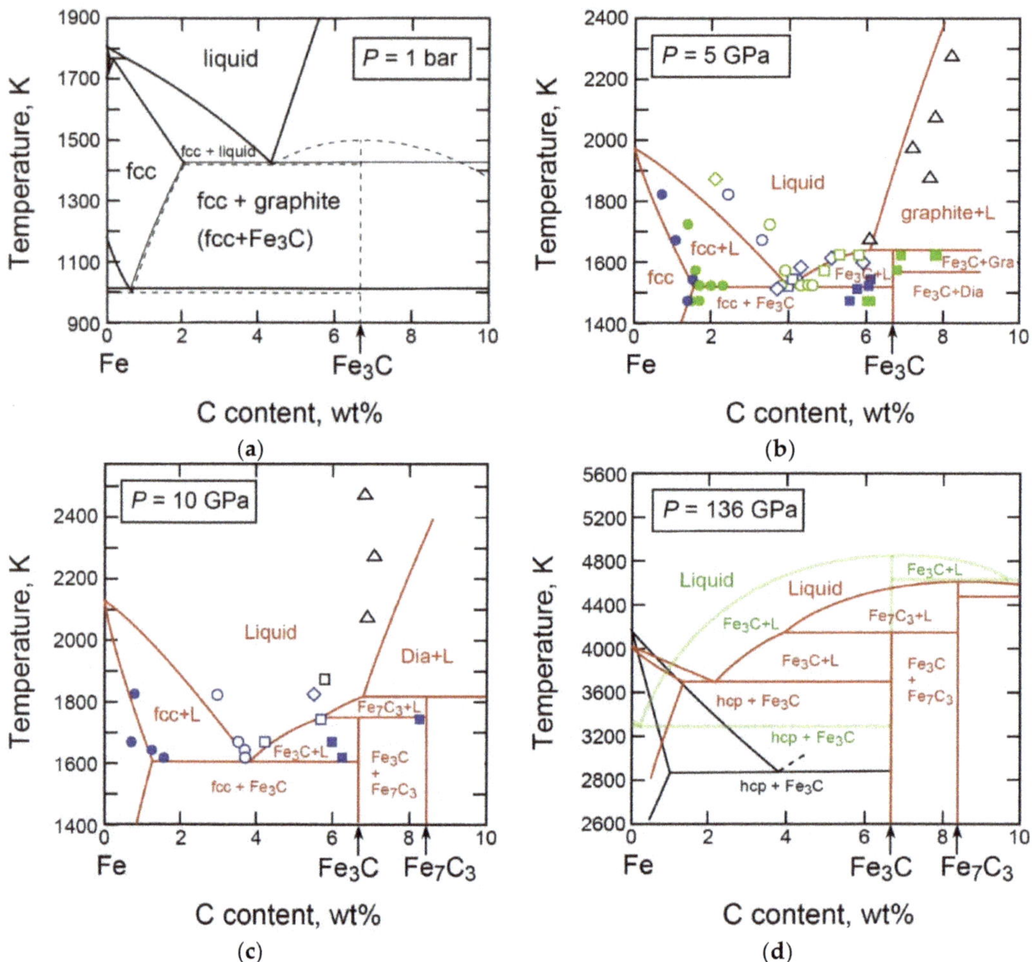

Figure 26. *Cont.*

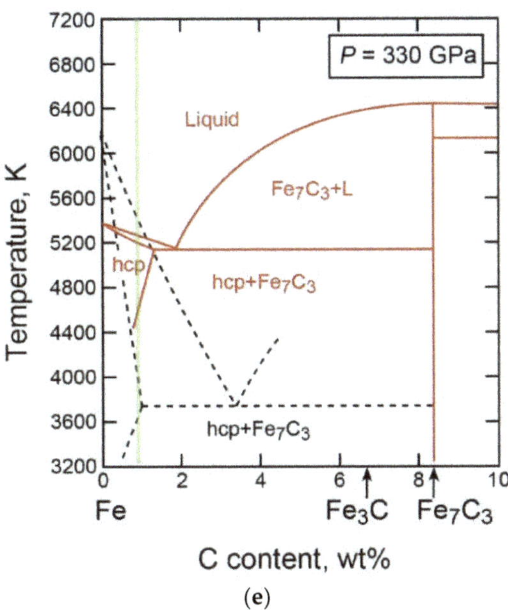

Figure 26. Eutectic relations in the Fe-C system at (**a**) 1 bar, (**b**) 5 GPa, (**c**) 10 GPa, (**d**) 136 GPa, and (**e**) 330 GPa. (**a**) is after Gustafson [152], and the red lines in (**b–e**) are calculated by Fei and Brosh [154]. The green and black lines in (**d**) are calculated by Wood [150] and are based on experiments by Mashino et al. [156], respectively (the dashed curve is inferred). The black dashed lines in (**e**) are inferred from thermodynamic calculations by [156]; the green line in (**e**) indicates the eutectic composition by [150]. In (**a**), the solid lines are equilibrium lines and dashed lines are metastable lines. Both phase relations are encountered in experiments [155]. In (**b,c**), the symbols are experimental data: blue [154]; green [153]; black [151]. The solid circles denote the compositions of solid iron coexisting with either melt or Fe_3C. The open circles denote the compositions of melt coexisting with solid iron. The solid squares represent the compositions of iron carbides (Fe_3C or Fe_7C_3). The open squares represent the compositions of melt coexisting with iron carbides. The open diamonds show that only melt was observed. The open triangles show the melt compositions coexisting with either graphite at 5 GPa or diamond at 10 GPa.

The key reaction is,

$$3Fe_3C = Fe_7C_3 + 2Fe \tag{R2}$$

which is an equilibrium univariant reaction. This was first inferred by Lord et al. [166] from the topology of high-pressure melting curves of Fe_3C and Fe_7C_3. Reaction (R2) was later examined in laser-heated DAC with in situ XRD by Liu et al. [157], who placed its boundary at about 150 GPa (Figure 29a). This observation is contradicted, however, by Tateno et al.'s [23] in situ XRD measurements of the formation of Fe_3C phase in their Fe sample in the DAC upon laser heating at about 340 GPa. More recent in situ XRD experiments also observed Fe_3C to pressures greater than 250 GPa [167] (Figure 29b). Mookherjee et al. [168] reported that Fe_3C was energetically stable at all pressures to the centre of the Earth from first-principles calculations on reaction (R2) at $T = 0$. On the other hand, Mashino et al. [156] reported the Fe_3C phase as a liquidus phase in their melting experiments in the Fe-C system to 203 GPa, but placed the possible occurrence of reaction (R2) at 255 GPa from chemical and textural analyses of recovered samples (Figure 29b). McGuire et al. [169] made thermodynamic calculations on the P-T locations of reaction (R2) using their newly established EoS for Fe_3C. The reaction is located at 87 GPa and 300 K and 251 GPa and 3000 K and its boundary is consistent with the results of both Tateno et al. [23] and Mashino et al. [156] (Figure 29b).

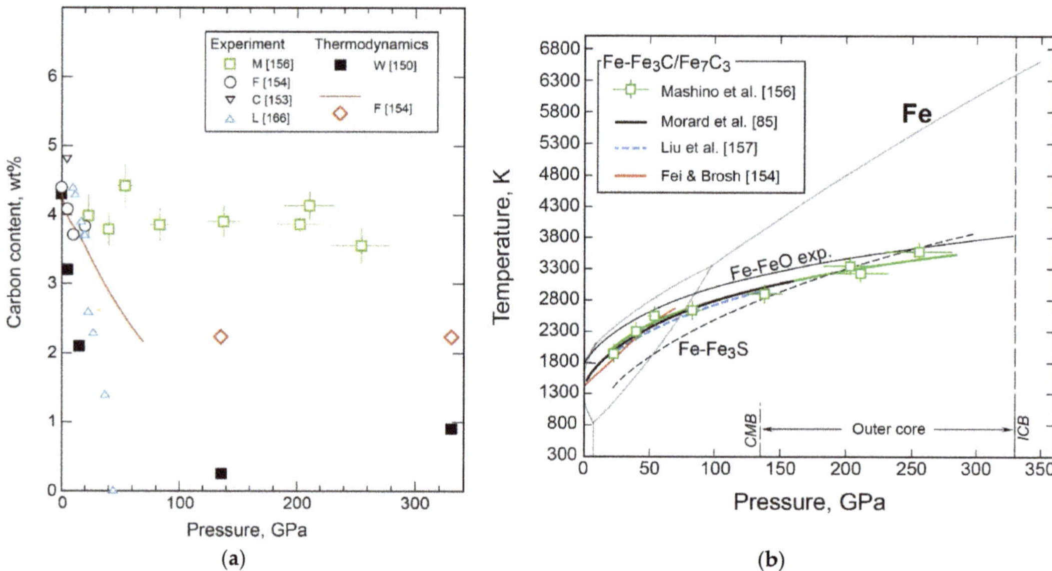

Figure 27. (**a**) Eutectic compositions in the Fe-Fe₃C/Fe₇C₃ system. Experimental data are plotted (Chabot et al. [153]; Lord et al. [166]; Fei and Brosh [154]; Mashino et al. [156]). Results of thermodynamic calculations by [150] and [154] are also shown. (**b**) Eutectic temperatures of the Fe-Fe₃C/Fe₇C₃ system (Mashino et al. [156]; Morard et al. [85]; Liu et al. [157]; Fei and Brosh [154]). Experimental data for the Fe-FeO system [86] and Fe-Fe₃S [122] are also shown for comparison. The phase relations for pure Fe is from [9].

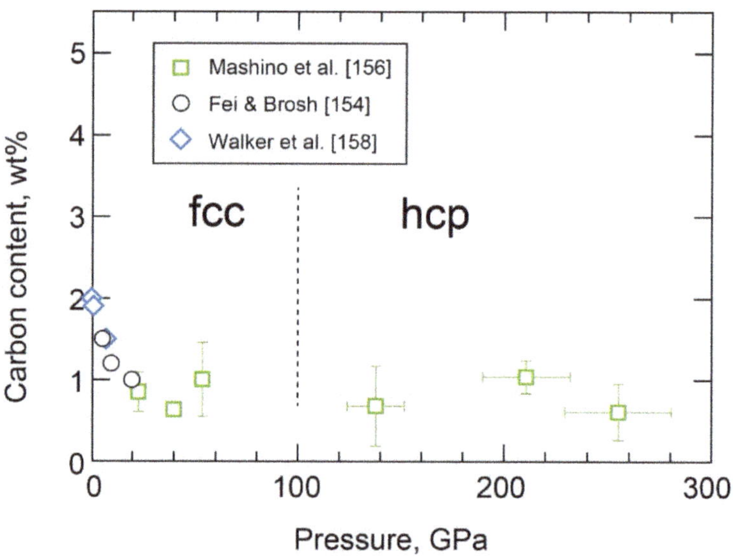

Figure 28. Carbon content in solid iron phases in the system Fe-C (Walker et al. [158]; Fei and Brosh [154]; Mashino et al. [156]). The figure is from [156].

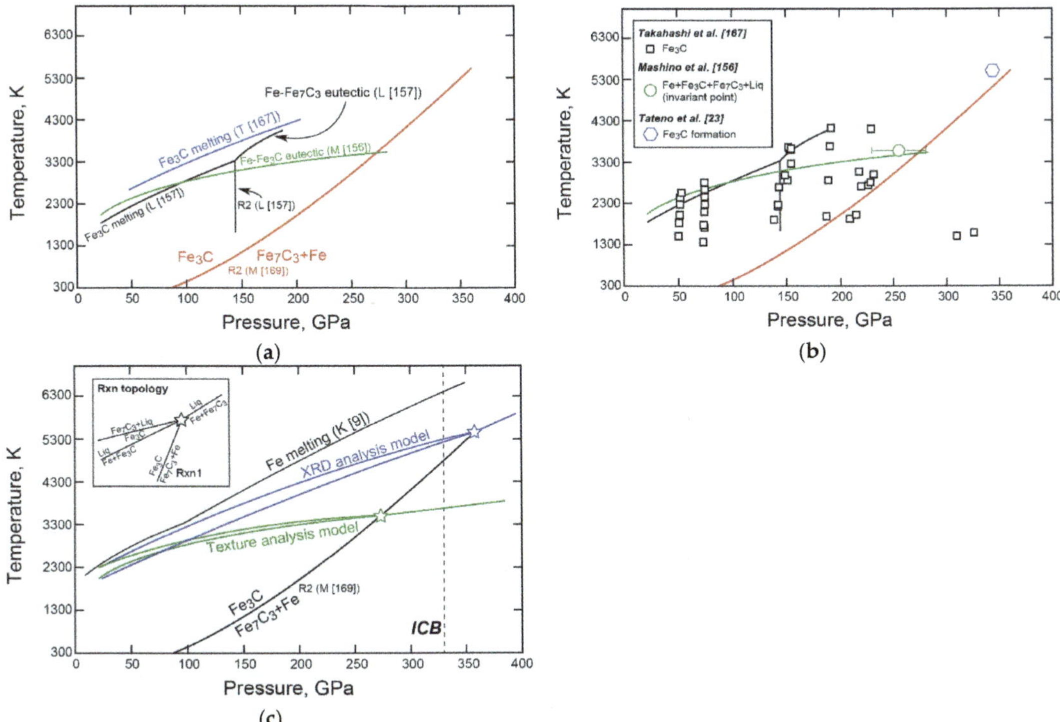

Figure 29. (**a**) Phase relations in the Fe-C system up to 400 GPa and 6000 K (Liu et al. [157]; Mashino et al. [156]; Takahashi et al. [167]; McGuire et al. [169]). The solid red line shows reaction $Fe_3C = Fe_7C_3 + Fe$ calculated in [169]. (**b**) Selected experimental data points from the literature. Mashino et al. [156] observed an invariant assemblage in their experimental run charge at the intersection of reaction $Fe_3C = Fe_7C_3 + Fe$ and the eutectic melting curve (green, [156]). The black lines are the observed phase boundaries by in situ XRD [157]. Experimental data of Takahashi et al. [167] are also plotted; their 300 K data are not shown. Tateno et al. [23] observed the formation of Fe_3C in their Fe sample at 5520 K and 344 GPa. (**c**) Inferred phase diagram for the Fe-C system. The stars denote two possible cases of the invariant point where Fe, Fe_3C, Fe_7C_3, and liquid coexist. The in situ XRD analysis-based work [23,167] forms the invariant point at 360 GPa and 5500 K (blue star) whereas the texture analysis-based model [156] placed it at 275 GPa and 3500 K (green star). The melting curve for pure Fe is taken from Komabayashi [9]. Rxn, reaction. The figures were modified after McGuire et al. [169].

An invariant point occurs on the reaction boundary (R2) where Fe, Fe_3C, Fe_7C_3, and liquid are stable, which places constraints on the liquidus temperature of a C-rich outer core [166,169]. From the invariant point, melting reactions originate: $Fe_3C = Fe_7C_3 + \text{liquid}$, $Fe + Fe_3C = \text{liquid}$, and $Fe + Fe_7C_3 = \text{liquid}$ (Figure 29c). Two possible *P-T* locations for the invariant point were predicted from existing experimental data combined with reaction (R2) determined in [169] (Figure 29c): 360 GPa and 5500 K based on XRD studies [23,167] and 275 GPa and 3500 K based on a texture analysis-based study [156]. If the XRD-based model is the case, Fe_3C is stable over the outer core pressure conditions and the Fe-Fe_3C subsystem may be relevant for an outer core composition near the eutectic point. In contrast, if the invariant point constrained by the textural analysis of DAC experiments [156] is the case, the subsolidus system would be Fe-Fe_7C_3.

- **Thermodynamic Model of the System**

  Thermodynamic models of the system at 1 bar revealed that liquids show negative nonideal mixing [152,170,171]. Wood [150] and more recently Fei and Brosh [154] established high-pressure thermodynamic models for melting relations based on available experimental data. Both models predicted the eutectic points moving towards the Fe side with increasing pressure but became fairly constant above 136 GPa (Figure 27a).

  At a core pressure of 136 GPa, the eutectic composition and temperature are not in agreement between the model by Fei and Brosh [154] and the experimental data by Mashino et al. [156] (Figure 26d). The disagreement could be reduced by refining the mixing model used in [154].

  On the other hand, Fei and Brosh [154] and Mashino et al. [156] agreed on the stability of Fe$_3$C which is decomposed between 136 and 330 GPa. As discussed above, Fe$_3$C might be stable at the inner core conditions and its stability is related with the liquidus temperatures of the Fe-C system (Figure 29c). Constraining the liquid properties combined with the thermodynamics of the subsolidus reaction 3Fe$_3$C = Fe$_7$C$_3$ + 2Fe [169] will provide a holistic understanding of the system.

- **Density and Velocity of Fe-C Liquid and Solid in the** Core

  Figure 30a shows the density of Fe$_3$C and Fe$_7$C$_3$ solid phases calculated from the EoS [169,172]. The experimental *P-V-T* data were collected to 117 GP and 2100 K for Fe$_3$C and 72 GPa and 1973 K for Fe$_7$C$_3$ and therefore Figure 30a is based on large extrapolations. Nevertheless, Fe$_3$C is less dense at any temperature than the inner core while the density of Fe$_7$C$_3$ may match the inner core PREM when the temperature is as low as 3700 ± 500 K.

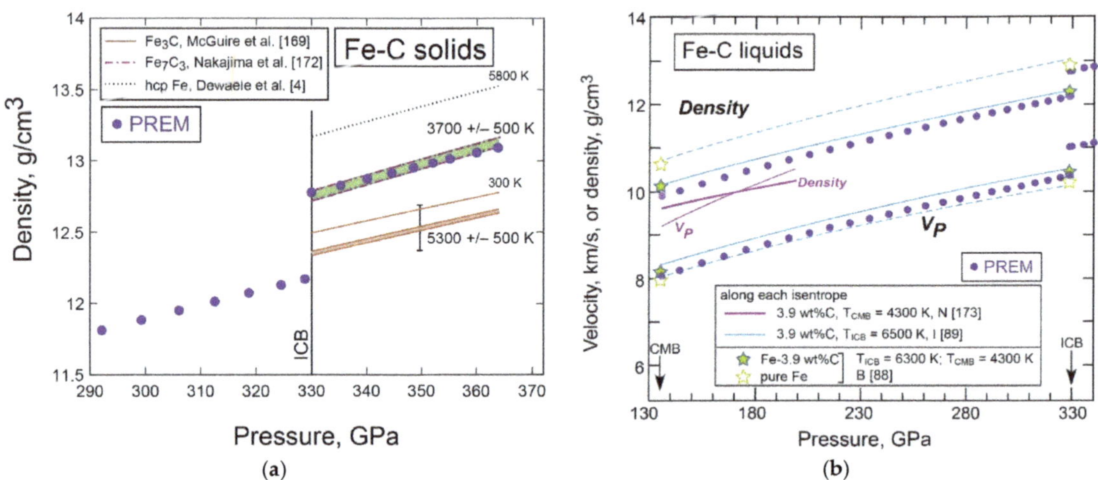

Figure 30. (**a**) Density versus pressure relation for the solid inner core with PREM and candidate iron carbides. The ICB pressure is indicated. Solid Fe$_3$C is shown at 300 K and 5300 ± 500 K in the solid orange line [169]. Solid Fe$_7$C$_3$ at 3700 ± 500 K is shown in the dot-dash line [172]. Solid hcp iron at 5800 K [4] is shown in the short-dash line. The uncertainty bar attached to the density of Fe$_3$C is propagated from the uncertainties in the EoS parameters. The figures are taken from [169]. (**b**) Density and *Vp* for Fe-C liquids over the outer core pressure range together with the PREM data. Results from the first-principles calculations are shown (Badro et al. [88]; Ichikawa and Tsuchiya [89]). Experimental measurements of Fe-3.9C using inelastic X-ray scattering are also plotted [173]. Isentropic profiles of liquid Fe are compared (dashed blue lines, Ichikawa et al. [50]) from Figure 3.

Comparison of the density and *Vp* of Fe-C liquids with the outer core profiles is made in Figure 30b. Results of first-principles calculations for Fe-3.9C show moderate reductions in density but great increases in *Vp* [88,89], which means there is no "best-representing" solution for the Fe-C system. The *Vp* of an Fe-C liquid with the same composition (Fe-3.9C) was examined in laser-heated DAC with synchrotron inelastic X-ray scattering to 70 GPa and 2800 K, followed by an evaluation of its thermal EoS from their *Vp* measurements [173]. Nakajima et al. [173] then calculated the *Vp* and density of liquid Fe-3.9C under core *P-T* conditions and compared them with the PREM (Figure 30b). Their plots are for a similar temperature to those of the first-principles calculations and therefore one can directly compare the data. The experimental determination [173] shows a greater reduction in density and increase in *Vp* than the theoretical results. This indicates that the properties of the Fe-C liquid are very sensitive to the carbon content. Nakajima et al. [173] estimated the carbon content required to account for the PREM values; 1.2–0.9 wt% C for velocity and 3.8–2.9 wt% for density, depending on the temperature and uncertainty of data extrapolation. They concluded that carbon cannot be a predominant light element in the outer core. However, a 1–2 wt% C in the outer core may crystallise a carbon-bearing hcp phase that could match the inner core PREM density as discussed above. As such, although the addition of carbon alone to iron cannot account for the properties of the outer core, the presence of carbon, particularly in the inner core, cannot be ruled out.

3.2.5. Fe-H

The Fe-H system is probably the most challenging system for an experimental study among the Fe-light element systems reviewed in this paper as the characterization of hydrogen dissolved in phases is very difficult. This is because Fe solid phases at 1 bar show little hydrogen solubility in the order of <0.05 at% (Figure 31a) and therefore the precise quantitative measurements of the amount of hydrogen in phases require in situ analysis [174] or very rapid (i.e., anvil breaking) decompression to turn dissolved hydrogen into bubbles which leave traces in the sample [175]. The hydrogen solubility in the Fe phases is conventionally expressed with a coefficient x, in iron hydride, FeHx. The x value drastically increases with increasing pressure (Figure 31b).

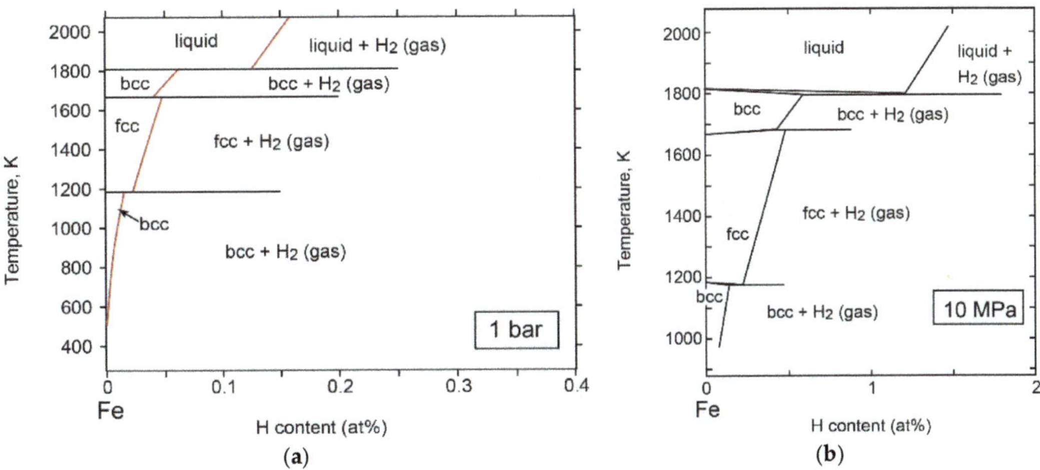

Figure 31. Phase relations for the Fe-H system at (a) 1 bar and (b) 10 MPa [176]. Note that the red lines in (a) show observed limit of H solubility and are not necessarily equilibrium phase boundaries. In (b) the H solubility in Fe phases is increased compared to at 1 bar.

- **The *T-X* Relations**

At 1 bar, the iron phases show negligible hydrogen solubility [176] and therefore, the phase assemblage in the Fe-H system is always $FeHx + H_2$ (gas) (Figure 31a). The hydrogen solubility, namely the x value, increases with increasing pressure (Figure 31b). Phase relations at 10 MPa shown in Figure 31b also show that the transitions between the bcc and fcc phases occur with loops [176]. The x value further increases with pressure and would reach 1, which means that stoichiometric FeH becomes stable at about 10 GPa [177] (Figure 32a). Fukai [178] suggested that the *T-X* relation at high pressures (e.g., 100 GPa) would be characterised by the presence of an extensive solid solution between Fe and FeH (x = 1) which covers a wide compositional range and the melting would occur over a loop (Figure 32b). This prediction was based on experimental observations that the x value became constant above 10 GPa [179,180]. However, Pepin et al. [181] reported new FeHx phases with x = 2 and 3 at 67 and 86 GPa, respectively, observed upon laser heating in DAC, which was later supported by Hirose et al. [182] to 127 GPa. As such, the compositional range of the solid-solution is expanding as the system is explored further.

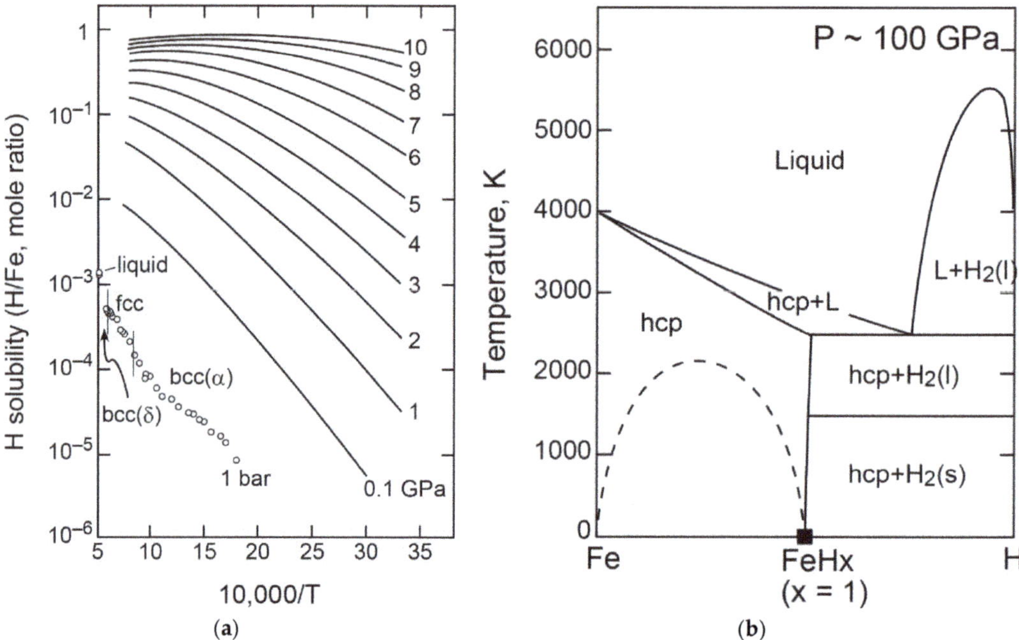

Figure 32. (a) Solubility of hydrogen in iron as a function of temperature and pressure. Experimental data for 1 bar are shown as open circles. The solid curves represent the results of thermodynamic calculations by Fukai and Suzuki [177]. The figure is taken from Fukai and Suzuki [177]. (b) A predicted phase diagram of the Fe-H system at ~100 GPa [178]. L, metallic liquid; H_2(s), solid hydrogen; H_2(l) liquid hydrogen. The figure was modified after Fukai [178].

As predicted by Fukai [178], the presence of solid FeHx solution under high pressure may result in a melting loop (Figure 32b). Sakamaki et al. [179], however, could not resolve the solidus and liquidus in their multi-anvil experiments to 21 GPa in which temperature precision was high, which suggests that the melting loop might be very narrow. A very narrow melting loop could partition hydrogen equally between solid and liquid and therefore could not account for the density contrast between the inner core and outer core.

- *P-T* **Relations**

The subsolidus phase relations of the Fe-H system can be compared with those of pure Fe. Notable differences include that a double hcp (dhcp) structure is stabilized instead of hcp under high pressure (Figure 33) and fcc FeH would be stable to high pressures. Previously dhcp, fcc, and liquid were expected to form a triple point (e.g., [179]), which is comparable to the invariant point in pure Fe where fcc, hcp, and liquid coexist. This was, however, recently challenged by observations of fcc FeH at higher pressures than previously expected [183–185]. Kato et al. [185] observed the formation of fcc FeH at 1000 K above 57 GPa, which implies that the fcc-dhcp boundary with a positive dP/dT slope constrained at low pressures to 21 GPa would bend back towards low temperature with increasing pressure (Figure 33). As a consequence, this topology hinders the formation of the triple point and the phase diagram looks very different from that for pure Fe. The high-pressure dhcp-fcc transition with a negative dP/dT slope can be tested by thermodynamic analysis. For example, the high-pressure phase, fcc must be denser than the dhcp phase upon transition.

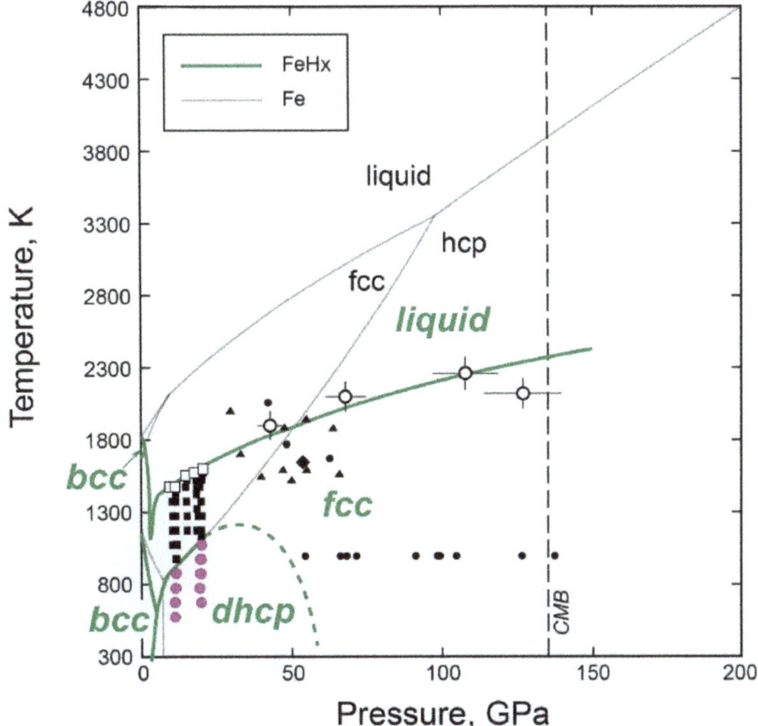

Figure 33. *P-T* phase relations for FeHx (x~1) in comparison with Fe [9]. The symbols are experimental observations: purple circles, dhcp; solid square, fcc; open square, liquid (Sakamaki et al. [179]); triangles, fcc (Thompson et al. [184]); diamond, fcc (Narygina et al. [183]); black solid circle, fcc (Kato et al. [185]); large open circle, liquid (Hirose et al. [182]).

The melting temperature of FeHx under hydrogen-saturated conditions is dramatically reduced with increasing pressure to 3 GPa [174,179] (Figure 33). This can be explained by the enhanced hydrogen incorporation into the phases; the x value in solid phases drastically increases with pressure, and iron liquid shows a greater concentration of hydrogen than the coexisting solid (Figure 31b). As a consequence, the melting temperature reduces by 500 K at 3 GPa [178,186]. The melting points above 5 GPa were reported by [179] in a multi-anvil apparatus and more recently in DAC by [182]. Hirose et al. [182] conducted laser-heated DAC experiments on the Fe-C-H system and they selected recovered samples which did not show a significant amount of carbon (0.2–0.3 wt% C) in Fe-H liquids and discussed the melting of FeHx at high pressures (Figure 33).

- **Density and Velocity of Fe-H Liquid and Solid in the Core**

Figure 34a shows relationships between the *Vp* and density (Birch plot) of solid FeHx and hcp Fe determined by experiment at 300 K or high temperatures and by first-principles calculation at $T = 0$ [180,187–190]. Both experiment and theory show that the addition of hydrogen to solid Fe increases both the *Vp* and shear wave velocity (*Vs*). The experimental data suggest that the *Vp* and density of the inner core may be explained by the addition of 0.2–0.3 wt% hydrogen [180]. There seems to be no solution for the *Vs* as the *Vs* of iron needs to be slowed down by the addition of a light element to Fe to match the inner core profile (Figure 34a), although the possibility that anharmonic effects could reduce the *Vs* under the inner core conditions cannot be ruled out [180].

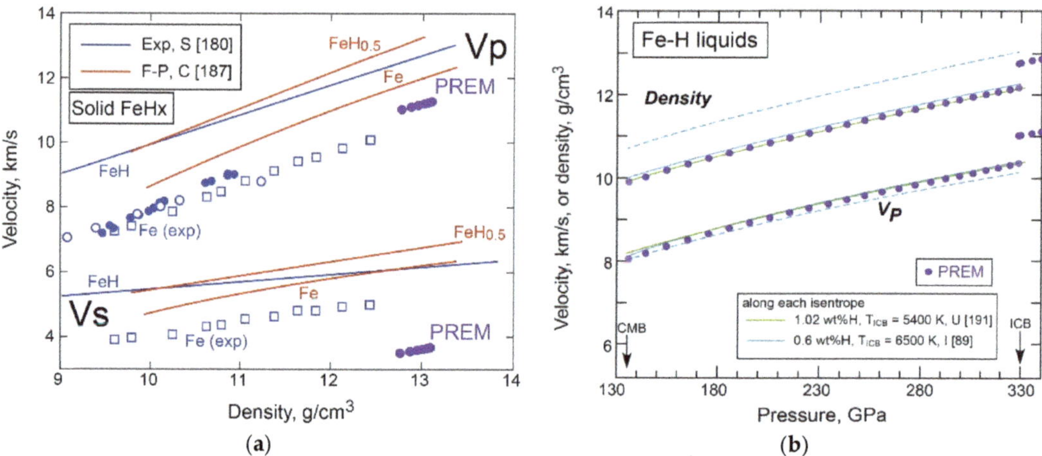

Figure 34. (**a**) *Vp*-density relations (Birch plot) for solid FeHx with the PREM data for the inner core. Experimentally constrained data using inelastic X-ray scattering are for FeH (x~1) by Shibazaki et al. [180] and results of first-principles calculations are for Fe and FeH$_{0.5}$ at $T = 0$ by Caracas [187]. The experimental data for pure hcp Fe are also shown for comparison [188–190]. (**b**) Density and *Vp* for Fe-H liquids over the outer core pressure range together with the PREM data. First-principles calculations were employed by Umemoto et al. [191] and Ichikawa and Tsuchiya [89]. Isentropic profiles of liquid Fe are compared (dashed blue lines, Ichikawa et al. [50]) from Figure 3.

Figure 34b compares the *Vp* and density of Fe-H liquids from first-principles calculations [89,191] with those of the outer core PREM. About 1 wt% hydrogen could match the calculated properties with the PREM profiles of the outer core.

3.3. Ternary Systems with Light Elements

Below I will review two ternary systems and associated topics.

3.3.1. Fe-Ni-Si

Phase relations in the Fe-Ni-Si system were experimentally investigated in DAC with in situ synchrotron XRD [58,110,192].

- **The fcc-hcp Transitions**

Komabayashi et al. [192] examined the *P-T* locations of the fcc-hcp transitions in Fe-5Ni-4Si in the internally resistive-heated DAC. Their results are compared with existing data for pure Fe [14] and the binary Fe-Ni [57] and Fe-Si [99] systems (Figure 35). Note that all the fcc–hcp transitions in those four different systems were examined in the internally heated DAC, which enables us to make a precise comparison. The addition of Ni reduces the transition temperature [55,57] whereas the Si incorporation has the opposite effect [97,99]. From these observations, one can expect that the simultaneous addition of Ni and Si will not greatly move the boundary from the case of pure Fe, as the effects of Ni and Si would cancel out.

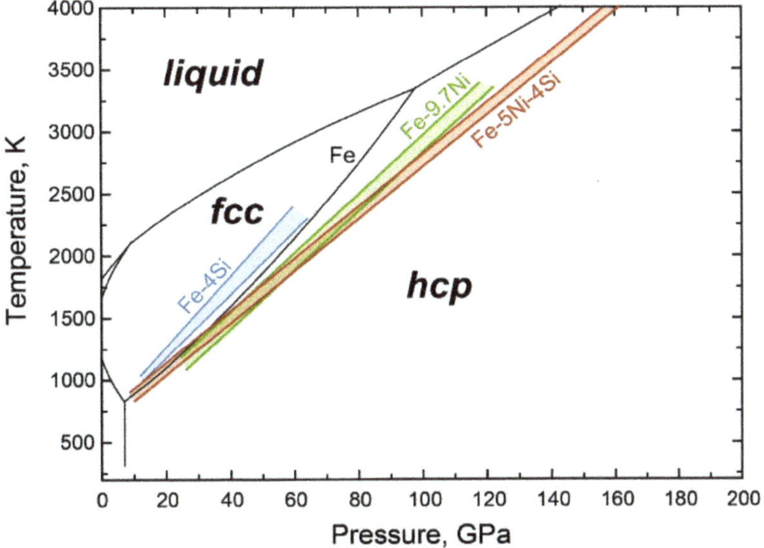

Figure 35. Comparison of the fcc-hcp transition boundaries in four different compositions: Fe [9,14], Fe-9.7Ni [57], Fe-4Si [99], and Fe-5Ni-4Si [192].

The transition in the Fe-Ni-Si system occurs at 15 GPa and 1000 K, similar to that for pure Fe. The Clausius–Clapeyron slope is however, 0.0480 GPa/K, which is larger than the reported slopes for Fe (0.0394 GPa/K, [14]), Fe-9.7Ni (0.0426 GPa/K, [57]), and Fe-4Si (0.0394 GPa/K, [99]), stabilising the fcc structure towards high pressure (Figure 35). Above 95 GPa, the boundary of the reaction hcp + fcc → fcc in Fe-5Ni-4Si is placed at a lower temperature than the hcp → hcp + fcc boundary in Fe-9.7Ni (Figure 35). This is even more important because Fe-5Ni-4Si contains less Ni than Fe-9.7Ni. As such, the simultaneous addition of Ni and Si has an anomalous effect on the transition pressure and temperature, which is greatly stabilising the fcc structure under high pressure.

The d*P*/d*T* slopes of the fcc-hcp transition boundaries in different systems and related properties, which include the volume change (ΔV) and entropy change (ΔS) upon transition, were summarized in [192]. The d*P*/d*T* slopes and ΔV were directly obtained from the experiments while ΔS were calculated through the Clausius–Clapeyron equation, $\Delta P/\Delta T = \Delta S/\Delta V$. Komabayashi et al. [192] showed that the increased slope in Fe-5Ni-4Si relative to pure Fe is because of a significantly small ΔV. The addition of Ni to Fe-Si changes ΔS little, but reduces ΔV, which leads to the increased d*P*/d*T* slope in Fe-5Ni-4Si. As such, the thermodynamic properties of the fcc-hcp transition in Fe-5Ni-4Si cannot readily be explained by a combination of the Fe-Ni and Fe-Si systems. The origin of the enlarged fcc stability may lie in the mixing properties of the phases. The triple point, where the fcc, hcp, and liquid phases coexist in Fe-5Ni-4Si is placed at 145 GPa and 3750 K (Figure 36).

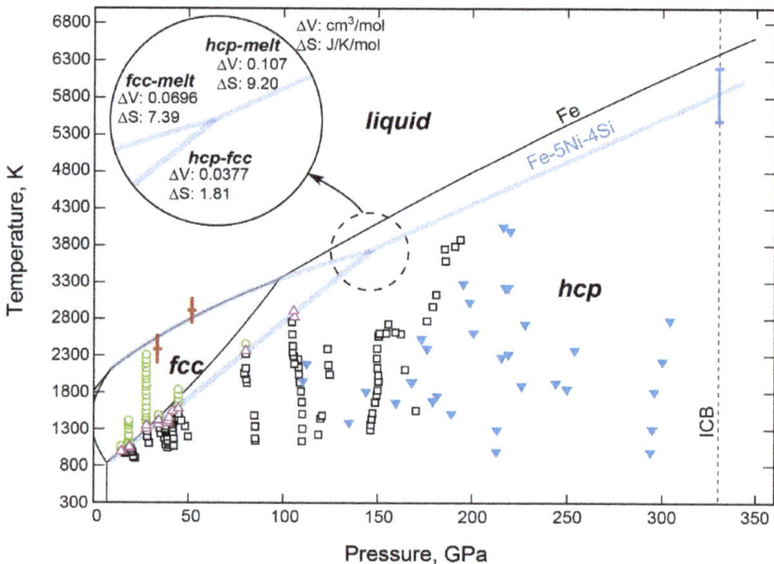

Figure 36. A phase diagram for Fe-5Ni-4Si (blue thick line). Experimental data plotted are: hcp in Fe-5Ni-4Si (open black square, Komabayashi et al. [192]) and Fe-4.8Ni-4Si (inverted blue triangle, Sakai et al. [58]), hcp+fcc in Fe-5Ni-4Si (purple triangle, [192]), fcc in Fe-5Ni-4Si (green circle, [192]), and melting of Fe-5Ni-10Si (red cross, Morard et al. [110]). Komabayashi et al.'s [192] data were based on internally resistive-heated DAC and laser-heated DAC. Inset: the thermodynamics of the triple point where the hcp, fcc, and liquid phases coexist.

- **The Melting Temperatures**

Figure 36 shows a phase diagram of Fe–5Ni–4Si reporting the fcc–hcp boundaries together with phase relations in pure Fe [9]. Experimental constraints on melting in Fe-5Ni-10Si under high pressure were also shown as the crosses in Figure 36 [110]. Since the melting temperature in Fe–Ni–Si is not greatly different from that in Fe up to 50 GPa, Komabayashi et al. [192] assumed the same fcc melting curve for Fe-5Ni-4Si. As the melting loop was not resolved in the previous works, they assumed a narrow melting interval, namely, between solidus and liquidus, and expressed it as a single thick line (Figure 36).

The triple point where the fcc, hcp, and liquid phases coexist is located at 100 GPa and 3400 K for pure Fe [9]. Assuming the same fcc melting curve as for pure Fe, the triple point for Fe-5Ni-4Si is located at 145 GPa and 3750 K. The melting curve of the hcp phase was obtained from the Clausius-Clapeyron relation at the triple point (inset of Figure 36) with the thermodynamics of fcc melting [9]. The phase diagram obtained is consistent with earlier laser-heated DAC experiments in Fe-4.8Ni-4Si to 304 GPa and 2780 K [58].

The melting temperature of the Fe-5Ni-4Si hcp phase at the ICB pressure is estimated to be 5850 ± 350 K, which is 550 K lower than the pure Fe melting temperature. This is consistent with shock wave measurements in Fe-8Ni-10Si [193] although their measurements did not address the structure of the phases. As mentioned above, the melting temperature of Fe-Ni-Si alloy is not very different from that of pure Fe when the alloy structure is fcc [110]. Due to the shift of the triple point towards the high pressure, the melting temperature of the hcp alloy should be largely reduced. As such, the phase relations in the Fe-Ni-Si system cannot readily be inferred from those in the binary Fe-Ni and Fe-Si systems, and therefore the melting temperature as well as the melting relations at 330 GPa needs to be directly constrained.

3.3.2. Fe-Si-O

The phase relations of the Fe-Si-O system at 1 bar are characterized by the presence of two-liquid immiscibility in the Fe-O-rich portion. The two-liquid field occurs due to the immiscibility between the metallic Fe-rich liquid and ionic FeO-rich liquid as was seen in Section 3.2.1.

- **Phase Relations and Mixing Properties for Liquids**

Phase relations of the Fe-rich portion under high pressure were constrained by laser-heated DAC experiments [194,195]. Hirose et al. [194] showed that the liquidus phases were SiO_2 for their experimental samples with compositions from Fe-12.1Si-8.3O to Fe-3.8Si-4.4O and the coexisting liquids were depleted in silicon or oxygen on the basis of chemical analysis of recovered samples. This implies that the eutectic point(s) should be placed near the Fe-Si or Fe-O axes (Figure 37). Arveson et al. [195] also conducted laser-heated DAC experiments on Fe-9Si-3O. They however reported a texture which they took as evidence for the presence of liquid immiscibility just above the solidus over a pressure range from 13 to 78 GPa. The temperature interval for immiscibility was rather constant over their experimental pressure range (Figure 38), which would be aligned with that in the Fe-FeO system over the 15 to 21 GPa pressure range explored by Tsuno et al. [77].

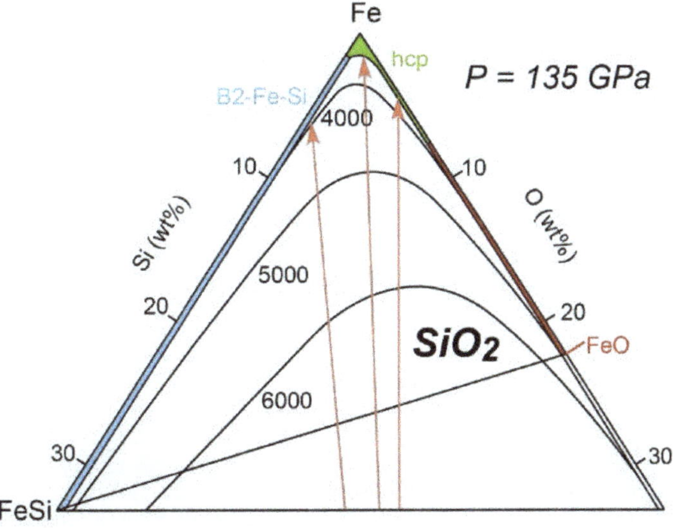

Figure 37. Liquidus phase relations in Fe-FeSi-FeO at about 135 GPa (Hirose et al. [194]) which are characterized by the presence of a large liquidus field with SiO_2. The arrows show the directions of liquid compositions moving away from SiO_2 upon crystallization, found in their experiments. Green, blue, and red areas are the liquidus fields with hcp Fe, CsCl type (B2) Fe-Si phase, and FeO, respectively. The figure is from [194].

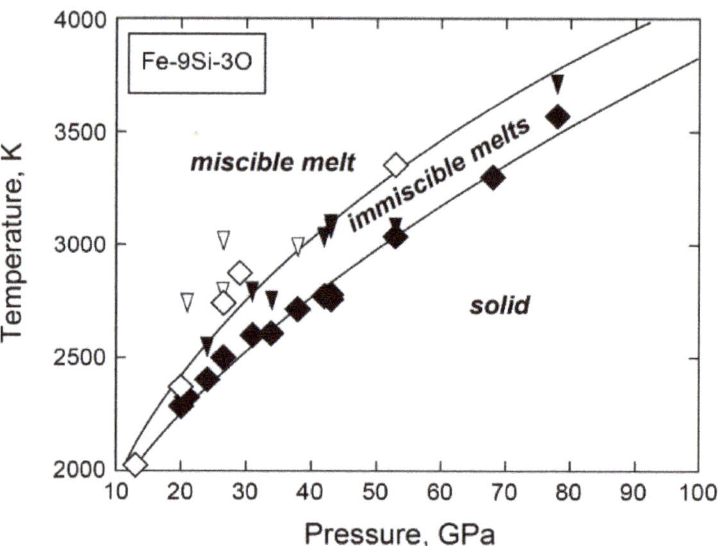

Figure 38. A proposed *P-T* diagram for Fe-9Si-3O showing a miscibility region based on textural observation of laser-heated DAC samples (Arveson et al. [195]). The open and solid diamonds are temperatures for closure of immiscibility and for solidus, respectively. The open and solid inverted triangles denote the maximum experimental temperatures for miscible and immiscible regions, respectively.

Arveson et al. [195] also employed first-principles molecular dynamics simulations. They used three indirect criteria for phase separation as there is no direct measure for it with these types of simulations: atom trajectories, oxygen clustering, and evolution of atom-atom coordination numbers. One of the key results was that most oxygens were clustered with each other via O-Fe/Si bonding under *P-T* conditions where their experiments observed immiscibility, supporting their interpretation of the texture showing immiscibility. However, this was challenged by Huang et al. [108] who also employed first-principles calculations, but with longer simulation durations for 10, 20, and 29 ps in contrast to 6.6–14.8 ps in Arveson et al. [195]. Huang et al. [108] observed full mixing in the longest simulation runs. They also concluded that the atoms of Fe, Si, and O mix ideally under the core conditions.

Huang et al. [108] also challenged the scenario proposed by Hirose et al. [194] that SiO_2 is being crystallised from the outer core at the CMB. Huang et al. [108] reversed the experiment by investigating a two-phase simulation containing solid SiO_2 in contact with liquid Fe and observed that SiO_2 and Fe phases mixed in a short time scale (less than 10 ps) above 4100 K at the CMB pressure, which is a relevant temperature for the outer core.

In summary, the discussion points here are whether the mixing property of Fe-Si-O liquids is ideal or nonideal, and whether the outer core is at its liquidus temperature for SiO_2 crystallisation or at a greater temperature at the CMB. Meanwhile, the enlarged SiO_2 liquidus field proposed by [194] has not been tested by first-principles calculation.

4. Conclusions

As reviewed above, the phase relations of every binary system have not fully been understood under high *P-T* conditions. The maximum *P-T* conditions achieved by experiment for some systems are still away from those corresponding to the centre of the Earth. Nevertheless, we can draw a significant conclusion that any of the binary systems reviewed here by itself cannot account for all the properties of Earth's core. While future research needs to resolve the existing discrepancies raised for each binary system, we should start rigorous and meticulous study on the ternary systems.

Funding: This research was funded by the European Research Council (ERC) Consolidator Grant, grant number #647723.

Acknowledgments: The author thanks Simone Anzellini and Daniel Errandonea for the invitation to this special volume. Chris McGuire and Sam Thompson are thanked for their discussions. The anonymous reviewers are acknowledged for their thorough reading and comments on the manuscript which improved the quality of the paper.

Conflicts of Interest: The author declares no conflict of interest. The funders had no role in the design of the study; in the collection, analyses, or interpretation of data; in the writing of the manuscript, or in the decision to publish the results.

References

1. Bancroft, D.; Peterson, E.L.; Minshall, S. Polymorphism of iron at high pressure. *J. Appl. Phys.* **1956**, *27*, 291–298. [CrossRef]
2. Anderson, O.L. *Equations of State of Solids for Geophysics and Ceramic Science*; Oxford University Press: New York, NY, USA, 1995; p. 405.
3. Jackson, I.; Rigden, S.M. Analysis of *P-V-T* data: Constraints on the thermoelastic properties of high-pressure minerals. *Phys. Earth Planet. Inter.* **1996**, *96*, 85–112. [CrossRef]
4. Dewaele, A.; Loubeyre, P.; Occelli, F.; Mezouar, M.; Dorogokupets, P.I.; Torrent, M. Quasihydrostatic equation of state of iron above 2 Mbar. *Phys. Rev. Lett.* **2006**, *97*, 215504. [CrossRef] [PubMed]
5. Alfè, D.; Price, G.D.; Gillan, M.J. Thermodynamics of hexagonal-close-packed iron under Earth's core conditions. *Phys. Rev. B* **2001**, *64*, 045123. [CrossRef]
6. Anderson, O.L.; Oda, H.; Isaak, D.G. A model for the computation of thermal expansivity at high compression and high temperatures: MgO as an example. *Geophys. Res. Lett.* **1992**, *19*, 1987–1990. [CrossRef]
7. Callen, H.B. *Thermodynamics*; John Wiley & Sons Inc.: Hoboken, NJ, USA, 1960; p. 376.
8. Dinsdale, A.T. SGTE data for pure elements. *Calphad* **1991**, *15*, 317–425. [CrossRef]
9. Komabayashi, T. Thermodynamics of melting relations in the system Fe-FeO at high pressure: Implications for oxygen in the Earth's core. *J. Geophys. Res.* **2014**, *119*. [CrossRef]
10. Komabayashi, T. Thermodynamics of the system Fe-Si-O under high pressure and temperature and its implications for Earth's core. *Phys. Chem. Miner.* **2020**, *47*. [CrossRef]
11. Uchida, T.; Wang, Y.; Rivers, M.L.; Sutton, S.R. Stability field and thermal equation of state of ε-iron determined by synchrotron X-ray diffraction in a multianvil apparatus. *J. Geophys. Res.* **2001**, *106*, 21709–21810. [CrossRef]
12. Shen, G.; Mao, H.-K.; Hemley, R.J.; Duffy, T.S.; Rivers, M.L. Melting and crystal structure of iron at high pressures and temperatures. *Geophys. Res. Lett.* **1998**, *25*, 373–376. [CrossRef]
13. Boehler, R. Temperatures in the Earth's core from melting-point measurements of iron at high static pressures. *Nature* **1993**, *363*, 534–536. [CrossRef]
14. Komabayashi, T.; Fei, Y.; Meng, Y.; Prakapenka, V. In-situ X-ray diffraction measurements of the γ-ε transition boundary of iron in an internally-heated diamond anvil cell. *Earth Planet. Sci. Lett.* **2009**, *282*, 252–257. [CrossRef]
15. Anzellini, S.; Dewaele, A.; Mezouar, M.; Loubeyre, P.; Morard, G. Melting of iron at Earth's inner core boundary based on fast X-ray diffraction. *Science* **2013**, *340*, 464–466. [CrossRef] [PubMed]
16. Mao, H.K.; Bell, P.M.; Hadidiacos, C. Experimental phase relations of iron to 360 kbar, 1400 °C, determined in an internally heated diamond-anvil apparatus. In *High-Pressure Research in Mineral Physics*; Manghnani, M.H., Syono, Y., Eds.; American Geophysical Union: Washington, DC, USA, 1987; pp. 135–138.
17. Funamori, N.; Yagi, T.; Uchida, T. High-pressure and high-temperature in situ x-ray diffraction study of iron to above 30 GPa using MA8-type apparatus. *Geophys. Res. Lett.* **1996**, *23*, 953–956. [CrossRef]
18. Kubo, A.; Ito, E.; Katsura, T.; Shinmei, T.; Yamada, H.; Nishikawa, O.; Song, M.; Funakoshi, K. In situ X-ray observation of iron using Kawai-type apparatus equipped with sintered diamond: Absence of β phase up to 44 GPa and 2100 K. *Geophys. Res. Lett.* **2003**, *30*. [CrossRef]
19. Morelli, A.; Dziewonski, A.M.; Woodhouse, J.H. Anisotropy of the inner core inferred from PKIKP travel times. *Geophys. Res. Lett.* **1986**, *13*, 1545–1548. [CrossRef]
20. Woodhouse, J.H.; Giardini, D.; Li, X.-D. Evidence for inner core anisotropy from free oscillations. *Geophys. Res. Lett.* **1986**, *13*, 1549–1552. [CrossRef]
21. Creager, K.C. Anisotropy of the inner core from differential travel times of the phases PKP and PKIKP. *Nat. Cell Biol.* **1992**, *356*, 309–314. [CrossRef]
22. Song, X.; Helmberger, D.V. Anisotropy of Earth's inner core. *Geophys. Res. Lett.* **1993**, *20*, 2591–2594. [CrossRef]
23. Tateno, S.; Hirose, K.; Ohishi, Y.; Tatsumi, Y. The structure of iron in Earth's inner core. *Science* **2010**, *330*, 359–361. [CrossRef]
24. Stixrude, L. Structure of Iron to 1 Gbar and 40 000 K. *Phys. Rev. Lett.* **2012**, *108*, 055505. [CrossRef]

25. Sinmyo, R.; Hirose, K.; Ohishi, Y. Melting curve of iron to 290 GPa determined in a resistance-heated diamond-anvil cell. *Earth Planet. Sci. Lett.* **2019**, *510*, 45–52. [CrossRef]
26. Kuwayama, Y.; Hirose, K.; Sata, N.; Ohishi, Y. Phase relations of iron and iron-nickel alloys up to 300 GPa: Implications for composition and structure of the Earth's inner core. *Earth Planet. Sci. Lett.* **2008**, *273*, 379–385. [CrossRef]
27. Brown, M.J.; McQueen, R.G. Phase transitions, Grueneisen parameter, and elasticity for shocked iron between 77 GPa and 400 GPa. *J. Geophys. Res.* **1986**, *91*, 7485–7494. [CrossRef]
28. Boehler, R. The phase diagram of iron to 430 kbar. *Geophys. Res. Lett.* **1986**, *13*, 1153–1156. [CrossRef]
29. Belonoshko, A.B.; Ahuja, R.; Johansson, B. Stability of the body-centred-cubic phase of iron in the Earth's inner core. *Nature* **2003**, *424*, 1032–1034. [CrossRef] [PubMed]
30. Mikhaylushkin, A.S.; Simak, S.I.; Dubrovinsky, L.; Dubrovinskaia, N.; Johansson, B.; Abrikosov, I.A. Pure Iron Compressed and Heated to Extreme Conditions. *Phys. Rev. Lett.* **2007**, *99*, 165505. [CrossRef] [PubMed]
31. Luo, W.; Johansson, B.; Eriksson, O.; Arapan, S.; Souvatzis, P.; Katsnelson, M.I.; Ahuja, R. Dynamical stability of body center cubic iron at the Earth's core conditions. *Proc. Natl. Acad. Sci. USA* **2010**, *107*, 9962–9964. [CrossRef] [PubMed]
32. Belonoshko, A.B.; Lukinov, T.; Fu, J.; Zhao, J.; Davis, S.; Simak, S.I. Stabilization of body-centred cubic iron under inner-core conditions. *Nat. Geosci.* **2017**, *10*, 312–316. [CrossRef]
33. Nguyen, J.H.; Holmes, N.C. Melting of iron at the physical conditions of the Earth's core. *Nature* **2004**, *427*, 339–342. [CrossRef]
34. Yoo, C.S.; Holmes, N.C.; Ross, M.; Webb, D.J.; Pike, C. Shock temperatures and melting of iron at Earth core conditions. *Phys. Rev. Lett.* **1993**, *70*, 3931–3934. [CrossRef] [PubMed]
35. Turneaure, S.J.; Sharma, S.M.; Gupta, Y.M. Crystal Structure and Melting of Fe Shock Compressed to 273 GPa: In Situ X-ray Diffraction. *Phys. Rev. Lett.* **2020**, *125*, 215702. [CrossRef] [PubMed]
36. Alfè, D.; Gillan, M.J.; Price, G.D. The melting curve of iron at the pressures of the Earth's core from ab initio calculations. *Nature* **1999**, *401*, 462–464. [CrossRef]
37. Belonoshko, A.B.; Ahuja, R.; Johansson, B. Quasi–Ab InitioMolecular Dynamic Study of Fe Melting. *Phys. Rev. Lett.* **2000**, *84*, 3638–3641. [CrossRef] [PubMed]
38. Alfè, D. Temperature of the inner-core boundary of the Earth: Melting of iron at high pressure from first-principles coexistence simulations. *Phys. Rev. B* **2009**, *79*, 060101. [CrossRef]
39. Laio, A.; Bernard, S.; Chiarotti, G.L.; Scandolo, S.; Tosatti, E. Physics of iron at Earth's core conditions. *Science* **2000**, *287*, 1027–1030. [CrossRef]
40. Jackson, J.M.; Sturhahn, W.; Lerche, M.; Zhao, J.Y.; Toellner, T.S.; Alp, E.E.; Sinogeikin, S.V.; Bass, J.D.; Murphy, C.A.; Wicks, J.K. Melting of compressed iron by monitoring atomic dynamics. *Earth Planet. Sci. Lett.* **2013**, *362*, 143–150. [CrossRef]
41. Morard, G.; Boccato, S.; Rosa, A.D.; Anzellini, S.; Miozzi, F.; Henry, L.; Garbarino, G.; Mezouar, M.; Harmand, M.; Guyot, F.; et al. Solving Controversies on the Iron Phase Diagram Under High Pressure. *Geophys. Res. Lett.* **2018**, *45*, 11074–11082. [CrossRef]
42. Guillermet, A.F.; Gustafson, P. An assessment of the thermodynamic properties and the (p, T) phase diagram of iron. *High Temp. High Press.* **1985**, *16*, 591–610.
43. Saxena, S.K.; Dubrovinsky, L.S. Thermodynamics of iron phases at high pressures and temperatures. In *Properties of Earth and Planetary Materials*; Manghnani, M.H., Yagi, T., Eds.; American Geophysical Union: Washington, DC, USA, 1998; pp. 271–279.
44. Tsujino, N.; Nishihara, Y.; Nakajima, Y.; Takahashi, E.; Funakoshi, K.; Higo, Y. Equation of state of gamma-Fe: Reference density for planetary cores. *Earth Planet. Sci. Lett.* **2013**, *375*, 244–253. [CrossRef]
45. Alfè, D.; Price, G.D.; Gillan, M.J. Iron under Earth's core conditions: Liquid-state thermodynamics and high-pressure melting curve from ab initio calculations. *Phys. Rev. B* **2002**, *65*, 165118. [CrossRef]
46. Tiwari, G.P. Modification of Richard's rule and correlation between entropy of fusion and allotropic behaviour. *Met. Sci.* **1978**, *12*, 317–320. [CrossRef]
47. Anderson, W.W.; Ahrens, T.J. An equation of state for liquid iron and implications for the Earth's core. *J. Geophys. Res.* **1994**, *99*, 4273–4284. [CrossRef]
48. Kuwayama, Y.; Morard, G.; Nakajima, Y.; Hirose, K.; Baron, A.Q.R.; Kawaguchi, S.I.; Tsuchiya, T.; Ishikawa, D.; Hirao, N.; Ohishi, Y. Equation of state of liquid iron under extreme conditions. *Phys. Rev. Lett.* **2020**, *124*, 165701. [CrossRef] [PubMed]
49. Wagle, F.; Steinle-Neumann, G. Liquid iron equation of state to the terapascal regime from ab initio simulations. *J. Geophys. Res. Solid Earth* **2019**, *124*, 3350–3364. [CrossRef]
50. Ichikawa, H.; Tsuchiya, T.; Tange, Y. The *P-V-T* equation of state and thermodynamic properties of liquid iron. *J. Geophys. Res. Solid Earth* **2014**, *119*, 240–252. [CrossRef]
51. Dziewonski, A.M.; Anderson, D.L. Preliminary reference Earth model. *Phys. Earth Planet. Inter.* **1981**, *25*, 297–356. [CrossRef]
52. Fei, Y.W.; Murphy, C.; Shibazaki, Y.; Shahar, A.; Huang, H.J. Thermal equation of state of hcp-iron: Constraint on the density deficit of Earth's solid inner core. *Geophys. Res. Lett.* **2016**, *43*, 6837–6843. [CrossRef]
53. Li, J.; Fei, Y. Experimental constraints on core composition. In *Treatise on Geochemistry Update*; Holland, H.D., Turekian, K.K., Eds.; Elsevier Ltd.: Amsterdam, The Netherlands, 2007; pp. 1–31.
54. Lin, J.-F.; Heinz, D.L.; Campbell, A.J.; Devine, J.M.; Mao, W.L.; Shen, G. Iron-Nickel alloy in the Earth's core. *Geophys. Res. Lett.* **2002**, *29*. [CrossRef]
55. Mao, W.L.; Campbell, A.J.; Heinz, D.L.; Shen, G. Phase relations of Fe–Ni alloys at high pressure and temperature. *Phys. Earth Planet. Inter.* **2006**, *155*, 146–151. [CrossRef]

56. Dubrovinsky, L.; Dubrovinskaia, N.; Narygina, O.; Kantor, I.; Kuznetzov, A.; Prakapenka, V.B.; Vitos, L.; Johansson, B.; Mikhaylushkin, A.S.; Simak, S.I.; et al. Body-centered cubic iron-nickel alloy in Earth's core. *Science* **2007**, *316*, 1880–1883. [CrossRef] [PubMed]
57. Komabayashi, T.; Hirose, K.; Ohishi, Y. In situ X-ray diffraction measurements of the fcc–hcp phase transition boundary of an Fe–Ni alloy in an internally heated diamond anvil cell. *Phys. Chem. Miner.* **2012**, *39*, 329–338. [CrossRef]
58. Sakai, T.; Ohtani, E.; Hirao, N.; Ohishi, Y. Stability field of the hcp-structure for Fe, Fe-Ni, and Fe-Ni-Si alloys up to 3 Mbar. *Geophys. Res. Lett.* **2011**, *38*, 09302. [CrossRef]
59. Tateno, S.; Hirose, K.; Komabayashi, T.; Ozawa, H.; Ohishi, Y. The structure of Fe-Ni alloy in Earth's inner core. *Geophys. Res. Lett.* **2012**, *39*. [CrossRef]
60. Zhang, D.Z.; Jackson, J.M.; Zhao, J.Y.; Sturhahn, W.; Alp, E.E.; Hu, M.Y.; Toellner, T.S.; Murphy, C.A.; Prakapenka, V.B. Temperature of Earth's core constrained from melting of Fe and $Fe_{0.9}Ni_{0.1}$ at high pressures. *Earth Planet. Sci. Lett.* **2016**, *447*, 72–83. [CrossRef]
61. Torchio, R.; Boccato, S.; Miozzi, F.; Rosa, A.D.; Ishimatsu, N.; Kantor, I.; Sevelin-Radiguet, N.; Briggs, R.; Meneghini, C.; Irifune, T.; et al. Melting Curve and Phase Relations of Fe-Ni Alloys: Implications for the Earth's Core Composition. *Geophys. Res. Lett.* **2020**, *47*. [CrossRef]
62. Zou, G.; Mao, H.K.; Bell, P.M.; Virgo, D. High pressure experiments on the iron oxide wüstite ($Fe_{1-x}O$). *Carnegie Inst. Wash. Yearb.* **1980**, *79*, 374–376.
63. Jeanloz, R.; Ahrens, T.J. Equations of state of FeO and CaO. *Geophys. J. R. Astron. Soc.* **1980**, *62*, 505–528. [CrossRef]
64. Knittle, E.; Jeanloz, R. High-Pressure metallization of FeO and implications for the Earth core. *Geophys. Res. Lett.* **1986**, *13*, 1541–1544. [CrossRef]
65. Knittle, E.; Jeanloz, R.; Mitchell, A.C.; Nellis, W.J. Metallization of $Fe_{0.94}O$ at elevated pressures and temperatures observed by shock-wave electrical resistivity measurements. *Solid State Commun.* **1986**, *59*, 513–515. [CrossRef]
66. Fei, Y.W.; Mao, H.K. In-situ determination of the NiAs phase of FeO at high-pressure and Temperature. *Science* **1994**, *266*, 1678–1680. [CrossRef] [PubMed]
67. Kondo, T.; Ohtani, E.; Hirao, N.; Yagi, T.; Kikegawa, T. Phase transitions of (Mg,Fe)O at megabar pressures. *Phys. Earth Planet. Inter.* **2004**, *143*, 201–213. [CrossRef]
68. Ozawa, H.; Hirose, K.; Tateno, S.; Sata, N.; Ohishi, Y. Phase transition boundary between B1 and B8 structures of FeO up to 210 GPa. *Phys. Earth Planet. Inter.* **2010**, *179*, 157–163. [CrossRef]
69. Ozawa, H.; Takahashi, F.; Hirose, K.; Ohishi, Y.; Hirao, N. Phase transition of FeO and stratification in Earth's outer core. *Science* **2011**, *334*, 792–794. [CrossRef] [PubMed]
70. Ohta, K.; Cohen, R.E.; Hirose, K.; Haule, K.; Shimizu, K.; Ohishi, Y. Experimental and theoretical evidence for pressure-induced metallization in FeO with rocksalt-type structure. *Phys. Rev. Lett.* **2012**, *108*, 026403. [CrossRef] [PubMed]
71. Ohta, K.; Hirose, K.; Shimizu, K.; Ohishi, Y. High-pressure experimental evidence for metal FeO with normal NiAs-type structure. *Phys. Rev. B* **2010**, *82*, 174120. [CrossRef]
72. Fischer, R.A.; Campbell, A.J.; Lord, O.T.; Shofner, G.A.; Dera, P.; Prakapenka, V.B. Phase transition and metallization of FeO at high pressures and temperatures. *Geophys. Res. Lett.* **2011**, *38*, L24301. [CrossRef]
73. Fischer, R.A.; Campbell, A.J.; Shofner, G.A.; Lord, O.T.; Dera, P.; Prakapenka, V.B. Equation of state and phase diagram of FeO. *Earth Planet. Sci. Lett.* **2011**, *304*, 496–502. [CrossRef]
74. Fischer, R.A.; Campbell, A.J. High-pressure melting of wüstite. *Am. Mineral.* **2010**, *95*, 1473–1477. [CrossRef]
75. Seagle, C.T.; Heinz, D.L.; Campbell, A.J.; Prakapenka, V.B.; Wanless, S.T. Melting and thermal expansion in the Fe-FeO system at high pressure. *Earth Planet. Sci. Lett.* **2008**, *265*, 655–665. [CrossRef]
76. Lindsley, D.H. Pressure-temperature relations in the system $FeO-SiO_2$. *Year Book Carnegie Inst. Wash.* **1966**, *65*, 226–230.
77. Tsuno, K.; Ohtani, E.; Terasaki, H. Immiscible two-liquid regions in the Fe–O–S system at high pressure: Implications for planetary cores. *Phys. Earth Planet. Inter.* **2007**, *160*, 75–85. [CrossRef]
78. Ohtani, E.; Ringwood, A.; Hibberson, W. Composition of the core, II. Effect of high pressure on solubility of FeO in molten iron. *Earth Planet. Sci. Lett.* **1984**, *71*, 94–103. [CrossRef]
79. Wriedt, H. The Fe-O (iron-oxygen) system. *J. Phase Equilibria Diffus.* **1991**, *12*, 170–200. [CrossRef]
80. Sundman, B. An assessment of the Fe-O system. *J. Phase Equilibria Diffus.* **1991**, *12*, 127–140. [CrossRef]
81. Kowalski, M.; Spencer, P.J. Thermodynamic reevaluation of the Cr-O, Fe-O and Ni-O systems-remodeling of the liquid, bcc and fcc phases. *Calphad* **1995**, *19*, 229–243. [CrossRef]
82. Frost, D.J.; Asahara, Y.; Rubie, D.C.; Miyajima, N.; Dubrovinsky, L.S.; Holzapfel, C.; Ohtani, E.; Miyahara, M.; Sakai, T. Partitioning of oxygen between the Earth's mantle and core. *J. Geophys. Res. Space Phys.* **2010**, *115*. [CrossRef]
83. Darken, L.S.; Gurry, R.W. The system iron-oxygen. II. Equilibrium and thermodynamics of liquid oxide and other phases. *J. Am. Chem. Soc.* **1946**, *68*, 798–816. [CrossRef]
84. Ringwood, A.E.; Hibberson, W. The system Fe-FeO revisited. *Phys. Chem. Miner.* **1990**, *17*, 313–319. [CrossRef]
85. Morard, G.; Andrault, D.; Antonangeli, D.; Nakajima, Y.; Auzende, A.L.; Boulard, E.; Cervera, S.; Clark, A.; Lord, O.T.; Siebert, J.; et al. Fe-FeO and $Fe-Fe_3C$ melting relations at Earth's core-mantle boundary conditions: Implications for a volatile-rich or oxygen-rich core. *Earth Planet. Sci. Lett.* **2017**, *473*, 94–103. [CrossRef]
86. Oka, K.; Hirose, K.; Tagawa, S.; Kidokoro, Y.; Nakajima, Y.; Kuwayama, Y.; Morard, G.; Coudurier, N.; Fiquet, G. Melting in the Fe-FeO system to 204 GPa: Implications for oxygen in Earth's core. *Am. Mineral.* **2019**, *104*, 1603–1607. [CrossRef]

87. Hattori, T.; Kinoshita, T.; Narushima, T.; Tsuji, K.; Katayama, Y. Pressure-induced structural change of liquid CdTe up to 23.5 GPa. *Phys. Rev. B* **2006**, *73*, 054203. [CrossRef]

88. Badro, J.; Cote, A.S.; Brodholt, J.P. A seismologically consistent compositional model of Earth's core. *Proc. Natl. Acad. Sci. USA* **2014**, *111*, 7542–7545. [CrossRef] [PubMed]

89. Ichikawa, H.; Tsuchiya, T. Ab Initio Thermoelasticity of Liquid Iron-Nickel-Light Element Alloys. *Minerals* **2020**, *10*, 59. [CrossRef]

90. Alfè, D.; Gillan, M.; Price, G. Composition and temperature of the Earth's core constrained by combining ab initio calculations and seismic data. *Earth Planet. Sci. Lett.* **2002**, *195*, 91–98. [CrossRef]

91. Dobson, D.P.; Crichton, W.A.; Bouvier, P.; Vočadlo, L.; Wood, I.G. The equation of state of CsCl-structured FeSi to 40 GPa: Implications for silicon in the Earth's core. *Geophys. Res. Lett.* **2003**, *30*. [CrossRef]

92. Lin, J.F.; Heinz, D.L.; Campbell, A.J.; Devine, J.M.; Shen, G.Y. Iron-silicon alloy in Earth's core? *Science* **2002**, *295*, 313–315. [CrossRef] [PubMed]

93. Lin, J.F.; Scott, H.P.; Fischer, R.A.; Chang, Y.Y.; Kantor, I.; Prakapenka, V.B. Phase relations of Fe-Si alloy in Earth's core. *Geophys. Res. Lett.* **2009**, *36*. [CrossRef]

94. Kuwayama, Y.; Hirose, K. Phase relations in the system Fe-FeSi at 21 GPa. *Am. Mineral.* **2004**, *89*, 273–276. [CrossRef]

95. Fischer, R.A.; Campbell, A.J.; Reaman, D.M.; Miller, N.A.; Heinz, D.L.; Dera, P.; Prakapenka, V.B. Phase relations in the Fe-FeSi system at high pressures and temperatures. *Earth Planet. Sci. Lett.* **2013**, *373*, 54–64. [CrossRef]

96. Fischer, R.A.; Campbell, A.J.; Caracas, R.; Reaman, D.M.; Heinz, D.L.; Dera, P.; Prakapenka, V.B. Equations of state in the Fe-FeSi system at high pressures and temperatures. *J. Geophys. Res. Solid Earth* **2014**, *119*, 2810–2827. [CrossRef]

97. Tateno, S.; Kuwayama, Y.; Hirose, K.; Ohishi, Y. The structure of Fe-Si alloy in Earth's inner core. *Earth Planet. Sci. Lett.* **2015**, *418*, 11–19. [CrossRef]

98. Ozawa, H.; Hirose, K.; Yonemitsu, K.; Ohishi, Y. High-pressure melting experiments on Fe-Si alloys and implications for silicon as a light element in the core. *Earth Planet. Sci. Lett.* **2016**, *456*, 47–54. [CrossRef]

99. Komabayashi, T.; Pesce, G.; Morard, G.; Antonangeli, D.; Sinmyo, R.; Mezouar, M. Phase transition boundary between fcc and hcp structures in Fe-Si alloy and its implications for terrestrial planetary cores. *Am. Mineral.* **2019**, *104*, 94–99. [CrossRef]

100. Fischer, R.A.; Campbell, A.J.; Caracas, R.; Reaman, D.M.; Dera, P.; Prakapenka, V.B. Equation of state and phase diagram of Fe-16Si alloy as a candidate component of Earth's core. *Earth Planet. Sci. Lett.* **2012**, *357*, 268–276. [CrossRef]

101. Asanuma, H.; Ohtani, E.; Sakai, T.; Terasaki, H.; Kamada, S.; Hirao, N.; Sata, N.; Ohishi, Y. Phase relations of Fe-Si alloy up to core conditions: Implications for the Earth inner core. *Geophys. Res. Lett.* **2008**, *35*. [CrossRef]

102. Kuwayama, Y.; Sawai, T.; Hirose, K.; Sata, N.; Ohishi, Y. Phase relations of iron–silicon alloys at high pressure and high temperature. *Phys. Chem. Miner.* **2009**, *36*, 511–518. [CrossRef]

103. Vočadlo, L.; Alfè, D.; Gillan, M.J.; Wood, I.G.; Brodholt, J.P.; Price, G.D. Possible thermal and chemical stabilization of body-centred-cubic iron in the Earth's core. *Nature* **2003**, *424*, 536–539. [CrossRef]

104. Belonoshko, A.B.; Rosengren, A.; Burakovsky, L.; Preston, D.L.; Johansson, B. Melting of Fe and $Fe_{0.9375}Si_{0.0625}$ at Earth's core pressures studied using ab initio molecular dynamics. *Phys. Rev. B* **2009**, *79*, 220102. [CrossRef]

105. Andrault, D.; Bolfan-Casanova, N.; Ohtaka, O.; Fukui, H.; Arima, H.; Fialin, M.; Funakoshi, K. Melting diagrams of Fe-rich alloys determined from synchrotron in situ measurements in the 15–23 GPa pressure range. *Phys. Earth Planet. Inter.* **2009**, *174*, 181–191. [CrossRef]

106. Morard, G.; Siebert, J.; Andrault, D.; Guignot, N.; Garbarino, G.; Guyot, F.; Antonangeli, D. The Earth's core composition from high pressure density measurements of liquid iron alloys. *Earth Planet. Sci. Lett.* **2013**, *373*, 169–178. [CrossRef]

107. Ohnuma, I.; Abe, S.; Shimenouchi, S.; Omori, T.; Kainuma, R.; Ishida, K. Experimental and thermodynamic studies of the Fe-Si binary system. *ISIJ Int.* **2012**, *52*, 540–548. [CrossRef]

108. Huang, D.Y.; Badro, J.; Brodholt, J.; Li, Y.G. Ab Initio molecular dynamics investigation of molten Fe-Si-O in Earth's core. *Geophys. Res. Lett.* **2019**, *46*, 6397–6405. [CrossRef]

109. Asanuma, H.; Ohtani, E.; Sakai, T.; Terasaki, H.; Kamada, S.; Kondo, T.; Kikegawa, T. Melting of iron–silicon alloy up to the core–mantle boundary pressure: Implications to the thermal structure of the Earth's core. *Phys. Chem. Miner.* **2009**, *37*, 353–359. [CrossRef]

110. Morard, G.; Andrault, D.; Guignot, N.; Siebert, J.; Garbarino, G.; Antonangeli, D. Melting of Fe-Ni-Si and Fe-Ni-S alloys at megabar pressures: Implications for the core-mantle boundary temperature. *Phys. Chem. Miner.* **2011**, *38*, 767–776. [CrossRef]

111. Lord, O.T.; Wann, E.T.H.; Hunt, S.A.; Walker, A.M.; Santangeli, J.; Walter, M.J.; Dobson, D.P.; Wood, I.G.; Vočadlo, L.; Morard, G.; et al. The NiSi melting curve to 70 GPa. *Phys. Earth Planet. Inter.* **2014**, *233*, 13–23. [CrossRef]

112. Yamasaki, M.; Banno, S. Zoning pattern of extremely fractionated plagioclase crystallized from simple system. *Bull. Volcanol. Soc. Jpn.* **1972**, *2*, 18–25. (In Japanese with English abstract)

113. Wade, J.; Wood, B.J. Core formation and the oxidation state of the Earth. *Earth Planet. Sci. Lett.* **2005**, *236*, 78–95. [CrossRef]

114. Antonangeli, D.; Siebert, J.; Badro, J.; Farber, D.L.; Fiquet, G.; Morard, G.; Ryerson, F.J. Composition of the Earth's inner core from high-pressure sound velocity measurements in Fe-Ni-Si alloys. *Earth Planet. Sci. Lett.* **2010**, *295*, 292–296. [CrossRef]

115. Rubie, D.C.; Frost, D.J.; Mann, U.; Asahara, Y.; Nimmo, F.; Tsuno, K.; Kegler, P.; Holzheid, A.; Palme, H. Heterogeneous accretion, composition and core-mantle differentiation of the Earth. *Earth Planet. Sci. Lett.* **2011**, *301*, 31–42. [CrossRef]

116. Badro, J.; Brodholt, J.P.; Piet, H.; Siebert, J.; Ryerson, F.J. Core formation and core composition from coupled geochemical and geophysical constraints. *Proc. Natl. Acad. Sci. USA* **2015**, *112*, 12310–12314. [CrossRef] [PubMed]

117. Antonangeli, D.; Morard, G.; Paolasini, L.; Garbarino, G.; Murphy, C.A.; Edmund, E.; Decremps, F.; Fiquet, G.; Bosak, A.; Mezouar, M.; et al. Sound velocities and density measurements of solid hcp-Fe and hcp-Fe-Si (9 wt%) alloy at high pressure: Constraints on the Si abundance in the Earth's inner core. *Earth Planet. Sci. Lett.* **2018**, *482*, 446–453. [CrossRef]

118. Nakajima, Y.; Kawaguchi, S.I.; Hirose, K.; Tateno, S.; Kuwayama, Y.; Sinmyo, R.; Ozawa, H.; Tsutsui, S.; Uchiyama, H.; Baron, A.Q.R. Silicon-depleted present-day Earth's outer core revealed by sound velocity measurements of liquid fe-si alloy. *J. Geophys. Res. Solid Earth* **2020**, *125*, e2020JB019399. [CrossRef]

119. Hirose, K.; Labrosse, S.; Hernlund, J. Composition and state of the core. *Annu. Rev. Earth Planet. Sci.* **2013**, *41*, 657–691. [CrossRef]

120. Campbell, A.J.; Seagle, C.T.; Heinz, D.L.; Shen, G.Y.; Prakapenka, V.B. Partial melting in the iron–sulfur system at high pressure: A synchrotron X-ray diffraction study. *Phys. Earth Planet. Inter.* **2007**, *162*, 119–128. [CrossRef]

121. Ozawa, H.; Hirose, K.; Suzuki, T.; Ohishi, Y.; Hirao, N. Decomposition of Fe_3S above 250 GPa. *Geophys. Res. Lett.* **2013**, *40*, 4845–4849. [CrossRef]

122. Mori, Y.; Ozawa, H.; Hirose, K.; Sinmyo, R.; Tateno, S.; Morard, G.; Ohishi, Y. Melting experiments on $Fe–Fe_3S$ system to 254 GPa. *Earth Planet. Sci. Lett.* **2017**, *464*, 135–141. [CrossRef]

123. Tateno, S.; Ozawa, H.; Hirose, K.; Suzuki, T.; I-Kawaguchi, S.; Hirao, N. Fe_2S: The most Fe-rich iron sulfide at the Earth's inner core pressures. *Geophys. Res. Lett.* **2019**, *46*, 11944–11949. [CrossRef]

124. Kamada, S.; Ohtani, E.; Terasaki, H.; Sakai, T.; Miyahara, M.; Ohishi, Y.; Hirao, N. Melting relationships in the $Fe–Fe_3S$ system up to the outer core conditions. *Earth Planet. Sci. Lett.* **2012**, *359–360*, 26–33. [CrossRef]

125. Pommier, A.; Laurenz, V.; Davies, C.J.; Frost, D.J. Melting phase relations in the Fe-S and Fe-S-O systems at core conditions in small terrestrial bodies. *ICARUS* **2018**, *306*, 150–162. [CrossRef]

126. Waldner, P.; Pelton, A.D. Thermodynamic modeling of the Fe-S system. *J. Phase Equilibria Diffus.* **2005**, *26*, 23–38. [CrossRef]

127. Chen, B.; Li, J.; Hauck, S.A. Non-ideal liquidus curve in the Fe-S system and Mercury's snowing core. *Geophys. Res. Lett.* **2008**, *35*. [CrossRef]

128. Fei, Y.W.; Li, J.; Bertka, C.M.; Prewitt, C.T. Structure type and bulk modulus of Fe_3S, a new iron-sulfur compound. *Am. Mineral.* **2000**, *85*, 1830–1833. [CrossRef]

129. Kamada, S.; Terasaki, H.; Ohtani, E.; Sakai, T.; Kikegawa, T.; Ohishi, Y.; Hirao, N.; Sata, N.; Kondo, T. Phase relationships of the Fe-FeS system in conditions up to the Earth's outer core. *Earth Planet. Sci. Lett.* **2010**, *294*, 94–100. [CrossRef]

130. Stewart, A.J.; Schmidt, M.W.; van Westrenen, W.; Liebske, C. Mars: A new core-crystallization regime. *Science* **2007**, *316*, 1323–1325. [CrossRef]

131. Chudinovskikh, L.; Boehler, R. Eutectic melting in the system Fe-S to 44 GPa. *Earth Planet. Sci. Lett.* **2007**, *257*, 97–103. [CrossRef]

132. Morard, G.; Andrault, D.; Guignot, N.; Sanloup, C.; Mezouar, M.; Petitgirard, S.; Fiquet, G. In situ determination of $Fe–Fe_3S$ phase diagram and liquid structural properties up to 65 GPa. *Earth Planet. Sci. Lett.* **2008**, *272*, 620–626. [CrossRef]

133. Morard, G.; Andrault, D.; Antonangeli, D.; Bouchet, J. Properties of iron alloys under the Earth's core conditions. *Comptes Rendus Geosci.* **2014**, *346*, 130–139. [CrossRef]

134. Li, J.; Fei, Y.; Mao, H.K.; Hirose, K.; Shieh, S.R. Sulfur in the Earth's inner core. *Earth Planet. Sci. Lett.* **2001**, *193*, 509–514. [CrossRef]

135. Seagle, C.T.; Campbell, A.J.; Heinz, D.L.; Shen, G.; Prakapenka, V.B. Thermal equation of state of Fe_3S and implications for sulfur in Earth's core. *J. Geophys. Res. Space Phys.* **2006**, *111*. [CrossRef]

136. Kamada, S.; Ohtani, E.; Terasaki, H.; Sakai, T.; Takahashi, S.; Hirao, N.; Ohishi, Y. Equation of state of Fe_3S at room temperature up to 2-megabars. *Phys. Earth Planet. Inter.* **2014**, *228*, 106–113. [CrossRef]

137. Chen, B.; Gao, L.L.; Funakoshi, K.; Li, J. Thermal expansion of iron-rich alloys and implications for the Earth's core. *Proc. Natl. Acad. Sci. USA* **2007**, *104*, 9162–9167. [CrossRef] [PubMed]

138. Thompson, S.; Komabayashi, T.; Breton, H.; Suehiro, S.; Glazyrin, K.; Pakhomova, A.; Ohishi, Y. Compression experiments to 126 GPa and 2500 K and thermal equation of state of Fe_3S: Implications for sulphur in the Earth's core. *Earth Planet. Sci. Lett.* **2020**, *534*, 116080. [CrossRef]

139. Hillert, M.; Staffansson, L.I. An analysis of the phase equilibria in the Fe-FeS system. *Metall. Mater. Trans. B* **1975**, *6*, 37–41. [CrossRef]

140. Sharma, R.C.; Chang, Y.A. Thermodynamics and phase relationships of transition metal-sulfur systems: Part III. Thermodynamic properties of the Fe-S liquid-phase and the calculation of the Fe-S phase diagram. *Metall. Trans. B* **1979**, *10*, 103–108. [CrossRef]

141. Guillermet, A.F.; Hillert, M.; Jansson, B.; Sundman, B. An assessment of the Fe-S system using a 2-sublattice model for the liquid-phase. *Metall. Trans. B* **1981**, *12*, 745–754. [CrossRef]

142. Chuang, Y.Y.; Hsieh, K.C.; Chang, Y.A. Thermodynamics and phase relationships of transition metal-sulfur systems: Part V. A reevaluation of the Fe-S System using an associated solution model for the liquid-phase. *Metall. Trans. B* **1985**, *16*, 277–285. [CrossRef]

143. Saxena, S.; Eriksson, G. Thermodynamics of Fe-S at ultra-high pressure. *Calphad* **2015**, *51*, 202–205. [CrossRef]

144. Umemoto, K.; Hirose, K.; Imada, S.; Nakajima, Y.; Komabayashi, T.; Tsutsui, S.; Baron, A.Q.R. Liquid iron-sulfur alloys at outer core conditions by first-principles calculations. *Geophys. Res. Lett.* **2014**, *41*, 6712–6717. [CrossRef]

145. Fu, J.; Cao, L.Z.; Duan, X.M.; Belonoshko, A.B. Density and sound velocity of liquid Fe-S alloys at Earth's outer core P-T conditions. *Am. Mineral.* **2020**, *105*, 1349–1354. [CrossRef]

146. Huang, H.J.; Wu, S.J.; Hu, X.J.; Wang, Q.S.; Wang, X.; Fei, Y.W. Shock compression of Fe-FeS mixture up to 204 GPa. *Geophys. Res. Lett.* **2013**, *40*, 687–691. [CrossRef]

147. Vočadlo, L.; Alfè, D.; Gillan, M.J.; Price, G.D. The properties of iron under core conditions from first principles calculations. *Phys. Earth Planet. Inter.* **2003**, *140*, 101–125. [CrossRef]

148. Grigorovich, V.K. The polymorphism of iron and the electron structure of iron alloys. *Izvt. Akad. Nauk. SSSR Met. JAN-FEB* **1969**, *1*, 53–68. (In Russian)

149. Shterenberg, L.E.; Slesarev, V.N.; Korsunskaya, I.A.; Kamenetskaya, D.S. The experimental study of the interaction between the melt, carbides and diamond in the iron-carbon system at high pressures. *High Temp. High Press.* **1975**, *7*, 517–522.

150. Wood, B.J. Carbon in the core. *Earth Planet. Sci. Lett.* **1993**, *117*, 593–607. [CrossRef]

151. Nakajima, Y.; Takahashi, E.; Suzuki, T.; Funakoshi, K. "Carbon in the core" revisited. *Phys. Earth Planet. Inter.* **2009**, *174*, 202–211. [CrossRef]

152. Gustafson, P. A Thermodynamic Evaluation of the Fe-C System. *Scand. J. Metall.* **1985**, *14*, 259–267. [CrossRef]

153. Chabot, N.L.; Campbell, A.J.; McDonough, W.F.; Draper, D.S.; Agee, C.B.; Humayun, M.; Watson, H.C.; Cottrell, E.; Saslow, S.A. The Fe-C system at 5 GPa and implications for Earth's core. *Geochim. Cosmochim. Acta* **2008**, *72*, 4146–4158. [CrossRef]

154. Fei, Y.W.; Brosh, E. Experimental study and thermodynamic calculations of phase relations in the Fe-C system at high pressure. *Earth Planet. Sci. Lett.* **2014**, *408*, 155–162. [CrossRef]

155. Benz, M.G.; Elliott, J.F. The austenite solidus and revised iron-carbon diagram. *Trans. Metall. Soc. AIME* **1961**, *221*, 323–331.

156. Mashino, I.; Miozzi, F.; Hirose, K.; Morard, G.; Sinmyo, R. Melting experiments on the Fe-C binary system up to 255 GPa: Constraints on the carbon content in the Earth's core. *Earth Planet. Sci. Lett.* **2019**, *515*, 135–144. [CrossRef]

157. Liu, J.; Lin, J.F.; Prakapenka, V.B.; Prescher, C.; Yoshino, T. Phase relations of Fe_3C and Fe_7C_3 up to 185 GPa and 5200 K: Implication for the stability of iron carbide in the Earth's core. *Geophys. Res. Lett.* **2016**, *43*, 12415–12422.

158. Walker, D.; Dasgupta, R.; Li, J.; Buono, A. Nonstoichiometry and growth of some Fe carbides. *Contrib. Mineral. Petrol.* **2013**, *166*, 935–957. [CrossRef]

159. Caracas, R. The influence of carbon on the seismic properties of solid iron. *Geophys. Res. Lett.* **2017**, *44*, 128–134. [CrossRef]

160. Yang, J.; Fei, Y.W.; Hu, X.J.; Greenberg, E.; Prakapenka, V.B. Effect of carbon on the volume of solid iron at high pressure: Implications for carbon substitution in iron structures and carbon content in the Earth's inner core. *Minerals* **2019**, *9*, 720. [CrossRef]

161. Gao, L.L.; Chen, B.; Wang, J.Y.; Alp, E.E.; Zhao, J.Y.; Lerche, M.; Sturhahn, W.; Scott, H.P.; Huang, F.; Ding, Y.; et al. Pressure-induced magnetic transition and sound velocities of Fe_3C: Implications for carbon in the Earth's inner core. *Geophys. Res. Lett.* **2008**, *35*. [CrossRef]

162. Gao, L.L.; Chen, B.; Zhao, J.Y.; Alp, E.E.; Sturhahn, W.; Li, J. Effect of temperature on sound velocities of compressed Fe_3C, a candidate component of the Earth's inner core. *Earth Planet. Sci. Lett.* **2011**, *309*, 213–220. [CrossRef]

163. Chen, B.; Lai, X.J.; Li, J.; Liu, J.C.; Zhao, J.Y.; Bi, W.L.; Alp, E.E.; Hu, M.Y.; Xiao, Y.M. Experimental constraints on the sound velocities of cementite Fe_3C to core pressures. *Earth Planet. Sci. Lett.* **2018**, *494*, 164–171. [CrossRef]

164. Prescher, C.; Dubrovinsky, L.; Bykova, E.; Kupenko, I.; Glazyrin, K.; Kantor, A.; McCammon, C.; Mookherjee, M.; Nakajima, Y.; Miyajima, N.; et al. High Poisson's ratio of Earth's inner core explained by carbon alloying. *Nat. Geosci.* **2015**, *8*, 220–223. [CrossRef]

165. Chen, B.; Li, Z.Y.; Zhang, D.Z.; Liu, J.C.; Hu, M.Y.; Zhao, J.Y.; Bi, W.L.; Alp, E.E.; Xiao, Y.M.; Chow, P.; et al. Hidden carbon in Earth's inner core revealed by shear softening in dense Fe_7C_3. *Proc. Natl. Acad. Sci. USA* **2014**, *111*, 17755–17758. [CrossRef]

166. Lord, O.T.; Walter, M.J.; Dasgupta, R.; Walker, D.; Clark, S.M. Melting in the Fe-C system to 70 GPa. *Earth Planet. Sci. Lett.* **2009**, *284*, 157–167. [CrossRef]

167. Takahashi, S.; Ohtani, E.; Sakai, T.; Kamada, S.; Ozawa, S.; Sakamaki, T.; Miyahara, M.; Ito, Y.; Hirao, N.; Ohishi, Y. Phase and melting relations of Fe_3C to 300 GPa and carbon in the core. In *Carbon in Earth's Interior*; Manning, C.E., Lin, J.-F., Mao, W.L., Eds.; AGU: Washington, DC, USA, 2020.

168. Mookherjee, M.; Nakajima, Y.; Steinle-Neumann, G.; Glazyrin, K.; Wu, X.A.; Dubrovinsky, L.; McCammon, C.; Chumakov, A. High-pressure behavior of iron carbide (Fe_7C_3) at inner core conditions. *J. Geophys. Res. Solid Earth* **2011**, *116*. [CrossRef]

169. McGuire, C.; Komabayashi, T.; Thompson, S.; Bromiley, G.; Greenberg, E.; Prakapenka, V.B. P-V-T measurements of Fe_3C to 117 GPa and 2100 K: Implications for stability of Fe_3C phase at core conditions. *Am. Mineral.* **2021**. [CrossRef]

170. Chipman, J. Thermodynamics and phase diagram of Fe-C System. *Metall. Trans.* **1972**, *3*, 55–64. [CrossRef]

171. Shubhank, K.; Kang, Y.B. Critical evaluation and thermodynamic optimization of Fe-Cu, Cu-C, Fe-C binary systems and Fe-Cu-C ternary system. *Calphad* **2014**, *45*, 127–137. [CrossRef]

172. Nakajima, Y.; Takahashi, E.; Sata, N.; Nishihara, Y.; Hirose, K.; Funakoshi, K.; Ohishi, Y. Thermoelastic property and high-pressure stability of Fe_7C_3: Implication for iron-carbide in the Earth's core. *Am. Mineral.* **2011**, *96*, 1158–1165. [CrossRef]

173. Nakajima, Y.; Imada, S.; Hirose, K.; Komabayashi, T.; Ozawa, H.; Tateno, S.; Tsutsui, S.; Kuwayama, Y.; Baron, A.Q.R. Carbon-depleted outer core revealed by sound velocity measurements of liquid iron-carbon alloy. *Nat. Commun.* **2015**, *6*, 9942. [CrossRef]

174. Fukai, Y.; Mori, K.; Shinomiya, H. The phase diagram and superabundant vacancy formation in Fe-H alloys under high hydrogen pressures. *J. Alloys Compd.* **2003**, *348*, 105–109. [CrossRef]

175. Okuchi, T. Hydrogen partitioning into molten iron at high pressure: Implications for Earth's core. *Science* **1997**, *278*, 1781–1784. [CrossRef]

176. Zinkevich, M.; Mattern, N.; Handstein, A.; Gutfleisch, O. Thermodynamics of Fe-Sm, Fe-H, and H-Sm systems and its application to the hydrogen-disproportionation-desorption-recombination (HDDR) process for the system $Fe_{17}Sm_2$-H_2. *J. Alloys Compd.* **2002**, *339*, 118–139. [CrossRef]
177. Fukai, Y.; Suzuki, T. Iron-water reaction under high-pressure and its implication in the evolution of the Earth. *J. Geophys. Res. Solid Earth Planets* **1986**, *91*, 9222–9230. [CrossRef]
178. Fukai, Y. Some properties of the Fe-H system at high pressures and temperatures, and their implications for the Earth's core. In *High-Pressure Research: Application to Earth and Planetary Sciences*; Syono, Y., Manghnani, M.H., Eds.; AGU: Washington, DC, USA, 1992; pp. 373–385.
179. Sakamaki, K.; Takahashi, E.; Nakajima, Y.; Nishihara, Y.; Funakoshi, K.; Suzuki, T.; Fukai, Y. Melting phase relation of FeHx up to 20 GPa: Implication for the temperature of the Earth's core. *Phys. Earth Planet. Inter.* **2009**, *174*, 192–201. [CrossRef]
180. Shibazaki, Y.; Ohtani, E.; Fukui, H.; Sakai, T.; Kamada, S.; Ishikawa, D.; Tsutsui, S.; Baron, A.Q.R.; Nishitani, N.; Hirao, N.; et al. Sound velocity measurements in dhcp-FeH up to 70 GPa with inelastic X-ray scattering: Implications for the composition of the Earth's core. *Earth Planet. Sci. Lett.* **2012**, *313*, 79–85. [CrossRef]
181. Pepin, C.M.; Dewaele, A.; Geneste, G.; Loubeyre, P.; Mezouar, M. New iron hydrides under high pressure. *Phys. Rev. Lett.* **2014**, *113*, 265504. [CrossRef] [PubMed]
182. Hirose, K.; Tagawa, S.; Kuwayama, Y.; Sinmyo, R.; Morard, G.; Ohishi, Y.; Genda, H. Hydrogen limits carbon in liquid iron. *Geophys. Res. Lett.* **2019**, *46*, 5190–5197. [CrossRef]
183. Narygina, O.; Dubrovinsky, L.S.; McCammon, C.A.; Kurnosov, A.; Kantor, I.Y.; Prakapenka, V.B.; Dubrovinskaia, N.A. X-ray diffraction and Mossbauer spectroscopy study of fcc iron hydride FeH at high pressures and implications for the composition of the Earth's core. *Earth Planet. Sci. Lett.* **2011**, *307*, 409–414. [CrossRef]
184. Thompson, E.C.; Davis, A.H.; Bi, W.; Zhao, J.; Alp, E.E.; Zhang, D.; Greenberg, E.; Prakapenka, V.B.; Campbell, A.J. High-pressure geophysical properties of fcc phase FeHx. *Geochem. Geophys. Geosyst.* **2018**, *19*, 305–314. [CrossRef]
185. Kato, C.; Umemoto, K.; Ohta, K.; Tagawa, S.; Hirose, K.; Ohishi, Y. Stability of fcc phase FeH to 137 GPa. *Am. Mineral.* **2020**, *105*, 917–921. [CrossRef]
186. Suzuki, T.; Akimoto, S.; Fukai, Y. The system iron enstatite water at high-pressures and Temperatures-formation of iron hydride and some geophysical implications. *Phys. Earth Planet. Inter.* **1984**, *36*, 135–144. [CrossRef]
187. Caracas, R. The influence of hydrogen on the seismic properties of solid iron. *Geophys. Res. Lett.* **2015**, *42*, 3780–3785. [CrossRef]
188. Fiquet, G.; Badro, J.; Guyot, F.; Requardt, H.; Krisch, M. Sound velocities in iron to 110 gigapascals. *Science* **2001**, *291*, 468–471. [CrossRef] [PubMed]
189. Antonangeli, D.; Komabayashi, T.; Occelli, F.; Borissenko, E.; Walters, A.C.; Fiquet, G.; Fei, Y.W. Simultaneous sound velocity and density measurements of hcp iron up to 93 GPa and 1100 K: An experimental test of the Birch's law at high temperature. *Earth Planet. Sci. Lett.* **2012**, *331*, 210–214. [CrossRef]
190. Murphy, C.A.; Jackson, J.M.; Sturhahn, W. Experimental constraints on the thermodynamics and sound velocities of hcp-Fe to core pressures. *J. Geophys. Res. Solid Earth* **2013**, *118*, 1999–2016. [CrossRef]
191. Umemoto, K.; Hirose, K. Liquid iron-hydrogen alloys at outer core conditions by first-principles calculations. *Geophys. Res. Lett.* **2015**, *42*, 7513–7520. [CrossRef]
192. Komabayashi, T.; Pesce, G.; Sinmyo, R.; Kawazoe, T.; Breton, H.; Shimoyama, Y.; Glazyrin, K.; Konopkova, Z.; Mezouar, M. Phase relations in the system Fe-Ni-Si to 200 GPa and 3900 K and implications for Earth's core. *Earth Planet. Sci. Lett.* **2019**, *512*, 83–88. [CrossRef]
193. Zhang, Y.J.; Sekine, T.; Lin, J.F.; He, H.L.; Liu, F.S.; Zhang, M.J.; Sato, T.; Zhu, W.J.; Yu, Y. Shock compression and melting of an Fe-Ni-Si alloy: Implications for the temperature profile of the Earth's core and the heat flux across the core-mantle boundary. *J. Geophys. Res. Solid Earth* **2018**, *123*, 1314–1327. [CrossRef]
194. Hirose, K.; Morard, G.; Sinmyo, R.; Umemoto, K.; Hernlund, J.; Helffrich, G.; Labrosse, S. Crystallization of silicon dioxide and compositional evolution of the Earth's core. *Nature* **2017**, *543*, 99–102. [CrossRef] [PubMed]
195. Arveson, S.M.; Deng, J.; Karki, B.B.; Lee, K.K.M. Evidence for Fe-Si-O liquid immiscibility at deep Earth pressures. *Proc. Natl. Acad. Sci. USA* **2019**, *116*, 10238–10243. [CrossRef] [PubMed]

Review

A Review of the Melting Curves of Transition Metals at High Pressures Using Static Compression Techniques

Paraskevas Parisiades

CNRS UMR 7590, Muséum National d'Histoire Naturelle, Institut de Minéralogie de Physique des Matériaux et de Cosmochimie, Sorbonne Université, 4 Place Jussieu, F-75005 Paris, France; paraskevas.parisiadis@upmc.fr

Abstract: The accurate determination of melting curves for transition metals is an intense topic within high pressure research, both because of the technical challenges included as well as the controversial data obtained from various experiments. This review presents the main static techniques that are used for melting studies, with a strong focus on the diamond anvil cell; it also explores the state of the art of melting detection methods and analyzes the major reasons for discrepancies in the determination of the melting curves of transition metals. The physics of the melting transition is also discussed.

Keywords: melting curves; laser-heated diamond anvil cell; extreme conditions; synchrotron radiation; transition metals; phase transitions

Citation: Parisiades, P. A Review of the Melting Curves of Transition Metals at High Pressures Using Static Compression Techniques. *Crystals* **2021**, *11*, 416. https://doi.org/10.3390/cryst11040416

Academic Editors: Daniel Errandonea and Simone Anzellini

Received: 10 March 2021
Accepted: 7 April 2021
Published: 13 April 2021

Publisher's Note: MDPI stays neutral with regard to jurisdictional claims in published maps and institutional affiliations.

1. Introduction

Transition metals are defined as those elements that have a partially filled d-electron sub-shell. The strong metallic bonding due to the delocalization of d-orbitals is responsible for a series of very interesting properties, such as high yield strength, corrosion and wear resistance, good ductility, easy alloy and metallic glass formation, paramagnetism, and high melting points and molar enthalpies of fusion, leading to a plethora of industrial applications [1–6].

In the field of high-pressure science, the melting of transition metals is of crucial importance from a geophysical point of view in order to define the structure and composition of planetary cores, with iron being by far the most abundant component of Earth's inner core, nickel being the second [7–11]. There is also a significant fundamental interest, and numerous efforts have been made to understand the phase diagrams, as well as the thermodynamic and microscopic processes of melting under pressure [12–18]. While the hcp-bcc-hcp-fcc phase sequence in transition metals can be understood by the progressive filling of the d-electron bands [19], the study of the melting behavior of transition metals has given rise to many controversies among the different experimental and theoretical studies.

The majority of experimental works on the melting of transition metals are conducted with static, mainly using a laser-heated diamond anvil cell (LH-DAC), or dynamic techniques, using shock wave (SW) compression. Despite the numerous technological advancements that have been made in both fields during the last two decades, reliable measurements remain very challenging, especially at the high end of P-T conditions. The diverse results for different experimental approaches concerning Ta [13,20–23], Fe [14,24–26], Mo [13,15,27–33], V [34,35], Ti [13,36], Zr [37,38], or Ni [39–41] clearly exhibit the need for a universally established methodology.

Static techniques are limited to a pressure of a few Mbar and temperatures up to 6000 K using infrared lasers (Section 3); however, new advancements in diamond anvil cell technology such as toroidal anvils or double-stage anvils [42–44] may offer the possibility to achieve even higher pressures in the close future. There are several methods to observe melting inside an LH-DAC, and the results tend to be very dependent on the di-agnostic. However, it is true that some convergence has been achieved between certain methodologies, especially in recent years, as it will be discussed in Section 4.

Dynamic compression offers the possibility to measure melting data at much higher pressures by generating strong shock waves with gas guns or lasers [45–47]. This range of pressures is crucial for the modeling of deep planetary interiors [48]. What is particularly challenging is the diagnostics of phase transitions in the short timescales of a dynamical process [49,50]. X-ray Free Electron Laser (XFEL) facilities such as the LCLS [51,52], or the HED instrument at European XFEL [53] are specifically designed to trigger materials at multi-Mbar pressures using ns pulses and a very high brilliance beam and are expected to greatly increase the quality of obtained data.

Theoretical data only show partial agreement with the experiment. The most widely used techniques are mainly based on molecular dynamics, mainly the superheating–supercooling (one-phase) hysteresis method [54] or the solid–liquid coexistence (two-phase) approach [55–57]. Other methods commonly used are the free energy approach [58] or the Z method [59,60]. Each method has a different way to calculate the melting temperature and has its own advantages and disadvantages.

This review aims to collect and try to unveil most of the discrepancies between different sets of experimental melting data for transition metals that occur in static experimental techniques. The behavior of melting curves, the physics of the melting phase transition, as well as the comparison between the melting curves of transition metals, alkali metals, alkaline earths, and rare earths will be discussed in Section 2. Section 3 will discuss the static experimental techniques, starting historically with large volume pressure (LVP) apparatuses (Section 3.1) and the resistively heated DAC (RH-DAC) (Section 3.2). The main focus of this review, however, is the LH-DAC, since it is the only static technique that can provide a range of pressures and temperatures wide enough to establish reliable melting curves for transition metals. The state of the art of LH-DAC technology will be briefly presented from a point of view concerning the melting curves (Section 3.3), and the reader can address a recent review [61] for more detailed information on the LH-DAC. As mentioned, the methodology, and principally the diagnostics of melting, comprises one of the main reasons for disagreement among the different studies; thus, the different experimental approaches and melting criteria will be thoroughly presented and compared (Section 4), while the various observed phenomena and the reasons for discrepancies between measurements will be discussed in Section 5. The concluding remarks of the manuscript contain a summary and some thoughts about future perspectives (Section 6).

2. Physics of the Melting Transition

2.1. Empirical Thermodynamic Models

Concerning transition metals, the metallic bonds tend to be weakest for elements that have nearly empty or nearly full valence shells and strongest for elements with half-filled valence shells. As a result, the melting point, boiling point, hardness, or enthalpy of fusion reach a maximum around group 6.

The melting line separates the solid and liquid phases in a pressure–temperature diagram, and these two variables are related by the Clausius–Clapeyron equation:

$$\frac{d\mathrm{T_m}}{d\mathrm{P}} = \frac{\mathrm{T}\Delta\mathrm{V}}{\mathrm{L}} = \frac{\Delta\mathrm{V}}{\Delta\mathrm{S}} \tag{1}$$

where $\mathrm{T_m}$ is the melting temperature, P the pressure, L the latent heat, and $\Delta\mathrm{V}$ and $\Delta\mathrm{S}$ the specific volume and entropy changes of the phase transition. This equation mainly states that the Gibbs free energies of the solid and the liquid are equal at the melting point at any pressure.

The melting curve of a metal can be expressed mathematically by the empirical Simon–Glatzel equation [62] to fit the data:

$$T_m = T_0 \left(\frac{P}{a} + 1 \right)^b \tag{2}$$

where T_0 is the melting point at ambient pressure, while a and b are fitting parameters. Sometimes a triple point is used in the place of T_0. As a monotonically increasing function, the Simon–Glatzel equation can mainly describe melting curves that rise with pressure. The tangent melting slope is defined by Equation (1) or by the derivative of Equation (2). For most metals, the density of the liquid is lower than that of the solid, while the entropy is higher, leading to mostly positive melting slopes (i.e., the melting temperature increases with pressure) [16,22,24,29,35–37,39,40,63]. The situation is much more intricate for alkali metals, however, whose phase diagrams reveal complex liquid melting-curve maxima and negative melting slopes [64–69], since these elements exhibit several density discontinuities and phase transitions already in the solid phase. In this case, the Kechin equation is more appropriate to describe the melting curve [70].

One of the most important models that has been proposed to explain melting, which can also be used to calculate the melting slope, is the Lindemann criterion [71,72]:

$$\partial \ln T_m / \partial \ln V_m = \frac{2}{3} - 2\gamma \tag{3}$$

where T_m, V_m, and γ are the melting temperature, the molar volume of the solid before melting, and the Grüneisen parameter, respectively. Equation (3) reveals that T_m increases with decreasing volume or increasing pressure. In the Lindemann melting criterion, thermal atomic vibrations increase with an increasing temperature, and melting is initiated when the vibrations become so large that the atoms invade the space of their nearest neighbors. For the sake of completion, the reader can also refer to the Born criterion [73], according to which melting occurs when the shear modulus vanishes and the crystal cannot further resist melting.

The Lindemann law is an empirical law based on investigations of simple gases at low pressures, and it is sometimes debated whether it can be used to describe the high-pressure melting curves of complex metals. In noble gases such as Ar, Kr, or Xe, the closed shell configuration (s^2p^6) is responsible for a steep melting curve, and something similar happens with noble metals such as Cu, Ag, or Au. However, in many cases the Lindemann law overestimates the melting curve of d-transition metals with partially filled cells, especially at higher pressures, no matter the experimental approach [30,37,74,75]. Moreover, the fact that the Lindemann law takes into account only the thermodynamic parameters of the solid phase, neglecting the liquid, can also lead to inaccurate predictions.

Often in calculations using Lindemann estimates, the Grüneisen parameter is considered volume independent. For better agreement with experimental data, γ needs to be expressed as an analytical expression of volume, with the formula $\gamma/\gamma_0 = (V/V_0)^q$ being often used. However, q is not constant, but it decreases with pressure, leading to discrepancies with the experimental data [76]. It has been argued that by reformulating the Lindemann law and modeling the Grüneisen parameter as a power series of the interatomic distance [30,71], the differences between theory and experiment can be reduced significantly, although there is still lack of total agreement. However, the experimental melting of Au was well reproduced by the Lindemann model using a volume-dependent Grüneisen parameter [77].

For a more in-depth analysis of empirical thermodynamic models regarding melting and taking into account also dynamic techniques, the reader can address references [78–82].

2.2. Electronic Structure and Phase Behavior of Transition Metals on Melting

Figure 1a plots together the melting lines of several transition metals, obtained in an LH-DAC [15,22,24,35–37,63,77,83,84]. The melting curves are normalized to the ambient melting temperature T_0 of each element. Since the different diagnostic methods for melting can yield different results (Section 4), the melting points in all the studies of Figure 1a have been determined using the same diagnostic (the appearance of a liquid diffuse scattering signal in synchrotron XRD, Section 4.2), except the study on Mo, which used also synchrotron XRD but defined melting by the appearance of microstructures in the quenched sample (Section 4.5). Figure 1a contains 3d, 4d, and 5d transition metals from groups 4 to 11. For the 3d (Ti, V, Fe, Ni) and 5d (Ta, Pt, Au) elements, increased electronegativity seems to lead to steeper melting lines, with Au and Pt exhibiting the steepest curves. However, this is not the case in 4d (Zr, Nb, Mo) elements, where the less electronegative Zr has the higher slope. Most melting curves are less steep at higher pressures, indicating that the volume change becomes less important with pressure.

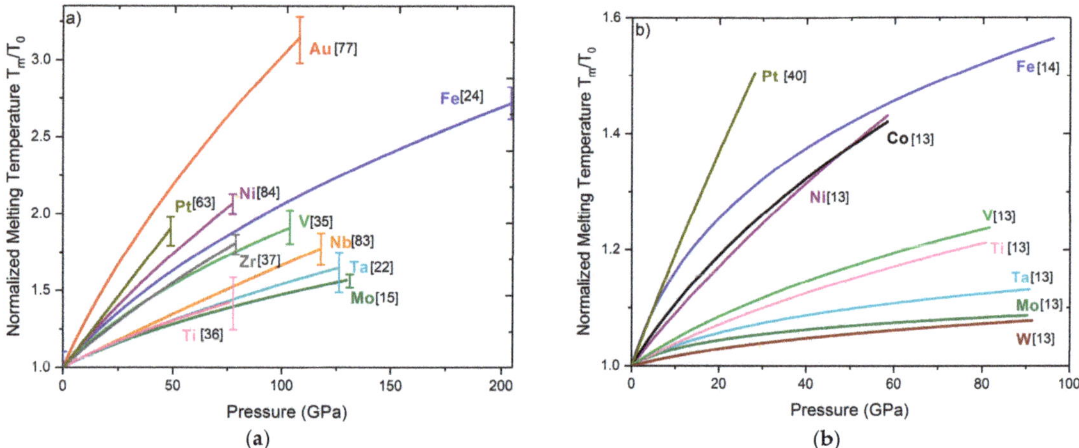

Figure 1. Normalized melting temperatures T_m/T_0 for several d-transition metals. All the melting curves in each graph were obtained with the same method: (**a**) XRD diagnostic (Section 4.2), (**b**) Speckle diagnostic (Section 4.1). The error bars in Figure 1a refer to the reported uncertainties in temperature measurement by using the XRD diagnostic. Datasheets can be found in the Supplemental Material.

The majority of transition metals have partially filled d-bands, which can be occupied by s- electrons via s->d electron scattering. The s-conduction electrons are less mobile since they have a higher effective mass, and this s-d electron transfer can be responsible for changes in the density of states (DOS) and the lowering of the melting curves in transition metals. It has been proposed that a small broadening of the liquid d-band (~1%) could lead to an increase in the stability of the liquid relative to the solid and thus suppress the melting slope significantly at higher pressures [85]. However, this was not found to be the case in Mo, where strong discrepancies between theory and experimental data still exist [15]. Another explanation for the low melting curves of metals with nearly half-filled d-bands such as Mo is the existence of Jann–Teller distortions which could create local structures in the liquid [86]. Elements with filled bands, and therefore a DOS that is less subject to change with melting, such as Cu or Al, have much steeper melting curves [40,85,87,88]. However, Al is characterized by sp^3 bonding electrons and can be drastically different from transition metals, where d-electron physics plays a dominant role.

A factor that could also affect the behavior of the melting curve of a material is the phase from which melting occurs. In fact, it has been proposed that bcc metals (a packing ratio of ~0.68) should have lower melting slopes than fcc or hcp metals (a packing ratio of ~0.74) [13]. It has been shown that the stability of a bcc structure is favored by a Fermi energy which falls in the minimum between two peaks of the DOS. According to Figure 1a, this seems to be a general rule, with bcc Mo exhibiting the lower slope, and fcc Au, Pt, or Ni the highest. For example, eightfold coordinated bcc Mo should undergo a smaller volume change than 12-fold coordinated Ni. Electron-band calculations have shown that hcp–fcc total energy differences are smaller than the corresponding bcc–fcc differences [89]; thus, the entropy change for Mo should be greater than for Ni or Fe. By taking all these into account together with the Clayperon criterion (Equation (1)), a smaller overall melting temperature can be expected.

Transitions within a metal phase diagram also affect the melting slope. For example, a flattening of a melting curve around the triple point fcc–hcp–liquid can be expected for Fe, because of the melting entropy differences between the two solid polymorphs [90].

Moreover, d-electron bonding can be responsible for the appearance of an icosahedral short-range order, which can be energetically favored in supercooled liquids and melts [91,92]. These short-range structures depend on the number of d-electrons and are more distorted in the liquids of early transition metals than those of late transition metals [17]. Icosahedral structures in liquids have been already observed in Ta [93], Zr [94], Ni [91,92], Ti [92], and Fe [91] and can act as impurities that lower the free energy and thus the melting slope [17].

Figure 1b shows the normalized melting curves obtained in an LH-DAC with the Speckle technique [13,14,40]. All melting curves appear a lot shallower compared to the data obtained with the XRD diagnostic (Figure 1a), and in all cases the normalized T_m has a lower value at the same pressure. With this method, the melting lines of Mo and W appear almost constant after ~50 GPa. It has been shown in many works [15,24,37,84] that the Speckle method often triggers the recrystallization of the sample instead of melting, therefore underestimating the melting temperature, as will be discussed in more detail in Section 5.6. However, it is interesting to see that melting from bcc, as in the case of W, Mo or Ta, yields the smoothest melting curves, as in the case of the XRD diagnostic. The datasheets for the plots of Figure 1a,b can be found in the Supplemental Material.

2.3. Comparison with Alkali Metals, Alkaline Earths, and Rare Earths

The melting curves of alkali metals behave rather differently than those of transition metals. These elements tend to have complex phase diagrams, with successive phase transitions in the solid phase and the presence of complex liquids in the liquid phase [64–69,95]. In many cases, alkali metals exhibit melting curve maxima, followed by negative slopes, where the molar volume of the liquid is expected to be less than that of the solid. Previous studies have shown abrupt changes in the coordination numbers and density, for example in liquid Cs [96]. The complexities in both crystal structure and melting have been attributed to the effects of s->d electronic transitions for the heavier alkali metals [97], but lighter elements such as Li and Na exhibit s-p orbital mixing [98]. It has been proposed that a possible explanation for the complex structures observed in alkali metals is the formation of an energy gap at the Brillouin zone boundary that can lower the kinetic energy of free electrons and thus stabilize such structures [99].

Alkaline earths also present several unusual structures and complex phase diagrams [100–102]. These phenomena could be explained by the sp->d electron transfer under compression [103]. It has been found that the changes in the melting slopes of Mg, Ca, and Sr are associated with the phase transitions observed at room temperature and with the increasing d-electron character of these elements [95]. Similar to transition metals, melting from a bcc phase leads to a less steep melting curve, as in the cases of Ca or Mg [95].

In rare earths, the application of pressure generally induces an s->d transition that increases the d-electron character of the conduction band. This electronic transition is responsible for the hcp->Sm-type->distorted hcp->fcc->distorted fcc series of structural transitions in the solid phase [104]. The highly localized f-electrons do not participate in bonding, since Y exhibits the same behavior even though its f-states are empty [105]. However, low symmetry structures have been observed under pressure for Pr, Nd, Sm, and Gd, which have been attributed to the delocalization of f-electrons [106–109]. The melting curves of rare earths seem to be affected by this delocalization, showing kinks or minima in the pressure values where it occurs [104].

3. Static Experimental Techniques

3.1. Large Volume Presses (LVP)

Large volume presses (LVP) such as the piston-cylinder or the multi-anvil press [110,111] are widely used for the synthesis of materials, especially when the production of large single crystals is essential [112,113]. Record temperatures of 4050 K [114] and pressures of 90 GPa [115] have been reported for multi-anvil presses, although for most conventional devices in laboratories and synchrotrons the maximum pressure is limited to 25 GPa and the maximum temperature to 2500 K. Therefore, the routinely obtainable P-T range in the LH-DAC (or RH-DAC) is much greater than that of a multi-anvil press, but the precision to which the melting temperature can be determined in a DAC is, in general, much lower.

Inside a multi-anvil press, the heating of the sample is carried out by an internal heating method, where a small heater and electrodes are placed inside a pressure transmitting medium (PTM). Electric power is supplied to the heater from an external power supply and through the conductive anvils. Several heaters can be used, the most prominent ones being Pt, Ta, Re, $LaCrO_3$, or graphite [116]. The sample temperature is measured by a thermocouple which is located close to the sample, and the pressure effect on the thermocouple's electromotive force has to be estimated [117–119]. For the most widely used W/WRe thermocouples, a variation of 35 °C has been reported at 15 GPa and 1800 °C [119]. Thus, the temperature determination with a thermocouple is very accurate and much superior to that obtained by spectral radiometry in an LH-DAC, which usually extends to hundreds of K (Section 3.3.3). The highest temperature that can be measured with a W/WRe thermocouple is 2300 °C. Higher temperatures are measured by extrapolating the relationship between temperature and applied power.

Pressure in a multi-anvil press is measured by the P-V-T equation of state (EoS) of a pressure standard which is placed inside the assembly. This material should ideally exhibit no phase transitions, have a low relatively bulk modulus, low yielding strength (so that the deviatoric stresses are easily released upon heating), chemical inertness, high melting temperature, low grain growth rate, and low X-ray cross-section (in the case of in situ XRD measurements). The most widely used pressure standards in a multi-anvil press are NaCl, Au, Pt, and MgO. From the known pressure scales, that of MgO seems to be the least controversial since it seems to be free from the free electron contribution to the thermal pressure [120–122]. In general, underestimations due to thermal pressure up to 3 GPa have been calculated, which is not dramatic compared to the LH-DAC (Section 3.3.3). The reader can refer to two very detailed reviews about the technical developments in the multi-anvil press [116,123].

The signature of melting in a large volume press can be evidenced either by differential thermal measurements [124] or electrical measurements [125–127] (Section 4.7). In the case of an LVP coupled with synchrotron radiation, in situ XRD [128,129] can be used as a reliable melting diagnostic (Section 4.2). Melting curves at low pressures determined in an LVP for gold [127], copper [130], and nickel [125] have been reported in the literature. Multi-anvil experiments provide an adequate method for investigating melting curves for pressures up to 25 GPa.

3.2. The Resistive Heating Diamond Anvil Cell (RH-DAC)

The resistively heated diamond anvil cell (RH-DAC) [131–138] is a complementary technique to laser heating, albeit less widely used because of the generally lower temperatures obtained and the complexity that lies in preparing the DAC. In resistive heating, samples are heated by conduction with the heat source outside the sample chamber, either by an external furnace (maximum temperature around 700 K) or by a small heater close to the diamond anvils (where much higher temperatures can be achieved). Various RH-DAC techniques have been proposed that can provide temperatures above 1200 K and can be used for the melting of metals, for example gold [139]. These techniques are based on (a) graphite heaters [135,136,138], (b) Mo wires [140], (c) W filaments [134], (d) Re gasket heating [137], or (e) metal strips placed directly in the sample chamber (internal resistive heating, [139,141–143]). The latter technique has been proven very effective, especially when the metal strip is the sample itself, and temperatures of almost 4000 K have been reached in a study of the Fe–Ni–Si system [143]. However, RH-DAC preparation, and especially the placing of metallic contacts on a micrometer sized sample, can be very challenging. At such high extreme conditions, spectral radiometry is needed for the temperature measurement.

Resistive heating methods have been extensively used to study the melting behavior and phase diagrams of alkali metals [64–66,68,69] and molecular systems [144–147], which generally have much lower melting temperatures than transition metals. The pressure range in an RH-DAC is limited compared to the LH-DAC, often because of the softening of the stress-bearing components, such as the gaskets and the diamond seats. The advantage of resistive heating is the significantly improved homogeneity of the temperature with respect to laser heating. Moreover, temperature control is independent and is not affected by any changes in the physical properties of the sample or the presence of phase transitions. Another advantage is that the temperature is measured with a thermocouple (up to ~2600 K), providing much smaller error bars with respect to spectral radiometry.

3.3. The Laser Heated Diamond Anvil Cell (LH-DAC)

The laser-heated diamond anvil cell was first presented in the pioneering work of Ming and Basset [148]. This technique takes advantage of the extreme hardness and the optical properties of diamond, which is transparent in a very wide wavelength range, from gamma and X-rays to mid-infrared, allowing not only laser irradiation but also coupling with various experimental techniques, both in synchrotron facilities and in laboratories. To demonstrate the versatility of the LH-DAC, one can refer to experimental works on X-ray diffraction (XRD) [149–157], X-ray absorption (XAS) [158–161], X-ray fluorescence (XRF) [162], Mössbauer spectroscopy (SMS) [163,164], inelastic X-ray scattering (IXS) [144–146], nuclear inelastic scattering (NIS) [165], nuclear magnetic resonance (NMR) [166], and Raman [167–171] and Brillouin [172–175] spectroscopies or the synthesis of novel materials [176–183]. Different types of diamond anvil cells exist for different applications, but especially for the LH-DAC, the angular opening of the cell is of crucial importance, especially when coupled with experimental techniques such as XRD or Raman/Brillouin spectroscopies, but also when an off-axis laser heating geometry is in place [61].

The principle of the LH-DAC has been thoroughly described in [61] and can be shown in Figure 2. The basic concept is based on a piston-cylinder mechanism, as the two opposing diamonds are approaching each other by the application of an external force which can be generated by screws or an external membrane. The sample is confined within a metallic gasket [184], usually made by stainless steel, rhenium, or tungsten, although composite gaskets such as amorphous boron-epoxy [185] or c-BN [186] also exist and can maximize the thickness of the sample chamber during the experiment. Alternatively, beryllium gaskets are used in the case both axial and radial access in the DAC is needed [187,188]. A pressure transmitting medium (PTM) is used to fill the sample chamber and offer the best hydrostatic conditions possible. The sample irradiation is performed with an IR laser source that accesses the sample by taking advantage of the transparency of the diamonds. The emitted thermal radiation passes through the diamonds and is guided to the entrance of a spectrometer, where it is analyzed in order to provide the temperature measurement (Section 3.3.3).

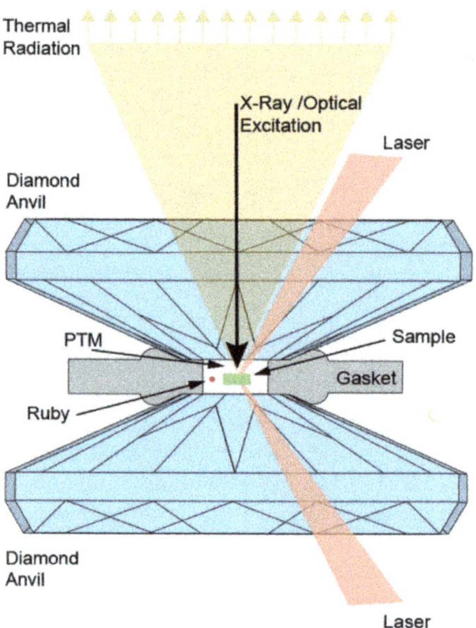

Figure 2. The principle of the laser-heated diamond anvil cell (LH-DAC).

3.3.1. Pressure Transmitting Medium (PTM)

Using a pressure transmitting medium in a melting experiment is of crucial importance for many reasons:

i. It introduces hydrostatic or quasi-hydrostatic pressure conditions, thus greatly reducing the deviatoric stresses and giving a better estimation of pressure [189], although shear stresses will always appear when the (pressure transmitting medium) PTM solidifies.

ii. It provides thermal insulation to the sample, since the diamond anvils are very good thermal conductors and can be a significant source of heat sink. It is practically impossible to laser-heat a sample without thermal insulation.

iii. It confines the liquid sample inside the sample chamber.

iv. It often prevents chemical reactions such as carbide formation, as it will be discussed further (Section 5.7) in more detail.

Several PTM have been used in diamond anvil cells, including liquids (methanol–ethanol mixture, methanol–ethanol–water mixture, silicon oil, Daphne 7373 or 7474, fluorinert), soft solids (CsCl, NaCl, KCl, KBr, LiF), hard solids (Al_2O_3, MgO, SiO_2), or condensed gases (He, Ne, Ar, N_2), He being by far the most hydrostatic of them all [190,191]. However, He is not used in high temperature studies, since it can escape the DAC in its gaseous form at high temperatures. Moreover, the choice of a PTM in melting curve ex-periments is mainly based on three factors: (a) the PTM has to stay insulating at high pressures and temperatures; (b) it has to be inert with the sample or the diamond anvils; and (c) it has a melting curve that is steeper than the material that is being studied (Section 5.3). By taking all this information into account, solid media such as NaCl, KCl, Al_2O_3, SiO_2, or MgO and noble gases such as Ar or Ne are the PTM that are mostly used in a laser heating melting experiment. Figure 3 gathers together all the experimentally established melting curves of the most commonly used PTM inside an LH-DAC [192–198].

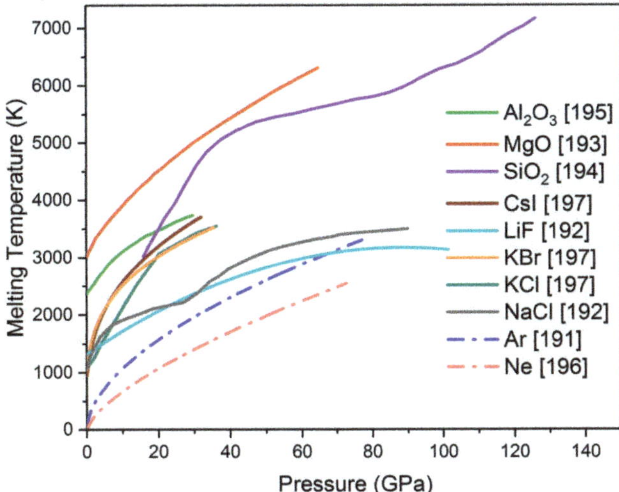

Figure 3. Experimentally defined melting curves of the most commonly used pressure transmitting media (PTMs) inside a laser-heated diamond anvil cell [192–198]. Solid lines: solid PTMs, dash-dotted lines: gas PTMs.

3.3.2. Lasers

All transition metals melt above 1200 K at ambient pressure, and this range of temperatures is easily accessible with laser heating. For the studies of metals (or semiconductors), solid state lasers of near-IR wavelengths such as Nd:YAG or Nd:YLF (λ = 1.053–1.070 µm) are preferred, since they are greatly absorbed by these materials [13,22,24,25,29,34,35,37,39,40,63,77,83,84,199–201]. The main mechanism lies in the photon-electron interaction between the electromagnetic field of the laser and the free electrons of the metal, inducing higher energy states in the conduction bands. The thermal energy transfer process takes place during the electron–phonon collisions; immediately after the collision, the electrons may change their directions, but the electron flux remains constant in any direction. However, some fraction of the excess electron energy is transferred to the phonons during the collision process, therefore increasing the temperature.

The small penetration depth of near-IR lasers makes it essential to use double-sided laser heating in order to reduce the thermal gradients from one side of the sample to the other, which can extend to many hundreds of K [95,202,203]. In general, defocusing the beam helps in obtaining a more homogenous laser spot, thus reducing the thermal gradients. In order to reduce the thermal gradients even further, a recent method [77,204] proposed encapsulating the sample inside a boron-doped diamond micro-oven.

CO_2 lasers have a much larger wavelength (λ = 10.6 µm) and thus a larger penetration depth, alleviating the need for two-sided laser heating in an LH-DAC; however, they are not well absorbed by metallic elements and are mainly used for the study of glasses, minerals, oxides, or optically transparent organic matter [205–207].

Most experimental works on the melting curve of transition metals have been carried out using continuous wave (CW) lasers; however, the use of pulsed lasers ("flash heating") [15,153,199] can potentially reduce the chemical reactions (oxidation, carbide formation, reactions between PTM and sample, Section 5.7), as well as any sample instabilities inside the DAC. This technique can become more powerful if the laser pulse is synchronized with a synchrotron radiation pulse and a fast detector.

3.3.3. P-T Metrology

The pressure in an LH-DAC can be measured by using the ruby or Sm^{2+}:SrB_4O_7 fluorescence methods [208–211], by the first-order Raman signal from the center of the diamond culet [212], or be derived from the thermal equation of state (EoS) of the PTM or the sample in an XRD experiment [37,84,90]. However, special care should be taken to take into account the thermal pressure P_{th}. It has been argued that the thermal pressure is strongly affected by the PTM used during the experiment [84]. The pressure is calculated empirically from XRD data using the following formula [84,90]:

$$P = P_{before} + \frac{P_{after} + \Delta P - P_{before}}{T_{max} - 300} \times (T - 300) \tag{4}$$

where P_{before} and P_{after} are the pressure before and after laser heating and T_{max} the maximum temperature reached during the experiment. ΔP is the pressure difference between the pressure at high temperature and the pressure measured after the heating cycle [90].

The temperature measurement in an LH-DAC is performed using spectral radiometry, where the raw intensity of the collected light is given by Planck's law, corrected for including the emissivity $\varepsilon(\lambda)$ in the grey body approximation:

$$Planck = I(\lambda, T, \varepsilon(\lambda)) = \varepsilon \frac{2\pi hc^2}{\lambda^5} \frac{1}{\exp(hc/\lambda kT) - 1} \tag{5}$$

In this equation, λ is the wavelength of the measured signal, $\varepsilon(\lambda)$ the emissivity, h the Planck constant, k the Boltzmann constant, and T the grey body temperature. In order to be able to fit the temperature, $\varepsilon(\lambda)$ is considered wavelength independent. Many complementary approaches exist for the temperature estimation using pyrometry, the Wien function and the two-color pyrometry being the most widely used [213,214].

The collected signal is analyzed by a spectrometer and sent to a CCD camera. Band pass filters are used to prevent reflections of the laser entering the spectrometer and thus saturating the CCD and perturbing the signal. The collection of the Planck radiation of the heated sample requires some sophisticated imagery and temperature measurement op-tics. Ideally, the temperature measurement and laser focusing optics should be independent. There are two main categories of optics to collect the Planck radiation: reflective (mirrors) or refractive (lenses). Reflective optics have the advantage or being almost free of chromatic aberrations, while refractive optics need the application of a numerical aperture to approach achromatic behavior, reducing the spatial resolution. However, the image quality is superior for refractive optics. Recent studies comparing the different optics have showed that the differences in the estimation of temperature are rather small [215,216].

Finally, special caution should be taken in the calibration of the optical system. By dividing the collected radiation with the system response, one can fit the spectra to the Planck radiation function. It has been argued that the absorbance of the diamonds should also be taken into account for the calibration in order to improve the accuracy of temperature measurements [217].

4. Melting Detection and Criteria

4.1. The Laser Speckle Method

One of the first diagnostics proposed to detect melting in an LH-DAC was the pioneering method of the laser speckle technique [13,14]. In this method, a visible laser (usually the 514.5 nm green line of an Argon laser) is applied to create interference patterns on the sample surface while the IR laser simultaneously heats the sample. Sometimes, visual observation of the sample surface has also been used [26,218]. Melting can be determined as the onset of convective motion with increasing temperature. Although it has been used extensively in numerous studies [13,14,16,21,40,41,201], the optical detection of melting based on speckle patterns has been questioned, especially after the development of the more recent techniques such as the in situ observation of a liquid signal in the synchrotron

XRD images (Section 4.2). It has been argued that the changes in the optical properties of the sample surface that are detected with this method could also be caused by structural changes in the PTM or the sample (such as a phase transition), or chemical reactions between the sample, the PTM, and/or the diamond anvils [22,219,220]. In many cases the speckle method was found to coincide with the onset of dynamic recrystallization of the sample rather than melting [22,24,36,37] since both the high temperature recrystallization and melting processes are endothermic and thus not easily distinguishable optically (Section 5.6). However, reliable results have been obtained for low reactivity transition metals such as Cu [16].

4.2. Appearance of the Liquid Diffuse Scattering Signal in the XRD Patterns

The technological advances in synchrotron facilities together with the technical developments in LH-DAC during the last 15 years have allowed the determination of the melting curves using both in situ XRD and XAS (Section 4.3) techniques. The focused X-ray beam in a high brilliance synchrotron radiation source can be reduced to a size much smaller than the size of the sample that is heated by the lasers, facilitating both the signal acquisition and the temperature determination. Concerning XRD, the melting criterion is based on the gradual appearance of a diffuse liquid X-ray scattering signal as the temperature is progressively increased [22,24,34–37,63,77,83,84,200,221], which can occur together with the disappearance of the Bragg peaks that dominate the signal of the solid phase. However, more often than not, the liquid diffuse signal appears before the complete disappearance of the Bragg peaks, indicating a partial melting of the sample. The Bragg reflections disappear progressively upon further increasing the temperature, until the XRD signal becomes fully liquid. The melting temperature is defined as the first temperature where the diffuse scattering is observed. The diffuse signal disappears on quenched diffraction pattern of the sample, i.e., when shutting off the heating (IR) lasers.

The diffuse scattering method has some remarkable advantages with respect to other methods, notably, the possibility to simultaneously detect chemical reactions and phase transitions in the sample since its structure is constantly triggered with in situ diffraction. However, it requires short exposure times and continuous monitoring of the temperature to be effective, as well as a very careful alignment of the X-ray beam, in order to ensure that the diffraction pattern is taken from the portion of the sample where the IR lasers are focused. The technique of using short exposure times (in the order of 1–2 s) has been first established for Pb and can potentially provide a more accurate determination of the various crystallographic changes in the sample [200], but the temperature fluctuations near the melting point happen in a much faster timescale (a few ms) (Section 5.1). This fact, combined with the small quantity of liquid signal that is usually observed with this method, requires very precise measurements and meticulous data analysis in order to properly detect the onset of melting. The large temperature error bars in many of the XRD melting studies [22,35,63], often calculated as the sum of the maximum error from spectral radiometry and the uncertainty on the detection of melting, clearly show that there are still several technical challenges to overcome.

Figure 4 presents an example of the solid–liquid transition upon laser heating in Zr [37] using XRD. The diffuse liquid scattering appears at a given temperature and progressively dominates the diffraction signal upon further heating, while at the same time, the solid contribution decreases, indicating that the quantity of liquid in the sample increases with temperature. In this experiment, the XRD patterns did not exhibit any parasitic phases, such as ZrC or ZrO_2.

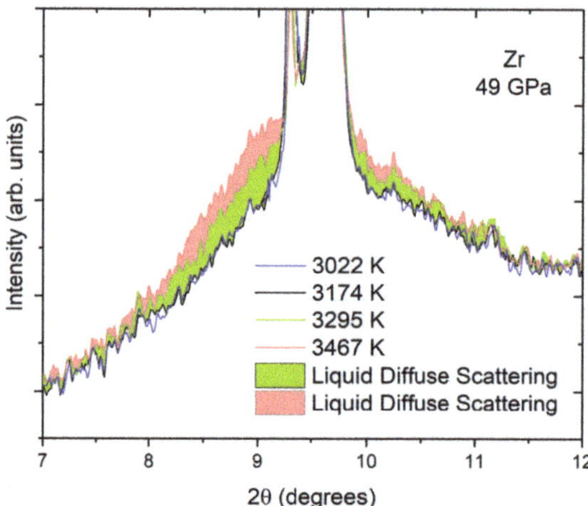

Figure 4. XRD patterns of a Zr sample in an LH-DAC [37]. The liquid diffuse scattering appears at melting as a background signal that increases in intensity with temperature as the quantity of the liquid increases. The light green liquid diffuse scattering background corresponds to 3295 K, the pink one to 3467 K. The patterns at 3022 K and 3174 K correspond to solid Zr.

Concerning the determination of the melting curves of transition metals, this technique has shown significant discrepancies with the speckle method, providing melting temperatures that can be 1000 K higher or even more at high pressures, Ta [21,22] and Fe [14,24] being characteristic examples. As mentioned, it has been argued that the speckle method often coincides with the onset of sub-solidus recrystallization, which can be observed in the diffraction patterns as rapidly moving single crystal spots that appear at high temperatures but before the onset of melting (Section 5.6).

4.3. Evolution of the XAS Spectrum

While XRD can probe the long-range structural changes in a solid–liquid transition, XAS is more sensitive to the local atomic environment since it is an element-selective technique. In particular, in an XAS experiment, the EXAFS (extended X-ray absorption fine structure) part of the signal provides information about the local atomic structure, while the XANES (X-ray absorption near edge structure) part of the signal provides information about the electronic structure.

In a melting experiment, the solid–liquid transition is defined by the disappearance of the shoulder on the XANES region of the XAS signal, as well as the flattening of the first few oscillations. Such changes have been observed in the 3d transition metals Fe [25] and Ni [39], as well as in Fe binary alloys [174,175]. Melting can be also identified by the "T-Scan method" [222], where the temperature dependence of the absorption coefficient at the shoulder region of the XANES signal follows a discontinuity attributed to the loss of long-range order.

XAS is therefore a multifaceted technique that can provide interesting information about melting, but also about the local, electronic, and magnetic configuration of a transition metal at extreme conditions. As in the melting studies with XRD, fast acquisition times are also essential in XAS, and therefore a polychromatic pink beam is often used to provide the maximum flux possible [223].

Figure 5 shows the X-ray absorption spectroscopy melting criterion in a recent work on Ni [39]. The melting criterion consists of the disappearance of the shoulder feature in point A and the flattening of the two oscillations in points B and C. For Ni, the XRD and XAS criteria have been found to give similar results [39,84], proposing that melting can be detected by probing either the long-range structural changes (XRD) or the local environment (XAS) of the sample. However, the studies on Fe comparing the two methods still remain controversial [24,25]. An argument was made that these discrepancies could rely on the fact that in some of the XAS studies, carbon-contaminated Fe samples were actually measured [90], leading to lower melting temperatures. Carbon contamination is discussed in Section 5.7.

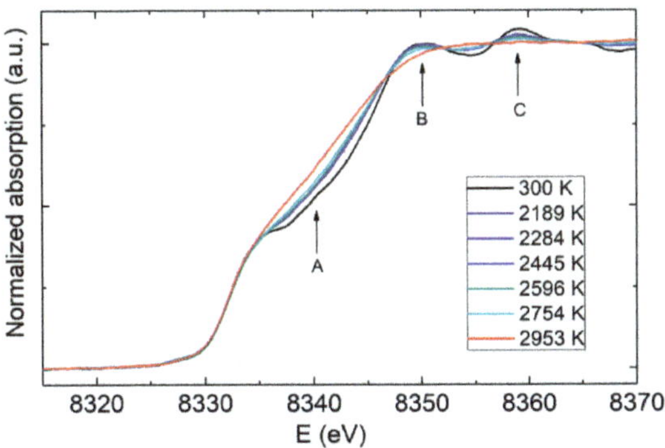

Figure 5. X-ray absorption near edge structure (XANES) spectra for Ni at different temperatures (reproduced from [39]). The melting can be determined by the disappearance of the shoulder in point A and the flattening of the first few oscillations (points B and C).

4.4. Temperature Plateaus

One of the most widely used methods for determining the melting temperature in an LH-DAC is the appearance of temperature plateaus, i.e., the temperature stays relatively stable near the melting point as the laser power continues to increase (in most cases in linear increments). This technique requires regular monitoring of the temperature (every few seconds) and is used often in laboratories when there is no possibility to obtain information about melting from an in situ technique, as is the case in a synchrotron experiment.

It has been proposed that after a certain point, the increasing laser power should only increase the volume of the melt, than raise the temperature of the molten material, given that the laser provides the latent heat of melting so that plateaus should normally be expected at any invariant melting point [128]. It is true that this method has provided relatively accurate results that have been verified in many cases with other melting diagnostics such as the detection of liquid diffuse scattering in an XRD or the disappearance of some XANES characteristic features in an XAS experiment [39,83,84,224]. An agreement of less than 100 K in the melting temperature was found between this method and the XRD method for Ni [84] and Nb [83], and a similar order of magnitude has been found for Ni measured with XANES [39]. However, in several works the temperature plateau was not observed at all or corresponded to a different temperature than the melting temperature T_m [22,36,37]. Figure 6 compares the Tm vs. laser power functions for different transition metals in order to clarify these differences.

It is clear from Figure 6 that this melting criterion is not applicable to all studies. Temperature plateaus can appear at $T=T_m$ (Figure 6a,c) $T<T_m$ (Figure 6b), or $T>T_m$ (Figure 6d). The trend can be different even for the same material, depending on the pressure or the heating run (Figure 6a,b). The range of temperature differences may also vary; for example, in Zr (Figure 6b, [37]), a melting using the temperature plateau criterion seems to appear at about 300 K before the XRD criterion, while for Ti (Figure 5d, [36]), melting was detected at a temperature almost 900 K higher.

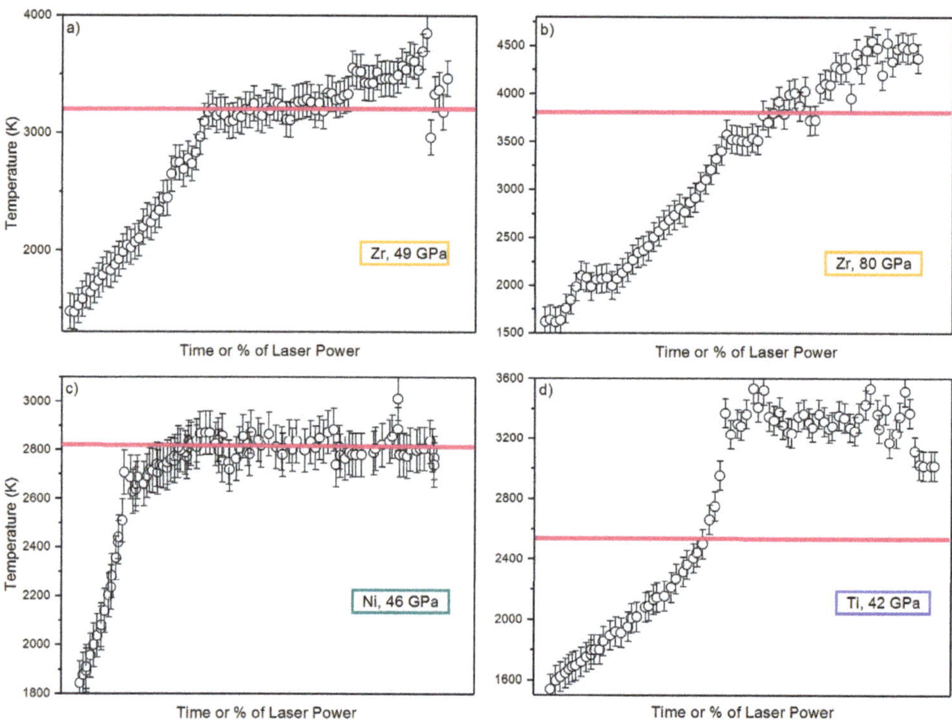

Figure 6. Comparison of temperature versus laser power curves in different samples and different heating runs: (**a**) Zr at 49 GPa [37], (**b**) Zr at 80 GPa [37], (**c**) Ni at 46 GPa [84], (**d**) Ti at 42 GPa [36]. In all of these works, the melting was determined by XRD liquid diffuse scattering (Section 4.2), and the melting temperature found at the given pressure is shown by the straight pink line.

It has been proposed that the latent heat of melting is insignificant with respect to the heat provided by the lasers [203]. In fact, there can be several reasons why the temperature–laser power relationship can change inside an LH-DAC:

i. Discontinuities in the reflectivity of the sample. The reflectivity cannot be a reliable indicator of melting since it is not an intrinsic property of materials.

ii. Increase in the conductivity of the PTM with the increasing heat provided by the lasers could also explain why the temperature is not always increased with laser power.

iii. The thickness of the PTM can change during an experimental run, which can affect the thermal insulation and the heating efficiency inside the LH-DAC. As a result, more laser power may be needed to heat the sample at a given temperature.

iv. A temperature plateau can appear because of the melting of the PTM (Section 5.3).

v. The melt may become mechanically unstable and flow, leading to sudden variations in temperature.

All of the above characteristics and parameters are very difficult to be calculated in situ in order to quantify their effect on the temperature of the sample. Therefore, the observation of temperature plateaus, although efficient in many cases, cannot be considered by itself a consistently reliable method, and often the use of a complementary diagnostic is advised.

4.5. Microstructure Formation on the Quench

A recent work [15] has demonstrated a new melting criterion by studying the microstructure formation of a Mo sample in a DAC using flash laser heating synchronized with X-ray diffraction and a fast detector. This approach is based on the observation that heating above the melting point followed by a quench could reduce the grain size of polycrystalline samples, sometimes to an nm scale, while quenching from annealing (i.e., heating below melting) generally increases the grain size [225]. The importance of using a short laser pulse duration (5–20 ms) lies in the fact that the changes in the microstructure of the sample happen in small steps, providing the opportunity to observe the changes in the diffraction patterns in a detailed way.

In this study [15] the sample was heated at different temperatures, measured by spectral radiometry, and then quenched. It has been observed that when quenched from a temperature below T_m, the diffraction single crystal spots were preferentially oriented in relation to the diffraction spots of MgO that has been used as the PTM. However, when quenched from a temperature above T_m, the diffraction images feature a fine-grained, randomly oriented microstructure (continuous Debye diffraction rings). With this technique, the authors were also able to track a microstructural transition in Mo that could be responsible for the lower melting temperatures reported in previous works [13,29,30].

4.6. Post-Heating Scanning Electron Microscopy (SEM)

The argument of obtained melting can be greatly reinforced by validating the experiment with post-heating characterization after the quenching of the sample, outside the LH-DAC. One of the existing possibilities is to study the surface of the heated spots with scanning electron microscopy (SEM) [199,226]. Drastic changes and bead-like features appear on the sample surface topography near the melting temperature. By careful examination of the SEM images in a recent work on Re and Mo, micrometer or sub-micrometer size recrystallization of the sample can be shown for $T<T_m$, while for $T>T_m$ the metal seems completely restructured with a boundary between the quenched liquid and the unmolten sample to a depth of several micrometers [199]. These observations are in contrast to the data acquired by the previous approach (Section 4.5) as the melting point found for Mo in [199] seemed to coincide with the recrystallization region in [15]. The discrepancies have been attributed to a microstructure transition that has been discovered for Mo [15].

A depth profile analysis of molten or unmolten regions can be also performed in a cross-section of the heated portion of the sample. The cross-section has to be prepared with a focused ion beam (FIB) [199,226]. In a recent work on Ni [39], the cross-section of the heated spot where the sample remained solid (Figure 7a) was found to be drastically different from the cross-section of the heated spot where the sample melted (Figure 7b). In fact, sharp boundaries extending to a few micrometers in depth were reported for the molten sample (Figure 7b). Energy dispersive X-ray spectroscopy (EDX) performed routinely on the samples did not show any chemical changes. In this work, KCl was used as a PTM, and as it is clear from Figure 7a,b, there are no significant differences in its morphology above or below the melting temperature of the sample.

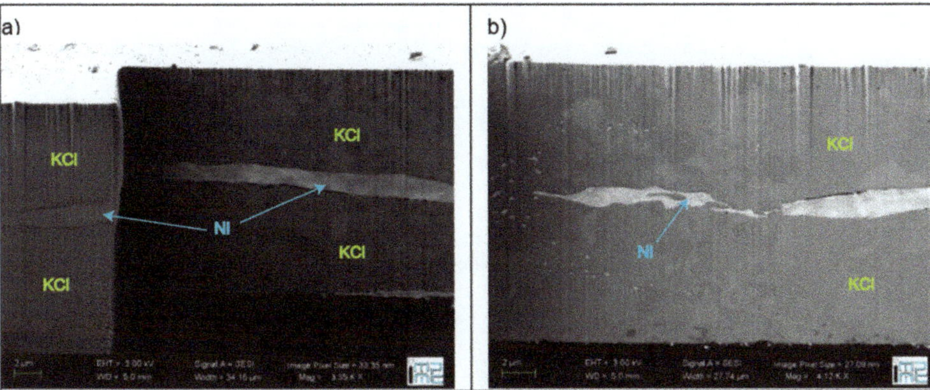

Figure 7. SEM image of the cross-section of a Ni sample heated (**a**) below melting and (**b**) above melting temperature (reproduced from [39]). In Figure 7a, cuts at different depths are shown so the three-dimensional structure of the sample assembly can be better visualized.

4.7. Changes in Resistivity

Phase transitions in a material can be often detected by changes in its physical properties. In a metal, the resistivity R generally increases with temperature T, but kinks will be observed near a phase transition region, for example, an increase in the slope of R versus T. The abrupt increase in a metal's resistivity with melting can be attributed to the loss of long-range order and thus the increased scattering of conduction electrons. Of course, such an increase in resistivity could also happen because of chemical contamination, but this possibility can be excluded by performing complementary diagnostics such as XRD, SEM, or EDS (Electron Dispersive Spectroscopy). Temperature or resistivity fluctuations can also be attributed to changes in the morphology of the sample upon melting [227].

The melting of several metals inside a DAC using this method has been obtained with good reproducibility by applying either resistive heating with graphite heaters and a four-point configuration [87,133] or laser heating [228]. For the LH-DAC, a split gasket method [229] has been employed, where two pre-indented steel gaskets were joined with a diamond filled epoxy cement that served as an insulator. The electrical leads soldered to the two gasket halves permitted the resistivity measurement. The LH-DAC resistivity method has given access to high pressures and temperatures. However, this kind of experiment remains very challenging, and the values of resistivity or thermal conductivity of metals in a DAC have been a subject of open debate [228,230,231].

4.8. Synchrotron Mössbauer Spectroscopy (SMS)

Recent technological developments in fast spectroradiometry and synchrotron technology have opened up the possibility of couple laser heating with synchrotron Mössbauer spectroscopy (SMS) to identify the melting of metals [164,232,233]. This method has mainly been demonstrated for ^{57}Fe [234,235] but can also be effective with other nuclear resonant isotopes. The diagnostic of melting relies on the dynamics of Fe atoms in a time window comparable to their nuclear lifetime. This is reflected in the reduction of the effective thickness of the sample and the collapse of the Mössbauer signal in the liquid phase. The ordering of atoms is irrelevant for SMS, in antithesis with other synchrotron techniques, for example XRD. In XRD, the scattering process is very fast and non-resonant, so that atomic motions become irrelevant. The Mössbauer signal, on the other hand, is very sensitive to the movement of the iron nuclei.

5. Sources of Controversies in the Melting Curves of Transition Metals from Static Experiments

5.1. Temperature Determination in the LH-DAC

The LH-DAC, although being the technique that can provide access to the highest temperatures and pressures possible in the static compression regime, could potentially yield error bars of a few hundreds of K in the estimation of temperature for several reasons:

i. The lasers can be unstable, especially near but below the melting temperature, and these fluctuations can alter the temperature within small time domains. It has been found that even a ~0.3% laser power fluctuation can lead to temperature fluctuations of up to 200 K [218,236].

ii. Spectral radiometry measurements can be complicated to perform and are strongly dependent on the wavelength region chosen to perform the Planck fit. The methodology for the temperature measurement inside a DAC using spectral radiometry, including the use of different optics, has been discussed in detail [213–217].

iii. Especially at lower temperatures (below 1400 K), there are not enough photons, and the Planck signal is difficult to detect. Larger acquisition times are needed which can lead to larger temperature error bars due to instabilities.

iv. It is crucial that the temperature measurement is taken from the same area of the sample that is heated by the lasers. In synchrotron radiation, this has been solved by perforating a polished mirror at the entrance of the spectrometer and aligning the X-rays to the hole using X-ray fluorescence [150,152,153,158]. Defocusing the IR lasers for a larger and more uniform heating spot can be also helpful.

5.2. Thermal Pressure Determination

In a melting experiment, the thermal pressure P_{th} has to be carefully estimated for every data point by the thermal equation of the sample or the PTM, but this can be a challenge in experiments where there is no possibility to perform in situ XRD. Pressure scales such as the fluorescence of the ruby or Sm^{2+}:SrB_4O_7 or the Raman of the diamond are not anymore valid at the high temperatures required for the melting of metals either because the signal is dampened or because the thermal radiation dominates the signal. In the absence of an XRD signal in order to determine the thermal EoS and therefore the pressure, the uncertainties can be minimized by measuring the pressure before and after the heating of the sample, but the error bar in the pressure determination can still be quite large.

5.3. Melting of the PTM

The selection of the PTM, apart for the reasons mentioned above (Section 3.3.1), is crucial for both temperature and melting determination, since the melting of the PTM prior to the melting of the sample could either hinder the fusion of the sample or give false signature of melting [22,84,198,224]. For example, spectral radiometry measurements can exhibit a temperature plateau (Section 4.4) with increasing laser power at the PTM melting temperature, confounding the experimental results. The pyrometry measurements are affected by the presence of molten PTM in the sample chamber, mainly because of changes to its optical properties (emissivity, absorption). Movements of the sample in the molten PTM and changes in spatial temperature distribution could underestimate the temperature by few hundreds of K. Moreover, the molten PTM could potentially give a diffuse signal in an XRD experiment, leading to a confusing interpretation of the data, since it could not be clear whether the liquid signal derives from the sample or the PTM. Therefore, the PTM has to be selected in a way that its melting curve is steeper and does not overlap with the melting curve of the sample.

5.4. Misalignments of the X-ray beam

During a melting experiment, it is crucial that the temperature is measured at exactly the same area of the sample that is triggered with the melt detection probe. This is especially important in Synchrotron experiments (for example XRD, XAS, or SMS), where the probe (i.e., the X-ray beam) is rather small, usually on the order of 2–3 microns. During laser heating, the optics may drift due to thermal expansion, and the X-rays are no longer triggering the same location of the sample where the temperature measurement is acquired.

The heating laser should also be very well aligned with the X-rays, and this is normally achieved by reducing the focusing of the laser in order to have a hotspot much larger than the X-ray beam (i.e., a laser spot of 10–20 microns for a 2–5 micron x-ray beam). However, in some cases, at the high temperatures required for melting the sample may move, meaning that the X-ray beam and the hotspot are no longer in the same place. Therefore, meticulous and continuous realignment of both the optics and the X-ray beam is required in order to obtain reliable measurements.

5.5. Additional Sources of Diffuse Scattering

The diffuse liquid scattering detected from XRD (Section 4.2) needs to be properly analyzed to avoid misinterpretations since the molten signal is often weak and can be masked by other factors. The Compton scattering arising from the diamonds in LH-DAC or RH-DAC can in some cases dominate the liquid diffraction signal, and it is therefore advised to be removed by subtracting the background signal of an empty DAC, before or after the experiment. Removing the Compton background may be less crucial for melting curves than it is for the determination of the liquid structure using a pair distribution function (PDF), but it can be important in cases where the molten region is too thin to create a measurable XRD signal. Thus, the quality of the obtained data can be improved using this technique [77]. As mentioned in Section 5.3, the melting of the PTM could also be a source of additional diffuse scattering, and the selection of the proper PTM is very important. Finally, extra care should be taken not to interpret glass formation as a liquid signal, given that many transition metals are known to form metallic glasses at high temperatures [1]. Usually, the possibility of glass formation can be eliminated by the absence of diffuse scattering in the quenched XRD pattern.

5.6. Fast Recrystallization for $T < T_m$

In static experiments, most controversial results come from the comparison of speckle (Section 4.1) and XRD (Section 4.2) methods in the LH-DAC, which are the two most widely used techniques to determine melting. It has been argued today that the main difference between the two methods relies on the phenomenon of "fast recrystallization" of the sample that occurs prior to melting [15,24,35–37,63,83,200]. The recrystallization process from a recent experiment on Zr [37] can be shown in Figure 8, where XRD patterns were collected every few seconds as the temperature of the sample was progressively in-creased. The diffraction patterns clearly show polycrystalline Zr spots that move and change in intensity with increasing temperature. This permanent reorientation of the crystal is due to movements in the surface of the sample, and it happens at temperatures of almost 1000 K before melting in some cases. The signature of melting comes much later with the appearance of liquid diffuse scattering. The quenched diffraction pattern (i.e., the one obtained after switching off the lasers) after melting shows a fine grain structure in good agreement with other methods using XRD [15].

Fast recrystallization and melting are both endothermic processes and cannot be distinguished easily by detecting the movements of the sample surface. Therefore, it is possible that earlier speckle works underestimated the melting curves of transition metals since they did not take into account the phenomenon of fast recrystallization. By using the complementary synchrotron technique of XAS, agreement of the melting temperature with XRD has been found in the case of Ni [39,84], proposing that melting detection can be performed either by probing the local atomic environment (XAS) or the long-range structural changes (XRD) in the sample. However, the works on Fe were controversial [24,25], and more experimental data need to become available to verify any possible convergence between these two methods.

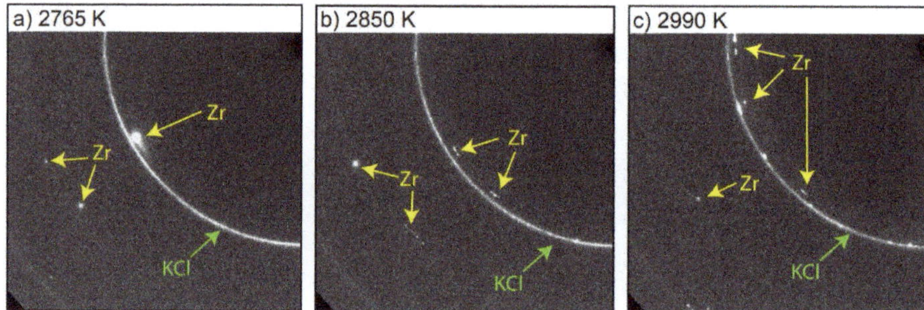

Figure 8. XRD 2D images of a Zr sample at different temperatures (P = 49 GPa) [37]. The movement of polycrystalline spots with temperature is related to movements in the sample surface. The actual melting temperature at this pressure point is 3235 K.

5.7. Chemical Reactions and Carbide Formation

At the very high temperatures that are generated inside an LH-DAC, especially when using CW lasers, carbon contamination of the sample from the diamond anvils is highly possible. Several transition metals have the ability to form carbides, notably, Fe (Fe_3C) [237] and Ta (TaC) [22], and carbide formation has been found to be the source of large discrepancies between different results, leading to an underestimation of the melting curve by many thousands degrees K [21,22]. Melting studies using in situ XRD can identify this problem and detect carbides or any other unwanted phases that are formed due to chemical reactions in the sample chamber, even in very small amounts. Raman spectroscopy is also a method that can be very sensitive to element detection, but the low Raman cross-section of metals, combined with the high fluorescence of the diamonds and the intense incandescent thermal radiation that dominates the signal, requires surface-enhanced and time-resolved techniques, making such experiments extremely complicated [168,238].

Carbon contamination may be caused by the diffusion of carbon from the gasket or anvil through the pressure medium and into the crystalline sample prior to melting. Choosing an inert PTM is therefore critical. Water in the PTM can also be a cause of chemical reactions; thus, a technique to reduce the humidity relies on heating the DAC before the experiment (for example, putting it in an oven for 1 h at 100 °C) [37,39,239].

Another way to minimize the chemical reactions is time constraining the heating duration to a few milliseconds with the use of a pulsed laser [15,199], but a recent work has detected carbon diffusion in both continued and pulsed laser heating [239]. Encapsulating the sample in a single crystal PTM such as MgO can also significantly reduce the lateral diffusion of unwanted elements in the sample chamber [15].

6. Concluding Remarks

The scope of this manuscript is to provide adequate information to the reader about the high-pressure melting curves of transition metals that are obtained with static compression techniques. In particular, the results from many static melting experiments have been collected and compared in detail, and the discussion about the thermodynamic models and the physics of the melting transition has tried to point out the main trends about the melting behavior of d-transition elements. This review focuses naturally on datasets obtained by the LH-DAC since it is the most widely used technique for melting experiments, offering access to a wide P-T region, but comparisons with other static methods (notably LVP and RH-DAC) were also made.

The advantages and drawbacks of the various experimental methods and melting criteria have been thoroughly discussed, as well as the main sources of controversies between the different datasets. The melting of transition metals is a very wide subject, and even though many significant technical developments have been carried out in the recent years concerning sample preparation [77,204], laser heating [164,204,239], pressure generation [42–44], temperature estimation [215,216], and synchrotron radiation [15,233], there is still lack of consensus between the different experimental techniques. All of this concentrated effort, however, could result in reducing some of the main problems that are present in the study of melting curves today, especially inside an LH-DAC. Obtaining data at pressures above 1 Mbar is crucially important since most of the melting curves use extrapolation to predict the behavior at higher pressures, which is not always reliable. More sophisticated sample assemblies and control of both the sample and PTM thickness could also result in better heating efficiency and insulation inside the DAC and help (together with fast temperature measurement techniques) to reduce temperature gradients. Fast diffraction measurements and flash heating could cut down (but probably not completely eliminate) chemical reactions. Finally, it may be very interesting to see if a sub-micrometer X-ray beam, as it is the case with new generation synchrotrons in the very near future, could yield more precise results. The same can be said for dynamic compression studies using time-resolved X-ray radiation in the new FEL facilities, which may potentially bridge the gap between differences in static and dynamic techniques.

Supplementary Materials: The following are available online at https://www.mdpi.com/article/10.3390/cryst11040416/s1, Table S1: Spreadsheets for all the data of Figure 1a. Only in situ XRD data are shown (i.e., not off-line obtained melting). Table S2: Spreadsheets for all the data of Figure 1b. Only in situ measurements are shown.

Funding: This research received no external funding.

Institutional Review Board Statement: Not Applicable.

Informed Consent Statement: Not Applicable.

Data Availability Statement: Data is contained within the article or Supplementary Material.

Acknowledgments: I wish to thank the three reviewers for their insightful comments and suggestions that have contributed to the improvement of the manuscript. I would also like to thank Silvia Boccato for help with Figures 5 and 7. I acknowledge the European Synchrotron Radiation Facility (ESRF) for provision of synchrotron radiation within Projects HC-2799 and HC-2554. Analysis of the recovered samples was performed with the help of Imène Estève at the Focused Ion Beam (FIB) and Scanning Electron Microscope (SEM) facility of the Institut de Minéralogie et de Physique des Milieux Condensés, supported by Région Ile de France grant SESAME 2006 I-07-593/R, INSU-CNRS, INP-CNRS, University Pierre et Marie Curie - Paris 6, and by the French National Research Agency (ANR) grant ANR-07-BLAN-0124-01.

Conflicts of Interest: The author declares no conflict of interest.

Abbreviations

The following abbreviations are used in this manuscript:

CW	Continuous Wave
DOS	Density of States
EDS	Electron Dispersive Spectroscopy
EoS	Equation of State
EXAFS	Extended X-ray Absorption Fine Structure
FIB	Focused Ion Beam
IR	Infrared
LH-DAC	Laser Heated Diamond Anvil Cell
LVP	Large Volume Press
PDF	Pair Distribution Function
P_{th}	Thermal Pressure
PTM	Pressure Transmitting Medium
RH-DAC	Resistively Heated Diamond Anvil Cell
SEM	Scanning Electron Microscopy
SMS	Synchrotron Mössbauer Spectroscopy
SW	Shock Wave
T_m	Melting Temperature
XANES	X-ray Absorption Near Edge Structure
XAS	X-ray Absorption Spectroscopy
XFEL	X-ray Free Electron Laser
XRD	X-ray Diffraction

References

1. Wang, W.H.; Dong, C.; Shek, C.H. Bulk metallic glasses. *Mater. Sci. Eng. R Rep* **2004**, *44*, 45–89. [CrossRef]
2. Motta, A.T.; Yilmazbayhan, A.; da Silva, M.J.G.; Comstock, R.J.; Was, G.S.; Busby, J.T.; Gartner, E.; Peng, Q.; Jeong, Y.H.; Park, J.Y. Zirconium alloys for supercritical water reactor applications: Challenges and possibilities. *J. Nucl. Mater.* **2007**, *371*, 61–75. [CrossRef]
3. Liu, X.; Chu, P.K.; Ding, C. Surface modification of titanium, titanium alloys, and related materials for biomedical applications. *Mater. Sci. Eng. R Rep.* **2004**, *47*, 49–121. [CrossRef]
4. Suryanarayana, C. Mechanical alloying and milling. *Prog. Mater. Sci.* **2001**, *46*, 1–184. [CrossRef]
5. Yap, C.Y.; Chua, C.K.; Dong, Z.L.; Liu, Z.H.; Zhang, D.Q.; Loh, L.E.; Sing, S.L. Review of selective laser melting: Materials and applications. *Appl. Phys. Rev.* **2015**, *2*, 041101. [CrossRef]
6. Boyer, R. An overview on the use of titanium in the aerospace industry. *Mater. Sci. Eng. A* **1996**, *213*, 103–114. [CrossRef]
7. Alfe, D.; Gillan, M.J.; Price, G.D. Temperature and composition of the Earth's core. *Contemp. Phys.* **2007**, *48*, 63–80. [CrossRef]
8. Allègre, C.J.; Poirier, J.-P.; Humler, E.; Hofmann, A.W. The chemical composition of the Earth. *Earth Planet. Sci. Lett.* **1995**, *134*, 515–526. [CrossRef]
9. Antonangeli, D.; Siebert, J.; Badro, J.; Farber, D.L.; Fiquet, G.; Morard, G.; Ryerson, F.J. Composition of the Earth's inner core from high-pressure sound velocity measurements in Fe–Ni–Si alloys. *Earth Planet. Sci. Lett.* **2010**, *295*, 292–296. [CrossRef]

10. Badro, J.; Côté, A.S.; Brodholt, J.P. A seismologically consistent compositional model of Earth's core. *Proc. Natl. Acad. Sci. USA* **2014**, *111*, 7542–7545. [CrossRef]

11. Hirose, K.; Labrosse, S.; Hernlund, J. Composition and State of the Core. *Annu. Rev. Earth Planet. Sci.* **2013**, *41*, 657–691. [CrossRef]

12. Landa, A.; Söderlind, P.; Ruban, A.V.; Peil, O.E.; Vitos, L. Stability in bcc Transition Metals: Madelung and Band-Energy Effects due to Alloying. *Phys. Rev. Lett.* **2009**, *103*, 235501. [CrossRef] [PubMed]

13. Errandonea, D.; Schwager, B.; Ditz, R.; Gessmann, C.; Boehler, R.; Ross, M. Systematics of transition-metal melting. *Phys. Rev. B* **2001**, *63*, 132104. [CrossRef]

14. Boehler, R. Temperatures in the Earth's core from melting-point measurements of iron at high static pressures. *Nat. Cell Biol.* **1993**, *363*, 534–536. [CrossRef]

15. Hrubiak, R.; Meng, Y.; Shen, G. Microstructures define melting of molybdenum at high pressures. *Nat. Commun.* **2017**, *8*, 14562. [CrossRef]

16. Japel, S.; Schwager, B.; Boehler, R.; Ross, M. Melting of Copper and Nickel at High Pressure: The Role of d Electrons. *Phys. Rev. Lett.* **2005**, *95*, 167801. [CrossRef]

17. Ross, M.; Boehler, R.; Errandonea, D. Melting of transition metals at high pressure and the influence of liquid frustration: The late metals Cu, Ni, and Fe. *Phys. Rev. B* **2007**, *76*, 184117. [CrossRef]

18. Alfe, D.; Adlo, L.V.; Price, G.D.; Gillan, M.J. Melting curve of materials: Theory versus experiments. *J. Phys. Condens. Matter* **2004**, *16*, S973–S982. [CrossRef]

19. Skriver, H.L. Crystal structure from one-electron theory. *Phys. Rev. B* **1985**, *31*, 1909–1923. [CrossRef]

20. Dai, C.; Hu, J.; Tan, H. Hugoniot temperatures and melting of tantalum under shock compression determined by optical pyrometry. *J. Appl. Phys.* **2009**, *106*, 043519. [CrossRef]

21. Errandonea, D.; Somayazulu, M.; Häusermann, D.; Mao, H.K. Melting of tantalum at high pressure determined by angle dispersive X-ray diffraction in a double-sided laser-heated diamond-anvil cell. *J. Phys. Condens. Matter* **2003**, *15*, 7635–7649. [CrossRef]

22. Dewaele, A.; Mezouar, M.; Guignot, N.; Loubeyre, P. High Melting Points of Tantalum in a Laser-Heated Diamond Anvil Cell. *Phys. Rev. Lett.* **2010**, *104*, 255701. [CrossRef] [PubMed]

23. Akin, M.C.; Nguyen, J.H.; Beckwith, M.A.; Chau, R.; Ambrose, W.P.; Fat'Yanov, O.V.; Asimow, P.D.; Holmes, N.C. Tantalum sound velocity under shock compression. *J. Appl. Phys.* **2019**, *125*, 145903. [CrossRef]

24. Anzellini, S.; Dewaele, A.; Mezouar, M.; Loubeyre, P.; Morard, G. Melting of Iron at Earth's Inner Core Boundary Based on Fast X-ray Diffraction. *Science.* **2013**, *340*, 464–466. [CrossRef] [PubMed]

25. Aquilanti, G.; Trapananti, A.; Karandikar, A.; Kantor, I.Y.; Marini, C.; Mathon, O.; Pascarelli, S.; Boehler, R. Melting of iron determined by X-ray absorption spectroscopy to 100 GPa. *Proc. Natl. Acad. Sci. USA* **2015**, *112*, 12042–12045. [CrossRef]

26. Williams, Q.; Jeanloz, R.; Bass, J.; Svendsen, B.; Ahrens, T.J. The Melting Curve of Iron to 250 Gigapascals: A Constraint on the Temperature at Earth's Center. *Science* **1987**, *236*, 181–182. [CrossRef] [PubMed]

27. Hixson, R.S.; Boness, D.A.; Shaner, J.W.; Moriarty, J.A. Acoustic Velocities and Phase Transitions in Molybdenum under Strong Shock Compression. *Phys. Rev. Lett.* **1989**, *62*, 637–640. [CrossRef] [PubMed]

28. Nguyen, J.H.; Akin, M.C.; Chau, R.; Fratanduono, D.E.; Ambrose, W.P.; Fat'Yanov, O.V.; Asimow, P.D.; Holmes, N.C. Molybdenum sound velocity and shear modulus softening under shock compression. *Phys. Rev. B* **2014**, *89*, 174109. [CrossRef]

29. Santamaría-Pérez, D.; Ross, M.; Errandonea, D.J.H.; Mukherjee, G.D.; Mezouar, M.; Boehler, R. X-ray diffraction measurements of Mo melting to 119 GPa and the high pressure phase diagram. *J. Chem. Phys.* **2009**, *130*, 124509. [CrossRef] [PubMed]

30. Errandonea, D. Improving the understanding of the melting behaviour of Mo, Ta, and W at extreme pressures. *Phys. B Condens. Matter* **2005**, *357*, 356–364. [CrossRef]

31. Errandonea, D.; Boehler, R.; Ross, M. Comment on "Molybdenum sound velocity and shear modulus softening under shock compression". *Phys. Rev. B* **2015**, *92*, 026101. [CrossRef]

32. Nguyen, J.H.; Akin, M.C.; Chau, R.; Fratanduono, D.E.; Ambrose, W.P.; Fat'yanov, O.V.; Asimow, P.D.; Holmes, N.C. Reply to "Comment on 'Molybdenum sound velocity and shear modulus softening under shock compres-sion'". *Phys. Rev. B* **2015**, *92*, 026102. [CrossRef]

33. Wang, J.; Coppari, F.; Smith, R.F.; Eggert, J.H.; Lazicki, A.E.; Fratanduono, D.E.; Rygg, J.R.; Boehly, T.R.; Collins, G.W.; Duffy, T.S. X-ray diffraction of molybdenum under shock compression to 450 GPa. *Phys. Rev. B* **2015**, *92*, 174114. [CrossRef]

34. Errandonea, D.; MacLeod, S.G.; Burakovsky, L.; Santamaria-Perez, D.; Proctor, J.E.; Cynn, H.; Mezouar, M. Melting curve and phase diagram of vanadium under high-pressure and high-temperature conditions. *Phys. Rev. B* **2019**, *100*, 094111. [CrossRef]

35. Zhang, Y.; Tan, Y.; Geng, H.Y.; Salke, N.P.; Gao, Z.; Li, J.; Sekine, T.; Wang, Q.; Greenberg, E.; Prakapenka, V.B.; et al. Melting curve of vanadium up to 256 GPa: Consistency between experiments and theory. *Phys. Rev. B* **2020**, *102*, 214104. [CrossRef]

36. Stutzmann, V.; Dewaele, A.; Bouchet, J.; Bottin, F.; Mezouar, M. High-pressure melting curve of titanium. *Phys. Rev. B* **2015**, *92*, 224110. [CrossRef]

37. Parisiades, P.; Cova, F.; Garbarino, G. Melting curve of elemental zirconium. *Phys. Rev. B* **2019**, *100*, 054102. [CrossRef]

38. Pigott, J.S.; Velisavljevic, N.; Moss, E.K.; Draganic, N.; Jacobsen, M.K.; Meng, Y.; Hrubiak, R.; Sturtevant, B.T. Experimental melting curve of zirconium metal to 37 GPa. *J. Phys. Condens. Matter* **2020**, *32*, 355402. [CrossRef]

39. Boccato, S.; Torchio, R.; Kantor, I.; Morard, G.; Anzellini, S.; Giampaoli, R.; Briggs, R.; Smareglia, A.; Irifune, T.; Pascarelli, S. The Melting Curve of Nickel Up to 100 GPa Explored by XAS: Melting curve of nickel up to 1 mbar. *J. Geophys. Res. Solid Earth* **2017**, *122*, 9921–9930. [CrossRef]

40. Errandonea, D. High-pressure melting curves of the transition metals Cu, Ni, Pd, and Pt. *Phys. Rev. B* **2013**, *87*, 054108. [CrossRef]

41. Lazor, P.; Shen, G.; Saxena, S.K. Laser-heated diamond anvil cell experiments at high pressure: Melting curve of nickel up to 700 kbar. *Phys. Chem. Miner.* **1993**, *20*, 86–90. [CrossRef]

42. Dewaele, A.; Loubeyre, P.; Occelli, F.; Marie, O.; Mezouar, M. Toroidal diamond anvil cell for detailed measurements under extreme static pressures. *Nat. Commun.* **2018**, *9*, 1–9. [CrossRef]

43. Dubrovinskaia, N.; Dubrovinsky, L.; Solopova, N.A.; Abakumov, A.; Turner, S.; Hanfland, M.; Bykova, E.; Bykov, M.; Prescher, C.; Prakapenka, V.B.; et al. Terapascal static pressure generation with ultrahigh yield strength nanodiamond. *Sci. Adv.* **2016**, *2*, e1600341. [CrossRef] [PubMed]

44. Dubrovinsky, L.; Dubrovinskaia, N.; Prakapenka, V.B.; Abakumov, A.M. Implementation of micro-ball nanodiamond anvils for high-pressure studies above 6 Mbar. *Nat. Commun.* **2012**, *3*, 1163. [CrossRef] [PubMed]

45. Swift, D.C.; Johnson, R.P. Quasi-isentropic compression by ablative laser loading: Response of materials to dynamic loading on nanosecond time scales. *Phys. Rev. E* **2005**, *71*, 066401. [CrossRef] [PubMed]

46. Kalita, P.; Brown, J.; Specht, P.; Root, S.; White, M.; Smith, J.S. Dynamic X-ray diffraction and nanosecond quantification of kinetics of formation of β -zirconium under shock compression. *Phys. Rev. B* **2020**, *102*, 060101. [CrossRef]

47. Eggert, J.H.; Hicks, D.G.; Celliers, P.M.; Bradley, D.K.; McWilliams, R.S.; Jeanloz, R.; Miller, J.E.; Boehly, T.R.; Collins, G.W. Melting temperature of diamond at ultrahigh pressure. *Nat. Phys.* **2009**, *6*, 40–43. [CrossRef]

48. Duffy, T.S.; Smith, R.F. Ultra-High Pressure Dynamic Compression of Geological Materials. *Front. Earth Sci.* **2019**, *7*, 23. [CrossRef]

49. Nguyen, J.H.; Holmes, N.C. Melting of iron at the physical conditions of the Earth's core. *Nat. Cell Biol.* **2004**, *427*, 339–342. [CrossRef] [PubMed]

50. Huser, G.; Koenig, M.; Benuzzi-Mounaix, A.; Henry, E.; Vinci, T.; Faral, B.; Tomasini, M.; Telaro, B.; Batani, D. Temperature and melting of laser-shocked iron releasing into an LiF window. *Phys. Plasmas* **2005**, *12*, 060701. [CrossRef]

51. Nagler, B.; Arnold, B.; Bouchard, G.; Boyce, R.F.; Callen, A.; Campell, M.; Curiel, R.; Galtier, E.; Garofoli, J.; Granados, E.; et al. The Matter in Extreme Conditions instrument at the Linac Coherent Light Source. *J. Synchrotron Radiat.* **2015**, *22*, 520–525. [CrossRef] [PubMed]

52. Glenzer, S.H.; Fletcher, L.B.; Galtier, E.; Nagler, B.; Alonso-Mori, R.; Barbrel, B.; Brown, S.B.; Chapman, D.A.; Chen, Z.; Curry, C.B.; et al. Matter under extreme conditions experiments at the Linac Coherent Light Source. *J. Phys. B At. Mol. Opt. Phys.* **2016**, *49*, 092001. [CrossRef]

53. Mason, P.; Banerjee, S.; Smith, J.; Butcher, T.; Phillips, J.; Höppner, H.; Möller, D.; Ertel, K.; De Vido, M.; Hollingham, I.; et al. Development of a 100 J, 10 Hz laser for compression experiments at the High Energy Density instrument at the European XFEL. *High Power Laser Sci. Eng.* **2018**, *6*, 65. [CrossRef]

54. Luo, S.-N.; Ahrens, T.J.; Çağın, T.; Strachan, A.; Goddard, W.A.; Swift, D.C. Maximum superheating and undercooling: Systematics, molecular dynamics simulations, and dynamic experiments. *Phys. Rev. B* **2003**, *68*, 134206. [CrossRef]

55. Morris, J.R.; Wang, C.Z.; Ho, K.M.; Chan, C.T. Melting line of aluminum from simulations of coexisting phases. *Phys. Rev. B* **1994**, *49*, 3109–3115. [CrossRef]

56. Belonoshko, A.B. Molecular dynamics of MgSiO₃ perovskite at high pressures: Equation of state, structure, and melting transition. *Geochim. Cosmochim. Acta* **1994**, *58*, 4039–4047. [CrossRef]

57. Gillan, M.J.; Alfe, D.; Brodholt, J.; Vočadlo, L.; Price, G.D. First-principles modelling of Earth and planetary materials at high pressures and temperatures. *Rep. Prog. Phys.* **2006**, *69*, 2365–2441. [CrossRef]

58. Cazorla, C.; Alfè, D.; Gillan, M.J. Constraints on the phase diagram of molybdenum from first-principles free-energy calculations. *Phys. Rev. B* **2012**, *85*, 064113. [CrossRef]

59. Belonoshko, A.B.; Rosengren, A. High-pressure melting curve of platinum from ab initio Z method. *Phys. Rev. B* **2012**, *85*, 174104. [CrossRef]

60. Belonoshko, A.B.; Davis, S.; Skorodumova, N.V.; Lundow, P.H.; Rosengren, A.; Johansson, B. Properties of the fcc Lennard-Jones crystal model at the limit of superheating. *Phys. Rev. B* **2007**, *76*, 064121. [CrossRef]

61. Anzellini, S.; Boccato, S. A Practical Review of the Laser-Heated Diamond Anvil Cell for University Laboratories and Syn-chrotron Applications. *Crystals* **2020**, *10*, 459. [CrossRef]

62. Simon, F.; Glatzel, G. Bemerkungen zur Schmelzdruckkurve. *Z. Anorg. Allg. Chem.* **1929**, *178*, 309–316. [CrossRef]

63. Anzellini, S.; Monteseguro, V.; Bandiello, E.; Dewaele, A.; Burakovsky, L.; Errandonea, D. In situ characterization of the high pressure–high temperature melting curve of platinum. *Sci. Rep.* **2019**, *9*, 1–10. [CrossRef]

64. Gregoryanz, E.; Degtyareva, O.; Somayazulu, M.; Hemley, R.J.; Mao, H.-K. Melting of Dense Sodium. *Phys. Rev. Lett.* **2005**, *94*, 185502. [CrossRef]

65. Gorelli, F.A.; De Panfilis, S.; Bryk, T.; Ulivi, L.; Garbarino, G.; Parisiades, P.; Santoro, M. Simple-to-Complex Transformation in Liquid Rubidium. *J. Phys. Chem. Lett.* **2018**, *9*, 2909–2913. [CrossRef] [PubMed]

66. Narygina, O.; McBride, E.E.; Stinton, G.W.; McMahon, M.I. Melting curve of potassium to 22 GPa. *Phys. Rev. B* **2011**, *84*, 054111. [CrossRef]

67. Guillaume, C.L.; Gregoryanz, E.; Degtyareva, O.; McMahon, M.I.; Hanfland, M.; Evans, S.; Guthrie, M.; Sinogeikin, S.V.; Mao, H.-K. Cold melting and solid structures of dense lithium. *Nat. Phys.* **2011**, *7*, 211–214. [CrossRef]
68. Decremps, F.; Ayrinhac, S.; Gauthier, M.; Antonangeli, D.; Morand, M.; Garino, Y.; Parisiades, P. Sound velocity and equation of state in liquid cesium at high pressure and high temperature. *Phys. Rev. B* **2018**, *98*, 184103. [CrossRef]
69. Ayrinhac, S.; Robinson, V.N.; Decremps, F.; Gauthier, M.; Antonangeli, D.; Scandolo, S.; Morand, M. High-pressure transformations in liquid rubidium. *Phys. Rev. Mater.* **2020**, *4*, 113611. [CrossRef]
70. Kechin, V.V. Melting curve equations at high pressure. *Phys. Rev. B* **2001**, *65*, 052102. [CrossRef]
71. Lindemann, F.A. The calculation of molecular vibration frequency. *Z. Phys.* **1910**, *11*, 609.
72. Wolf, G.H.; Jeanloz, R. Lindemann Melting Law: Anharmonic correction and test of its validity for minerals. *J. Geophys. Res. Space Phys.* **1984**, *89*, 7821–7835. [CrossRef]
73. Born, M. Thermodynamics of Crystals and Melting. *J. Chem. Phys.* **1939**, *7*, 591–603. [CrossRef]
74. Hieu, H.K. Melting of solids under high pressure. *Vacuum* **2014**, *109*, 184–186. [CrossRef]
75. Kushwah, S.; Tomar, Y.; Upadhyay, A. On the volume-dependence of the Grüneisen parameter and the Lindemann law of melting. *J. Phys. Chem. Solids* **2013**, *74*, 1143–1145. [CrossRef]
76. Wang, Z.; Lazor, P.; Saxena, S. A simple model for assessing the high pressure melting of metals: Nickel, aluminum and platinum. *Phys. B Condens. Matter* **2001**, *293*, 408–416. [CrossRef]
77. Weck, G.; Recoules, V.; Queyroux, J.-A.; Datchi, F.; Bouchet, J.; Ninet, S.; Garbarino, G.; Mezouar, M.; Loubeyre, P. Determination of the melting curve of gold up to 110 GPa. *Phys. Rev. B* **2020**, *101*, 014106. [CrossRef]
78. Lomonosov, I. Multi-phase equation of state for aluminum. *Laser Part. Beams* **2007**, *25*, 567–584. [CrossRef]
79. Khishchenko, K. Equation of state and phase diagram of tin at high pressures. *J. Phys. Conf. Ser.* **2008**, *121*, 022025. [CrossRef]
80. Kulyamina, E.Y.; Zitserman, V.Y.; Fokin, L.R. Titanium Melting Curve: Data Consistency Assessment, Problems and Achievements. *Tech. Phys.* **2018**, *63*, 369–373. [CrossRef]
81. Kulyamina, E.Y.; Zitserman, V.Y.; Fokin, L.R. Calculating the melting curves by the thermodynamic data matching method: Platinum-group refractory metals (Ru, Os, and Ir). *Tech. Phys.* **2017**, *62*, 68–74. [CrossRef]
82. Kulyamina, E.Y.; Zitserman, V.Y.; Fokin, L.R. Osmium: Melting curve and matching of high-temperature data. *High Temp.* **2015**, *53*, 151–154. [CrossRef]
83. Errandonea, D.; Burakovsky, L.; Preston, D.L.; MacLeod, S.G.; Santamaría-Perez, D.; Chen, S.; Cynn, H.; Simak, S.I.; McMahon, M.I.; Proctor, J.E.; et al. Experimental and theoretical confirmation of an orthorhombic phase transition in niobium at high pressure and temperature. *Commun. Mater.* **2020**, *1*, 1–11. [CrossRef]
84. Lord, O.T.; Wood, I.G.; Dobson, D.P.; Vočadlo, L.; Wang, W.; Thomson, A.R.; Wann, E.T.; Morard, G.; Mezouar, M.; Walter, M.J. The melting curve of Ni to 1 Mbar. *Earth Planet. Sci. Lett.* **2014**, *408*, 226–236. [CrossRef]
85. Ross, M.; Yang, L.H.; Boehler, R. Melting of aluminum, molybdenum, and the light actinides. *Phys. Rev. B* **2004**, *70*, 184112. [CrossRef]
86. Akella, J.; Kennedy, G.C. Melting of gold, silver, and copper-proposal for a new high-pressure calibration scale. *J. Geophys. Res. Space Phys.* **1971**, *76*, 4969–4977. [CrossRef]
87. Errandonea, D. The melting curve of ten metals up to 12 GPa and 1600 K. *J. Appl. Phys.* **2010**, *108*, 033517. [CrossRef]
88. Boehler, R.; Ross, M. Melting curve of aluminum in a diamond cell to 0.8 Mbar: Implications for iron. *Earth Planet. Sci. Lett.* **1997**, *153*, 223–227. [CrossRef]
89. Belonoshko, A.; Burakovsky, L.; Chen, S.P.; Johansson, B.; Mikhaylushkin, A.S.; Preston, D.L.; Simak, S.I.; Swift, D.C. Molybdenum at High Pressure and Temperature: Melting from Another Solid Phase. *Phys. Rev. Lett.* **2008**, *100*, 135701. [CrossRef]
90. Morard, G.; Boccato, S.; Rosa, A.D.; Anzellini, S.; Miozzi, F.; Henry, L.; Garbarino, G.; Mezouar, M.; Harmand, M.; Guyot, F.; et al. Solving Controversies on the Iron Phase Diagram Under High Pressure. *Geophys. Res. Lett.* **2018**, *45*. [CrossRef]
91. Schenk, T.; Holland-Moritz, D.; Simonet, V.; Bellissent, R.; Herlach, D.M. Icosahedral Short-Range Order in Deeply Undercooled Metallic Melts. *Phys. Rev. Lett.* **2002**, *89*, 075507. [CrossRef] [PubMed]
92. Lee, G.W.; Gangopadhyay, A.K.; Kelton, K.F.; Hyers, R.W.; Rathz, T.J.; Rogers, J.R.; Robinson, D.S. Difference in Icosahedral Short-Range Order in Early and Late Transition Metal Liquids. *Phys. Rev. Lett.* **2004**, *93*, 037802. [CrossRef] [PubMed]
93. Ross, M.; Boehler, R.; Japel, S. Melting of bcc transition metals and icosahedral clustering. *J. Phys. Chem. Solids* **2006**, *67*, 2178–2182. [CrossRef]
94. Jakse, N.; Pasturel, A. Local Order of Liquid and Supercooled Zirconium by Ab Initio Molecular Dynamics. *Phys. Rev. Lett.* **2003**, *91*, 195501. [CrossRef] [PubMed]
95. Errandonea, D. Phase behavior of metals at very high P–T conditions: A review of recent experimental studies. *J. Phys. Chem. Solids* **2006**, *67*, 2017–2026. [CrossRef]
96. Falconi, S.; Lundegaard, L.F.; Hejny, C.; McMahon, M.I. X-ray Diffraction Study of Liquid Cs up to 9.8 GPa. *Phys. Rev. Lett.* **2005**, *94*, 125507. [CrossRef]
97. Ross, M.; McMahan, A.K. Systematics of the $s \rightarrow d$ and $p \rightarrow d$ electronic transition at high pressure for the elements I through La. *Phys. Rev. B* **1982**, *26*, 4088–4093. [CrossRef]
98. Christensen, N.; Novikov, D. High-pressure phases of the light alkali metals. *Solid State Commun.* **2001**, *119*, 477–490. [CrossRef]
99. Degtyareva, V.F. Brillouin zone concept and crystal structures of sp metals under high pressure. *High Press. Res.* **2003**, *23*, 253–257. [CrossRef]

100. Olijnyk, H.; Holzapfel, W. Phase transitions in alkaline earth metals under pressure. *Phys. Lett. A* **1984**, *100*, 191–194. [CrossRef]

101. Winzenick, M.; Holzapfel, W.B. Structural study on the high-pressure phase strontium III. *Phys. Rev. B* **1996**, *53*, 2151–2154. [CrossRef]

102. Kenichi, T. High-pressure structural study of barium to 90 GPa. *Phys. Rev. B* **1994**, *50*, 16238–16246. [CrossRef]

103. Duthie, J.C.; Pettifor, D.G. Correlation between d-Band Occupancy and Crystal Structure in the Rare Earths. *Phys. Rev. Lett.* **1977**, *38*, 564–567. [CrossRef]

104. Errandonea, D.; Boehler, R.; Ross, M. Melting of the Rare Earth Metals and f-Electron Delocalization. *Phys. Rev. Lett.* **2000**, *85*, 3444–3447. [CrossRef] [PubMed]

105. Vohra, Y.K.; Olijnik, H.; Grosshans, W.; Holzapfel, W.B. Structural Phase Transitions in Yttrium under Pressure. *Phys. Rev. Lett.* **1981**, *47*, 1065–1067. [CrossRef]

106. Vohra, Y.; Akella, J.; Weir, S.; Smith, G.S. A new ultra-high pressure phase in samarium. *Phys. Lett. A* **1991**, *158*, 89–92. [CrossRef]

107. Akella, J.; Smith, G.S.; Jephcoat, A.P. High-pressure phase transformation studies in gadolinium to 106 GPa. *J. Phys. Chem. Solids* **1988**, *49*, 573–576. [CrossRef]

108. Akella, J.; Weir, S.T.; Vohra, Y.K.; Prokop, H.; Catledge, S.A.; Chesnut, G.N. High pressure phase transformations in neodymium studied in a diamond anvil cell using diamond-coated rhenium gaskets. *J. Phys. Condens. Matter* **1999**, *11*, 6515–6520. [CrossRef]

109. Grosshans, W.A.; Holzapfel, W.B. Atomic volumes of rare-earth metals under pressures to 40 GPa and above. *Phys. Rev. B* **1992**, *45*, 5171–5178. [CrossRef]

110. Frost, D.J.; Poe, B.T.; Trønnes, R.G.; Liebske, C.; Duba, A.; Rubie, D.C. A new large-volume multianvil system. *Phys. Earth Planet. Ineriors* **2004**, *143*, 507–514. [CrossRef]

111. Shcheka, S.S.; Wiedenbeck, M.; Frost, D.J.; Keppler, H. Carbon solubility in mantle minerals. *Earth Planet. Sci. Lett.* **2006**, *245*, 430–742. [CrossRef]

112. Nieto-Sanz, D.; Loubeyre, P.; Crichton, W.; Mezouar, M. X-ray study of the synthesis of boron oxides at high pressure: Phase diagram and equation of state. *Phys. Rev. B* **2004**, *70*, 214108. [CrossRef]

113. Horvath-Bordon, E.; Riedel, R.; Zerr, A.; McMillan, P.F.; Auffermann, G.; Prots, Y.; Bronger, W.; Kniep, R.; Kroll, P. High-pressure chemistry of nitride-based materials. *Chem. Soc. Rev.* **2006**, *35*, 987–1014. [CrossRef] [PubMed]

114. Zhou, X.; Ma, D.; Wang, L.; Zhao, Y.; Wang, S. Large-volume cubic press produces high temperatures above 4000 Kelvin for study of the refractory materials at pressures. *Rev. Sci. Instrum.* **2020**, *91*, 015118. [CrossRef] [PubMed]

115. Ito, E.; Yamazaki, D.; Yoshino, T.; Fukui, H.; Zhai, S.; Shatzkiy, A.; Katsura, T.; Tange, Y.; Funakoshi, K.-I. Pressure generation and investigation of the post-perovskite transformation in MgGeO3 by squeezing the Kawai-cell equipped with sintered diamond anvils. *Earth Planet. Sci. Lett.* **2010**, *293*, 84–89. [CrossRef]

116. Ito, E. Theory and Practice—Multianvil cells and high-pressure experimental methods. In *Mineral Physics*; Price, G.D., Ed.; Elsevier: Amsterdam, The Netherlands, 2007; Volume 2, pp. 197–230.

117. Getting, I.; Kennedy, G. Effect of pressure on the emf of chromel-alumel and platinum-platinum 10%rhodium thermocouples. *J. Appl. Phys.* **1970**, *41*, 4552–4562. [CrossRef]

118. Hanneman, R.E.; Strong, H.M. Pressure Dependence of the emf of Thermocouples to 1300 °C and 50 kbar. *J. Appl. Phys.* **1965**, *36*, 523. [CrossRef]

119. Li, J.; Hadidiacos, C.; Mao, H.-K.; Fei, Y.; Hemley, R.J. Behavior of thermocouples under high pressure in a multi-anvil apparatus. *High Press. Res.* **2003**, *23*, 389–401. [CrossRef]

120. Matsui, M.; Nishiyama, N. Comparison between the Au and MgO pressure calibration standards at high temperature: Au and mgo pressure calibration standards. *Geophys. Res. Lett.* **2002**, *29*, 6-1-6-4. [CrossRef]

121. Speziale, S.; Zha, C.-S.; Duffy, T.S.; Hemley, R.J.; Mao, H.-K. Quasi-hydrostatic compression of magnesium oxide to 52 GPa: Implications for the pressure-volume-temperature equation of state. *J. Geophys. Res. Space Phys.* **2001**, *106*, 515–528. [CrossRef]

122. Zha, C.-S.; Mao, H.-K.; Hemley, R.J. Elasticity of MgO and a primary pressure scale to 55 GPa. *Proc. Natl. Acad. Sci. USA* **2000**, *97*, 13494–13499. [CrossRef]

123. Liebermann, R.C. Multi-anvil, high pressure apparatus: A half-century of development and progress. *High Press. Res.* **2011**, *31*, 493–532. [CrossRef]

124. Lazicki, A.; Fei, Y.; Hemley, R.J. High-pressure differential thermal analysis measurements of the melting curve of lithium. *Solid State Commun.* **2010**, *150*, 625–627. [CrossRef]

125. Silber, R.E.; Secco, R.A.; Yong, W. Constant electrical resistivity of Ni along the melting boundary up to 9 GPa: Constant Ni Melting Resistivity to 9 GPa. *J. Geophys. Res. Solid Earth* **2017**, *122*, 5064–5081. [CrossRef]

126. Ezenwa, I.C.; Secco, R.A. Invariant electrical resistivity of Co along the melting boundary. *Earth Planet. Sci. Lett.* **2017**, *474*, 120–127. [CrossRef]

127. Berrada, M.; Secco, R.A.; Yong, W. Decreasing electrical resistivity of gold along the melting boundary up to 5 GPa. *High Press. Res.* **2018**, *38*, 367–376. [CrossRef]

128. Lord, O.T.; Walter, M.J.; Dasgupta, R.; Walker, D.; Clark, S.M. Melting in the Fe-C system to 70 GPa. *Earth Planet. Sci. Lett.* **2009**, *284*, 157–167. [CrossRef]

129. Sanloup, C.; Guyot, F.; Gillet, P.; Fiquet, G.; Hemley, R.J.; Mezouar, M.; Martinez, I. Structural changes in liquid Fe at high pressures and high temperatures from Synchrotron X-ray Diffraction. *Europhys. Lett.* **2000**, *52*, 151–157. [CrossRef]

130. Brand, H.; Dobson, D.P.; Vočadlo, L.; Wood, I.G. Melting curve of copper measured to 16 GPa using a multi-anvil press. *High Press. Res.* **2006**, *26*, 185–191. [CrossRef]

131. Fan, D.W.; Zhou, W.G.; Wei, S.Y.; Liu, Y.G.; Ma, M.N.; Xie, H.S. A simple external resistance heating diamond anvil cell and its application for synchrotron radiation X-ray diffraction. *Rev. Sci. Instrum.* **2010**, *81*, 5. [CrossRef]

132. Jenei, Z.; Cynn, H.; Visbeck, K.; Evans, W. High-temperature experiments using a resistively heated high-pressure membrane diamond anvil cell. *Rev. Sci. Instrum.* **2013**, *84*, 0951144. [CrossRef] [PubMed]

133. Weir, S.T.; Jackson, D.D.; Falabella, S.; Samudrala, G.; Vohra, Y.K. An electrical microheater technique for high-pressure and high-temperature diamond anvil cell experiments. *Rev. Sci. Instrum.* **2009**, *80*, 013905. [CrossRef]

134. Pasternak, S.; Aquilanti, G.; Pascarelli, S.; Poloni, R.; Canny, B.; Coulet, M.-V.; Zhang, L. A diamond anvil cell with resistive heating for high pressure and high temperature X-ray diffraction and absorption studies. *Rev. Sci. Instrum.* **2008**, *79*, 085103. [CrossRef]

135. Du, Z.; Miyagi, L.; Amulele, G.; Lee, K.K.M. Efficient graphite ring heater suitable for diamond-anvil cells to 1300 K. *Rev. Sci. Instrum.* **2013**, *84*, 024502. [CrossRef]

136. Dubrovinsky, L.S.; Saxena, S.K.; Lazor, P. High-pressure and high-temperature in situ X-ray diffraction study of iron and corundum to 68 GPa using an internally heated diamond anvil cell. *Phys. Chem. Miner.* **1998**, *25*, 434–441. [CrossRef]

137. Balzaretti, N.M.; Gonzalez, E.J.; Piermarini, G.J.; Russell, T.P. Resistance heating of the gasket in a gem-anvil high pressure cell. *Rev. Sci. Instrum.* **1999**, *70*, 4316–4323. [CrossRef]

138. Liermann, H.-P.; Merkel, S.; Miyagi, L.; Wenk, H.-R.; Shen, G.; Cynn, H.; Evans, W.J. Experimental method for in situ determination of material textures at simultaneous high pressure and high temperature by means of radial diffraction in the diamond anvil cell. *Rev. Sci. Instrum.* **2009**, *80*, 104501. [CrossRef] [PubMed]

139. Zha, C.-S.; Bassett, W.A. Internal resistive heating in diamond anvil cell for in situ X-ray diffraction and Raman scattering. *Rev. Sci. Instrum.* **2003**, *74*, 1255. [CrossRef]

140. Bassett, W.A.; Shen, A.H.; Bucknum, M.; Chou, I.-M. A new diamond anvil cell for hydrothermal studies to 2.5 GPa and from −190 to 1200 °C. *Rev. Sci. Instrum.* **1993**, *64*, 2340–2345. [CrossRef]

141. Weir, S.T.; Lipp, M.J.; Falabella, S.; Samudrala, G.; Vohra, Y.K. High pressure melting curve of tin measured using an internal resistive heating technique to 45 GPa. *J. Appl. Phys.* **2012**, *111*, 123529. [CrossRef]

142. Komabayashi, T.; Fei, Y.; Meng, Y.; Prakapenka, V. In-situ X-ray diffraction measurements of the γ-ε transition boundary of iron in an internally-heated diamond anvil cell. *Earth Planet. Sci. Lett.* **2009**, *282*, 252–257. [CrossRef]

143. Komabayashi, T.; Pesce, G.; Sinmyo, R.; Kawazoe, T.; Breton, H.; Shimoyama, Y.; Glzyrin, K.; Konôpková, Z.; Mezouar, M. Phase relations in the system Fe–Ni–Si to 200 GPa and 3900 K and implications for Earth's core. *Earth Planet. Sci. Lett.* **2019**, *512*, 83–88. [CrossRef]

144. Goncharov, A.F.; Hemley, R.J. Probing hydrogen-rich molecular systems at high pressures and temperatures. *Chem. Soc. Rev.* **2006**, *35*, 899–907. [CrossRef]

145. Smith, D.; Hakeem, M.A.; Parisiades, P.; Maynard-Casely, H.E.; Foster, D.; Eden, D.; Bull, D.J.; Marshall, A.R.L.; Adawi, A.M.; Howie, R.; et al. Crossover between liquidlike and gaslike behavior in C H_4 at 400 K. *Phys. Rev. E* **2017**, *96*, 052113. [CrossRef]

146. Ninet, S.; Datchi, F. High pressure–high temperature phase diagram of ammonia. *J. Chem. Phys.* **2008**, *128*, 154508. [CrossRef]

147. Datchi, F.; Loubeyre, P.; LeToullec, R. Extended and accurate determination of the melting curves of argon, helium, ice (H_2O), and hydrogen(H_2). *Phys. Rev. B* **2000**, *61*, 6535–6546. [CrossRef]

148. Ming, L.; Bassett, W.A. Laser heating in the diamond anvil press up to 2000 °C sustained and 3000 °C pulsed at pressures up to 260 kilobars. *Rev. Sci. Instrum.* **1974**, *45*, 1115–1118. [CrossRef]

149. Murakami, M. Post-Perovskite Phase Transition in $MgSiO_3$. *Science* **2004**, *304*, 855–858. [CrossRef] [PubMed]

150. Petitgirard, S.; Salamat, A.; Beck, P.; Weck, G.; Bouvier, P. Strategies for in situ laser heating in the diamond anvil cell at an X-ray diffraction beamline. *J. Synchrotron Radiat.* **2014**, *21*, 89–96. [CrossRef] [PubMed]

151. Prakapenka, V.B.; Kubo, A.; Kuznetsov, A.; Laskin, A.; Shkurikhin, O.; Dera, P.; Rivers, M.L.; Sutton, S.R. Advanced flat top laser heating system for high pressure research at GSECARS: Application to the melting behavior of germanium. *High Press. Res.* **2008**, *28*, 225–235. [CrossRef]

152. Anzellini, S.; Kleppe, A.K.; Daisenberger, D.; Wharmby, M.T.; Giampaoli, R.; Boccato, S.; Baron, M.A.; Miozzi, F.; Keeble, D.S.; Ross, A.; et al. Laser-heating system for high-pressure X-ray diffraction at the Extreme Conditions beamline I15 at Diamond Light Source. *J. Synchrotron Radiat.* **2018**, *25*, 1860–1868. [CrossRef]

153. Meng, Y.; Hrubiak, R.; Rod, E.; Boehler, R.; Shen, G. New developments in laser-heated diamond anvil cell with in situ synchrotron X-ray diffraction at High Pressure Collaborative Access Team. *Rev. Sci. Instrum.* **2015**, *86*, 072201. [CrossRef] [PubMed]

154. Liermann, H.-P.; Konôpková, Z.; Morgenroth, W.; Glazyrin, K.; Bednarčik, J.; McBride, E.E.; Petitgirard, S.; Delitz, J.T.; Wendt, M.; Bican, Y.; et al. The Extreme Conditions Beamline P02.2 and the Extreme Conditions Science Infrastructure at PETRA III. *J. Synchrotron Radiat.* **2015**, *22*, 908–924. [CrossRef] [PubMed]

155. Watanuki, T.; Shimomura, O.; Yagi, T.; Kondo, T.; Isshiki, M. Construction of laser-heated diamond anvil cell system for in situ X-ray diffraction study at SPring-8. *Rev. Sci. Instrum.* **2001**, *72*, 1289. [CrossRef]

156. Stan, C.V.; Beavers, C.M.; Kunz, M.; Tamura, N. X-ray Diffraction under Extreme Conditions at the Advanced Light Source. *Quantum Beam Sci.* **2018**, *2*, 4. [CrossRef]

157. Andrault, D.; Bolfan-Casanova, N.; Bouhifd, M.; Boujibar, A.; Garbarino, G.; Manthilake, G.; Mezouar, M.; Monteux, J.; Parisiades, P.; Pesce, G. Toward a coherent model for the melting behavior of the deep Earth's mantle. *Phys. Earth Planet. Inter.* **2017**, *265*, 67–81. [CrossRef]

158. Boehler, R.; Musshoff, H.G.; Ditz, R.; Aquilanti, G.; Trapananti, A. Portable laser-heating stand for synchrotron applications. *Rev. Sci. Instrum.* **2009**, *80*, 045103. [CrossRef]

159. Andrault, D.; Muñoz, M.; Bolfan-Casanova, N.; Guignot, N.; Perrillat, J.-P.; Aquilanti, G.; Pascarelli, S. Experimental evidence for perovskite and post-perovskite coexistence throughout the whole D″ region. *Earth Planet. Sci. Lett.* **2010**, *293*, 90–96. [CrossRef]

160. Kantor, I.; Marini, C.; Mathon, O.; Pascarelli, S. A laser heating facility for energy-dispersive X-ray absorption spectroscopy. *Rev. Sci. Instrum.* **2018**, *89*, 13111. [CrossRef]

161. Torchio, R.; Boccato, S.; Cerantola, V.; Morard, G.; Irifune, T.; Kantor, I. Probing the local, electronic and magnetic structure of matter under extreme conditions of temperature and pressure. *High Press. Res.* **2016**, *36*, 293–302. [CrossRef]

162. Andrault, D.; Petitgirard, S.; Nigro, G.L.; Devidal, J.; Veronesi, G.; Garbarino, G.; Mezouar, M. Solid-liquidiron partitioning in Earth's deep mantle. *Nature* **2012**, *487*, 354. [CrossRef] [PubMed]

163. Potapkin, V.; McCammon, C.; Glazyrin, K.; Kantor, A.; Kupenko, I.; Prescher, C.; Sinmyo, R.; Smirnov, G.V.; Chumakov, A.I.; Rüffer, R.; et al. Effect of iron oxidation state on the electrical conductivity of the Earth's lower mantle. *Nat. Commun.* **2013**, *4*, 1427. [CrossRef] [PubMed]

164. Kupenko, I.; Dubrovinsky, L.; Dubrovinskaia, N.; McCammon, C.; Glazyrin, K.; Bykova, E.; Ballaran, T.B.; Sinmyo, R.; Chumakov, A.I.; Potapkin, V.; et al. Portable double-sided laser-heating system for Mössbauer spectroscopy and X-ray diffraction experiments at synchrotron facilities with diamond anvil cells. *Rev. Sci. Instrum.* **2012**, *83*, 124501. [CrossRef]

165. McCammon, C.; Caracas, R.; Glazyrin, K.; Potapkin, V.; Kantor, A.; Sinmyo, R.; Prescher, C.; Kupenko, I.; Chumakov, A.; Dubrovinsky, L. Sound velocities of bridgmanite from density of states determined by nuclear inelastic scattering and first-principles calculations. *Prog. Earth Planet. Sci.* **2016**, *3*, 10. [CrossRef]

166. Meier, T.; Dwivedi, A.P.; Khandarkhaeva, S.; Fedotenko, T.; Dubrovinskaia, N.; Dubrovinsky, L. Table-top nuclear magnetic resonance system for high-pressure studies with in situ laser heating. *Rev. Sci. Instrum.* **2019**, *90*, 123901. [CrossRef] [PubMed]

167. Zinin, P.V.; Prakapenka, V.B.; Burgess, K.; Odake, S.; Chigarev, N.; Sharma, S.K. Combined laser ultrasonics, laser heating, and Raman scattering in diamond anvil cell system. *Rev. Sci. Instrum.* **2016**, *87*, 123908. [CrossRef]

168. Goncharov, A.F.; Crowhurst, J.C. Pulsed laser Raman spectroscopy in the laser-heated diamond anvil cell. *Rev. Sci. Instrum.* **2005**, *76*, 63905. [CrossRef]

169. Santoro, M.; Lin, J.-F.; Mao, H.-K.; Hemley, R.J. In situ high P-T Raman spectroscopy and laser heating of carbon dioxide. *J. Chem. Phys.* **2004**, *121*, 2780. [CrossRef]

170. Zhou, Q.; Ma, Y.; Cui, Q.; Cui, T.; Zhang, J.; Xie, Y.; Yang, K.; Zou, G. Raman scattering system for a laser heated diamond anvil cell. *Rev. Sci. Instrum.* **2004**, *75*, 2432–2434. [CrossRef]

171. Lin, J.-F.; Santoro, M.; Struzhkin, V.V.; Mao, H.-K.; Hemley, R.J. In situ high pressure-temperature Raman spectroscopy technique with laser-heated diamond anvil cells. *Rev. Sci. Instrum.* **2004**, *75*, 3302–3306. [CrossRef]

172. Murakami, M.; Asahara, Y.; Ohishi, Y.; Hirao, N.; Hirose, K. Development of in situ Brillouin spectroscopy at high pressure and high temperature with synchrotron radiation and infrared laser heating system: Application to the Earth's deep interior. *Phys. Earth Planet. Inter.* **2009**, *174*, 282–291. [CrossRef]

173. Sinogeikin, S.V.; Lakshtanov, D.L.; Nicholas, J.D.; Jackson, J.M.; Bass, J.D. High temperature elasticity measurements on oxides by Brillouin spectroscopy with resistive and IR laser heating. *J. Eur. Ceram. Soc.* **2005**, *25*, 1313–1324. [CrossRef]

174. Li, F.; Cui, Q.; He, Z.; Cui, T.; Gao, C.; Zhou, Q.; Zou, G. Brillouin scattering spectroscopy for a laser heated diamond anvil cell. *Appl. Phys. Lett.* **2006**, *88*, 203507. [CrossRef]

175. Zhang, J.S.; Bass, J.D.; Zhu, G. Single-crystal Brillouin spectroscopy with CO2 laser heating and variable q. *Rev. Sci. Instrum.* **2015**, *86*, 063905. [CrossRef] [PubMed]

176. Young, A.F.; Sanloup, C.; Gregoryanz, E.; Scandolo, S.; Hemley, R.J.; Mao, H.K. Synthesis of novel transition metal nitrides IrN_2 and OsN_2. *Phys. Rev. Lett.* **2006**, *96*, 155501. [CrossRef] [PubMed]

177. Solozhenko, V.L.; Kurakevych, O.O.; Andrault, D.; Le Godec, Y.; Mezouar, M. Ultimate Metastable Solubility of Boron in Diamond: Synthesis of Superhard Diamond like BC_5. *Phys. Rev. Lett.* **2009**, *102*, 015506. [CrossRef]

178. Nielsen, M.B.; Ceresoli, D.; Parisiades, P.; Prakapenka, V.B.; Yu, T.; Wang, Y.; Bremholm, M. Phase stability of the $SrMnO_3$ hexagonal perovskite system at high pressure and temperature. *Phys. Rev. B* **2014**, *90*, 214101. [CrossRef]

179. Crowhurst, J.C.; Goncharov, A.F.; Sadigh, B.; Zaug, J.; Aberg, D.; Meng, Y.; Prakapenka, V.B. Synthesis and characterization of nitrides of iridium and palladium. *J. Mater. Res.* **2008**, *23*, 1–5. [CrossRef]

180. Salamat, A.; Fischer, R.A.; Briggs, R.; McMahon, M.I.; Petitgirard, S. In situ synchrotron X-ray diffraction in the laser-heated diamond anvil cell: Melting phenomena and synthesis of new materials. *Coord. Chem. Rev.* **2014**, *277–278*, 15–30. [CrossRef]

181. Pépin, C.M.; Geneste, G.; Dewaele, A.; Mezouar, M.; Loubeyre, P. Synthesis of FeH_5: A layered structure with atomic hydrogen slabs. *Science* **2017**, *357*, 382–385. [CrossRef]

182. Drozdov, A.P.; Kong, P.P.; Minkov, V.S.; Besedin, S.P.; Kuzovnikov, M.A.; Mozaffari, S.; Balicas, L.; Balakirev, F.F.; Graf, D.E.; Prakapenka, V.B.; et al. Superconductivity at 250 K in lanthanum hydride under high pressures. *Nat. Cell Biol.* **2019**, *569*, 528–531. [CrossRef] [PubMed]

183. Guigue, B.; Marizy, A.; Loubeyre, P. Direct synthesis of pure H_3S from S and H elements: No evidence of the cubic superconducting phase up to 160 GPa. *Phys. Rev. B* **2017**, *95*, 020104. [CrossRef]

184. Block, S.; Weir, C.W.; Piermarini, G.J. High-Pressure Single-Crystal Studies of Ice VI. *Science* **1965**, *148*, 947–948. [CrossRef]

185. Lin, J.-F.; Shu, J.; Mao, H.-K.; Hemley, R.J.; Shen, G. Amorphous boron gasket in diamond anvil cell research. *Rev. Sci. Instrum.* **2003**, *74*, 4732–4736. [CrossRef]

186. Funamori, N.; Sato, T. A cubic boron nitride gasket for diamond-anvil experiments. *Rev. Sci. Instrum.* **2008**, *79*, 053903. [CrossRef]

187. Hemley, R.J.; Beckmann, R.; Bubeck, R.; Grassucci, R.; Penczek, P.; Verschoor, A.; Blobel, G.; Frank, J. X-ray Imaging of Stress and Strain of Diamond, Iron, and Tungsten at Megabar Pressures. *Science* **1997**, *276*, 1242–1245. [CrossRef]

188. Mao, H.-K.; Shu, J.; Shen, G.; Hemley, R.J.; Li, B.; Singh, A.K. Elasticity and rheology of iron above 220 GPa and the nature of the Earth's inner core. *Nat. Cell Biol.* **1998**, *396*, 741–743. [CrossRef]

189. Wu, T.-C.; Bassett, W.A. Deviatoric stress in a diamond anvil cell using synchrotron radiation with two diffraction geometries. *PAGEOPH* **1993**, *141*, 509–519. [CrossRef]

190. Klotz, S.; Chervin, J.-C.; Munsch, P.; Le Marchand, G. Hydrostatic limits of 11 pressure transmitting media. *J. Phys. D Appl. Phys.* **2009**, *42*, 075413. [CrossRef]

191. Dewaele, A.; Loubeyre, P. Pressurizing conditions in helium-pressure-transmitting medium. *High Press. Res.* **2007**, *27*, 419–429. [CrossRef]

192. Boehler, R.; Ross, M.; Söderlind, P.; Boercker, D.B. High-Pressure Melting Curves of Argon, Krypton, and Xenon: Deviation from Corresponding States Theory. *Phys. Rev. Lett.* **2001**, *86*, 5731–5734. [CrossRef]

193. Boehler, R.; Ross, M.; Boercker, D.B. Melting of LiF and NaCl to 1 Mbar: Systematics of Ionic Solids at Extreme Conditions. *Phys. Rev. Lett.* **1997**, *78*, 4589–4592. [CrossRef]

194. Kimura, T.; Ohfuji, H.; Nishi, M.; Irifune, T. Melting temperatures of MgO under high pressure by micro-texture analysis. *Nat. Commun.* **2017**, *8*, 15735. [CrossRef]

195. Andrault, D.; Morard, G.; Garbarino, G.; Mezouar, M.; Bouhifd, M.A.; Kawamoto, T. Melting behavior of SiO_2 up to 120 GPa. *Phys. Chem. Miner.* **2020**, *47*, 10. [CrossRef]

196. Shen, G.; Lazor, P. Measurement of melting temperatures of some minerals under lower mantle pressures. *J. Geophys. Res. Space Phys.* **1995**, *100*, 17699–17713. [CrossRef]

197. Santamaría-Pérez, D.; Mukherjee, G.D.; Schwager, B.; Boehler, R. High-pressure melting curve of helium and neon: Deviations from corresponding states theory. *Phys. Rev. B* **2010**, *81*, 214101. [CrossRef]

198. Boehler, R.; Ross, M.; Boercker, D.B. High-pressure melting curves of alkali halides. *Phys. Rev. B* **1996**, *53*, 556–563. [CrossRef] [PubMed]

199. Yang, L.; Karandikar, A.; Boehler, R. Flash heating in the diamond cell: Melting curve of rhenium. *Rev. Sci. Instrum.* **2012**, *83*, 63905. [CrossRef] [PubMed]

200. Dewaele, A.; Mezouar, M.; Guignot, N.; Loubeyre, P. Melting of lead under high pressure studied using second-scale time-resolved X-ray diffraction. *Phys. Rev. B* **2007**, *76*, 144106. [CrossRef]

201. Patel, N.N.; Sunder, M. High pressure melting curve of osmium up to 35 GPa. *J. Appl. Phys.* **2019**, *125*, 055902. [CrossRef]

202. Du, Z.; Amulele, G.; Benedetti, L.R.; Lee, K.K.M. Mapping temperatures and temperature gradients during flash heating in a diamond-anvil cell. *Rev. Sci. Instrum.* **2013**, *84*, 75111. [CrossRef] [PubMed]

203. Geballe, Z.M.; Jeanloz, R. Origin of temperature plateaus in laser-heated diamond anvil cell experiments. *J. Appl. Phys.* **2012**, *111*, 123518. [CrossRef]

204. Weck, G.; Datchi, F.; Garbarino, G.; Ninet, S.; Queyroux, J.-A.; Plisson, T.; Mezouar, M.; Loubeyre, P. Melting Curve and Liquid Structure of Nitrogen Probed by X-ray Diffraction to 120 GPa. *Phys. Rev. Lett.* **2017**, *119*, 235701. [CrossRef]

205. Zerr, A.; Diegeler, A.; Boehler, R. Solidus of Earth's Deep Mantle. *Science* **1998**, *281*, 243–246. [CrossRef]

206. Hearne, G.; Bibik, A.; Zhao, J. CO2laser-heated diamond-anvil cell methodology revisited. *J. Phys. Condens. Matter* **2002**, *14*, 11531–11535. [CrossRef]

207. Fiquet, G.; Andrault, D.; Itié, J.; Gillet, P.; Richet, P. X-ray diffraction of periclase in a laser-heated diamond-anvil cell. *Phys. Earth Planet. Inter.* **1996**, *95*, 1–17. [CrossRef]

208. Mao, H.K.; Bell, P.M.; Shaner, J.W.; Steinberg, D.J. Specific volume measurements of Cu, Mo, Pd, and Ag and calibration of the ruby R_1 fluorescence pressure gauge from 0.06 to 1 Mbar. *J. Appl. Phys.* **1978**, *49*, 3276–3283. [CrossRef]

209. Dewaele, A.; Torrent, M.; Loubeyre, P.; Mezouar, M. Compression curves of transition metals in the Mbar range: Experiments and projector augmented-wave calculations. *Phys. Rev. B* **2008**, *78*, 104102. [CrossRef]

210. Shen, G.; Wang, Y.; Dewaele, A.; Wu, C.; Fratanduono, D.E.; Eggert, J.; Klotz, S.; Dziubek, K.F.; Loubeyre, P.; Fat'Yanov, O.V.; et al. Toward an international practical pressure scale: A proposal for an IPPS ruby gauge (IPPS-Ruby2020). *High Press. Res.* **2020**, *40*, 299–314. [CrossRef]

211. Datchi, F.; Dewaele, A.; Loubeyre, P.; Letoullec, R.; Le Godec, Y.; Canny, B. Optical pressure sensors for high-pressure–high-temperature studies in a diamond anvil cell. *High Press. Res.* **2007**, *27*, 447–463. [CrossRef]

212. Akahama, Y.; Kawamura, H. Pressure calibration of diamond anvil Raman gauge to 410 GPa. *J. Phys. Conf. Ser.* **2010**, *215*, 012195. [CrossRef]

213. Benedetti, L.R.; Loubeyre, P. Temperature gradients, wavelength-dependent emissivity, and accuracy of high and very-high temperatures measured in the laser-heated diamond cell. *High Press. Res.* **2004**, *24*, 423–445. [CrossRef]

214. Benedetti, L.R.; Antonangeli, D.; Farber, D.L.; Mezouar, M. An integrated method to determine melting temperatures in high-pressure laser-heating experiments. *Appl. Phys. Lett.* **2008**, *92*, 141903. [CrossRef]

215. Mezouar, M.; Giampaoli, R.; Garbarino, G.; Kantor, I.; Dewaele, A.; Weck, G.; Boccato, S.; Svitlyk, V.; Rosa, A.D.; Torchio, R.; et al. Methodology for in situ synchrotron X-ray studies in the laser-heated diamond anvil cell. *High Press. Res.* **2017**, *37*, 170–180. [CrossRef]

216. Giampaoli, R.; Kantor, I.; Mezouar, M.; Boccato, S.; Rosa, A.D.; Torchio, R.; Garbarino, G.; Mathon, O.; Pascarelli, S. Measurement of temperature in the laser heated diamond anvil cell: Comparison between reflective and refractive optics. *High Press. Res.* **2018**, *38*, 250–269. [CrossRef]

217. Benedetti, L.R.; Guignot, N.; Farber, D.L. Achieving accuracy in spectroradiometric measurements of temperature in the laser-heated diamond anvil cell: Diamond is an optical component. *J. Appl. Phys.* **2007**, *101*, 013109. [CrossRef]

218. Jeanloz, R.; Kavner, A. Melting criteria and imaging spectroradiometry in laser-heated diamond-cell experiments. *Philos. Trans. R. Soc. A Math. Phys. Eng. Sci.* **1996**, *354*, 1279–1305. [CrossRef]

219. Belonoshko, A.; Ahuja, R.; Johansson, B. Molecular Dynamics Study of Melting and fcc-bcc Transitions in Xe. *Phys. Rev. Lett.* **2001**, *87*, 165505. [CrossRef] [PubMed]

220. Belonoshko, A.B.; Ahuja, R.; Johansson, B. Stability of the body-centred-cubic phase of iron in the Earth's inner core. *Nat. Cell Biol.* **2003**, *424*, 1032–1034. [CrossRef]

221. Andrault, D.; Morard, G.; Bolfan-Casanova, N.; Ohtaka, O.; Fukui, H.; Arima, H.; Guignot, N.; Funakoshi, K.; Lazor, P.; Mezouar, M. Study of partial melting at high-pressure using in situ X-ray diffraction. *High Press. Res.* **2006**, *26*, 267–276. [CrossRef]

222. Di Cicco, A.; Trapananti, A. Study of local icosahedral ordering in liquid and undercooled liquid copper. *J. Non-Cryst. Solids* **2007**, *353*, 3671–3678. [CrossRef]

223. Pascarelli, S.; Mathon, O.; Mairs, T.; Kantor, I.; Agostini, G.; Strohm, C.; Pasternak, S.; Perrin, F.; Berruyer, G.; Chappelet, P.; et al. The Time-resolved and Extreme-conditions XAS (TEXAS) facility at the European Synchrotron Radiation Facility: The energy-dispersive X-ray absorption spectroscopy beamline ID24. *J. Synchrotron Radiat.* **2016**, *23*, 353–368. [CrossRef]

224. Lord, O.T.; Wann, E.T.; Hunt, S.A.; Walker, A.M.; Santangeli, J.; Walter, M.J.; Dobson, D.P.; Wood, I.G.; Vočadlo, L.; Morard, G.; et al. The NiSi melting curve to 70 GPa. *Phys. Earth Planet. Inter.* **2014**, *233*, 13–23. [CrossRef]

225. Prakapenka, V.B.; Shen, G.; Rivers, M.L.; Sutton, S.R.; Dubrovinsky, L. Grain-size control in situ at high pressures and high temperatures in a diamond-anvil cell. *J. Synchrotron Radait.* **2005**, *12*, 560–565. [CrossRef] [PubMed]

226. Karandikar, A.; Boehler, R. Flash melting of tantalum in a diamond cell to 85 GPa. *Phys. Rev. B* **2016**, *93*, 054107. [CrossRef]

227. Sinmyo, R.; Hirose, K.; Ohishi, Y. Melting curve of iron to 290 GPa determined in a resistance-heated diamond-anvil cell. *Earth Planet. Sci. Lett.* **2019**, *510*, 45–52. [CrossRef]

228. Basu, A.; Field, M.R.; McCulloch, D.G.; Boehler, R. New measurement of melting and thermal conductivity of iron close to outer core conditions. *Geosci. Front.* **2020**, *11*, 565–568. [CrossRef]

229. Boehler, R. The phase diagram of iron to 430 kbar. *Geophys. Res. Lett.* **1986**, *13*, 1153–1156. [CrossRef]

230. Ohta, K.; Kuwayama, Y.; Hirose, K.; Shimizu, K.; Ohishi, Y. Experimental determination of the electrical resistivity of iron at Earth's core conditions. *Nat. Cell Biol.* **2016**, *534*, 95–98. [CrossRef]

231. Konôpková, Z.; McWilliams, R.S.; Gómez-Pérez, R.S.M.N.; Goncharov, A.F. Direct measurement of thermal conductivity in solid iron at planetary core conditions. *Nat. Cell Biol.* **2016**, *534*, 99–101. [CrossRef]

232. Aprilis, G.; Strohm, C.; Kupenko, I.; Linhardt, S.; Laskin, A.; Vasiukov, D.; Cerantola, V.; Koemets, E.; McCammon, C.; Kurnosov, A.; et al. Portable double-sided pulsed laser heating system for time-resolved geoscience and materials science applications. *Rev. Sci. Instrum.* **2017**, *88*, 084501. [CrossRef] [PubMed]

233. Zhang, D.; Jackson, J.M.; Zhao, J.; Sturhahn, W.; Alp, E.E.; Toellner, T.S.; Hu, M.Y. Fast temperature spectrometer for samples under extreme conditions. *Rev. Sci. Instrum.* **2015**, *86*, 013105. [CrossRef] [PubMed]

234. Zhang, D.; Jackson, J.M.; Zhao, J.; Sturhahn, W.; Alp, E.E.; Hu, M.Y.; Toellner, T.S.; Murphy, C.A.; Prakapenka, V.B. Temperature of Earth's core constrained from melting of Fe and Fe$_{0.9}$Ni$_{0.1}$ at high pressures. *Earth Planet. Sci. Lett.* **2016**, *447*, 72–83. [CrossRef]

235. Jackson, J.M.; Sturhahn, W.; Lerche, M.; Zhao, J.; Toellner, T.S.; Alp, E.E.; Sinogeikin, S.V.; Bass, J.D.; Murphy, C.A.; Wicks, J.K. Melting of compressed iron by monitoring atomic dynamics. *Earth Planet. Sci. Lett.* **2013**, *362*, 143–150. [CrossRef]

236. Jeanloz, R.; Heinz, D. Experiments at high temperature and pressure: Laser heating through the diamond cell. *J. Phys. Colloq.* **1984**, *45*, C8-83. [CrossRef]

237. Rouquette, J.; Dolejs, D.; Kantor, I.Y.; McCammon, C.A.; Frost, D.J.; Prakapenka, V.B.; Dubrovinsky, L.S. Iron-carbon interactions at high temperatures and pressures. *Appl. Phys. Lett.* **2008**, *92*, 121912. [CrossRef]

238. Ren, B.; Liu, G.-K.; Lian, X.-B.; Yang, Z.-L.; Tian, Z.-Q. Raman spectroscopy on transition metals. *Anal. Bioanal. Chem.* **2007**, *388*, 29–45. [CrossRef] [PubMed]

239. Aprilis, G.; Kantor, I.; Kupenko, I.; Cerantola, V.; Pakhomova, A.; Collings, I.E.; Torchio, R.; Fedotenko, T.; Chariton, S.; Bykov, M.; et al. Comparative study of the influence of pulsed and continuous wave laser heating on the mobilization of carbon and its chemical reaction with iron in a diamond anvil cell. *J. Appl. Phys.* **2019**, *125*, 095901. [CrossRef]

MDPI
St. Alban-Anlage 66
4052 Basel
Switzerland
Tel. +41 61 683 77 34
Fax +41 61 302 89 18
www.mdpi.com

Crystals Editorial Office
E-mail: crystals@mdpi.com
www.mdpi.com/journal/crystals